Pharmaceutical Process Development
Current Chemical and Engineering Challenges

RSC Drug Discovery Series

Editor-in-Chief
Professor David Thurston, *London School of Pharmacy, UK*

Series Editors:
Dr David Fox, *Pfizer Global Research and Development, Sandwich, UK*
Professor Salvatore Guccione, *University of Catania, Italy*
Professor Ana Martinez, *Instituto de Quimica Medica-CSIC, Spain*
Dr David Rotella, *Montclair State University, USA*

Advisor to the Board:
Professor Robin Ganellin, *University College London, UK*

Titles in the Series:
1: Metabolism, Pharmacokinetics and Toxicity of Functional Groups: Impact of Chemical Building Blocks on ADMET
2: Emerging Drugs and Targets for Alzheimer's Disease; Volume 1: Beta-Amyloid, Tau Protein and Glucose Metabolism
3: Emerging Drugs and Targets for Alzheimer's Disease; Volume 2: Neuronal Plasticity, Neuronal Protection and Other Miscellaneous Strategies
4: Accounts in Drug Discovery: Case Studies in Medicinal Chemistry
5: New Frontiers in Chemical Biology: Enabling Drug Discovery
6: Animal Models for Neurodegenerative Disease
7: Neurodegeneration: Metallostasis and Proteostasis
8: G Protein-Coupled Receptors: From Structure to Function
9: Pharmaceutical Process Development: Current Chemical and Engineering Challenges

How to obtain future titles on publication:
A standing order plan is available for this series. A standing order will bring delivery of each new volume immediately on publication.

For further information please contact:
Book Sales Department, Royal Society of Chemistry, Thomas Graham House, Science Park, Milton Road, Cambridge, CB4 0WF, UK
Telephone: +44 (0)1223 420066, Fax: +44 (0)1223 420247, Email: books@rsc.org
Visit our website at http://www.rsc.org/Shop/Books/

Pharmaceutical Process Development
Current Chemical and Engineering Challenges

Edited by

A. John Blacker
Institute of Process Research and Development, University of Leeds, Leeds, UK

Mike T. Williams
CMC Consultant

RSC Publishing

RSC Drug Discovery Series No. 9

ISBN: 978-1-84973-146-1
ISSN: 2041-3203

A catalogue record for this book is available from the British Library

Published by The Royal Society of Chemistry,
Thomas Graham House, Science Park, Milton Road,
Cambridge CB4 0WF, UK

Registered Charity Number 207890

For further information see our web site at www.rsc.org

Foreword

It is pleasing to contemplate a manufacture rising gradually from its first mean state by the successive labours of innumerable minds.
Samuel Johnson, The Rambler, 17 April 1750

In more than 45 years of learning and teaching chemistry I have encountered a rich array of students and colleagues whose pursuits have been animated by similar ideals. Some were driven by the intellectual stimulus and rewards of working at the frontier of knowledge. Others were driven by the urge to challenge existing tenets in the pursuit of greater understanding. Still others were inspired by the practicality of the subject and the desire to leave a beneficent legacy. The issue of practicality has loomed rather large recently as governments have sought a tangible return for very substantial investments. Evidence for the enhanced status of practicality can be gleaned from the enrichment of "standard" university programmes in chemistry with polymer chemistry, nanotechnology, material science and medicinal chemistry. Conspicuous for their absence from this short list are courses in process development; hence, there is little appreciation by university graduates of the complex and multidimensional science that transforms a reaction or sequence of reactions to a process. The absence of process development in academia can be attributed in part to the fact that most process development takes place in the industrial sphere under a veil of secrecy. However, a number of changes has taken place in the last 20 years that augur well for the future. Firstly, many companies encourage their employees to publish their work in open scientific journals, with the consequent revelation of the wealth of skill and creativity that underpins process research. Secondly, process chemists are often guests in university seminar programmes. Thirdly, the publication of high quality monographs written by experts in the field has aided dissemination. Finally, academic chemists are now aware of the many challenges in process development and they are better attuned to the benefits of exploitation.

RSC Drug Discovery Series No. 9
Pharmaceutical Process Development: Current Chemical and Engineering Challenges
Edited by A. John Blacker and Mike T. Williams
© Royal Society of Chemistry 2011
Published by the Royal Society of Chemistry, www.rsc.org

In *Pharmaceutical Process Development: Current Chemical & Chemical Engineering Challenges,* John Blacker and Mike Williams, expert practitioners, present a brief overview of the many facets of process development and how recent advances in synthetic organic chemistry, process technology and chemical engineering have impacted on the manufacture of pharmaceuticals. It is aimed at chemistry, engineering and pharmacy undergraduates, postgraduates and early to mid-career professionals in allied disciplines. In 15 concise chapters the book covers such diverse subjects as route selection and economics, the interface with medicinal chemistry, the impact of green chemistry, safety, the crucial role of physical organic measurements in gaining a deeper understanding of chemical behaviour, the role of the analyst, new tools and innovations in reactor design, purification and separation, solid state chemistry and its role in formulation. The book ends with an assessment of future trends and challenges.

Philip Kocienski

Preface

Anyone who has helped to plan an industrial synthesis tends to pity the poverty of the criteria that academic synthesis must meet.

Sir John Cornforth, *Aust. J. Chem.*, 1993, **46**, 157.

The industrialisation of any synthetic process is a challenging endeavour, and this is particularly so for pharmaceutical agents because of their relative complexity. In addition to being efficient and atom economic, reactions used on scale to produce pharmaceuticals need to meet exacting safety and environmental standards, while the efficiency of every operation in the work-up, isolation and purification of the reaction needs to be examined. One of the key goals of process research and development (R&D) scientists is to ensure that their processes are economic. Because the price of pharmaceuticals falls sharply when their patents expire, there is a common misconception that they are overpriced by their innovator companies during their patent lifetime. The reality is usually very different: the pharmaceutical development process is both extremely costly and risky. There is no law of economics stating that the launch of a product onto the market ensures that it will be profitable, and it has been estimated that over half of all drugs that reach the market fail to recoup their discovery and development costs. Hence, process R&D scientists are charged with driving down the cost of goods of drug candidates, to help ensure their commercial viability.

Pharmaceutical process R&D is therefore an exacting, multidisciplinary effort. It has, however, been a somewhat neglected discipline in the chemical curriculum, so we were delighted to have the opportunity to contribute this volume to the RSC Drug Discovery series. The aim of this book is to communicate to those interested in the field the fascinating, and interdependent nature of activities associated with producing drug substance. The book differs from the few others in this field in its organisation, which attempts to walk the reader logically through key aspects of the chemical R&D process. In the wide

RSC Drug Discovery Series No. 9
Pharmaceutical Process Development: Current Chemical and Engineering Challenges
Edited by A. John Blacker and Mike T. Williams
© Royal Society of Chemistry 2011
Published by the Royal Society of Chemistry, www.rsc.org

variety of chapters, contributed mainly by current practitioners from the pharmaceutical industry, the authors have taken the opportunity to discuss the latest technological developments and their impact upon the many changes taking place in the market. It is hoped the reader will appreciate the discussions in each chapter, which are frequently animated by contemporary examples of drug syntheses and processes.

We would like to thank all of the authors for their support of this project, and for the high quality of their contributions to this book. We would also like to thank Gwen Jones at the RSC for her patience and support in bringing this volume to completion at a difficult time for the industry when sites were closing, several authors were struggling with the search for new employment, and timelines drifted. If this book contributes to the growing awareness of the complexities and challenges of process R&D, and provides a useful resource for academic and industrial scientists and engineers, then all of our efforts will have been worthwhile.

A. J. Blacker and M. T. Williams

Contents

RSC Drug Discovery Series No. 9
Pharmaceutical Process Development: Current Chemical and Engineering Challenges
Edited by A. John Blacker and Mike T. Williams
© Royal Society of Chemistry 2011
Published by the Royal Society of Chemistry, www.rsc.org

Chapter 14   Technology Transfer of an Active Pharmaceutical Ingredient   317

Stephen McGhie and Stuart Young

Contributors

John H. Atherton, *Department of Chemical and Biological Sciences, University of Huddersfield, Queensgate, Huddersfield, HD1 3DH, UK.* Email: j.h.atherton@hud.ac.uk

Alexis Bertrand, *MSD Ltd (a subsidiary of Merck & Co., Inc), Hertford Road, Hoddesdon, Hertfordshire, EN11 9BU, UK.*

A. John Blacker, *University of Leeds, Institute of Process Research and Development, Leeds, LS2 9JT, UK.* E-mail: J.Blacker@leeds.ac.uk

Mike Butters, *Pharmaceutical Development, AstraZeneca, Avlon Works, Severn Rd, Bristol, BS10 7ZE, UK.* E-mail: ButtersM@cardiff.ac.uk

John S. Carey, *Reckitt Benckiser Pharmaceuticals, Dansom Lane, Hull, HU8 7DS, UK.* E-mail: john.carey@rb.com

David J. Dale, *The Briars, Clavertye, Elham, Kent, CT4 6YE, UK.* E-mail: daviddale50@hotmail.co.uk

Pieter D. de Koning, *Pfizer Ltd, Research Active Pharmaceutical Ingredients, Ramsgate Road, Sandwich, Kent CT13 9NJ, UK.* E-mail: Pieter.de.koning@pfizer.com

Robert Docherty, *Pharmaceutical Sciences, Pfizer Global R&D, Ramsgate Road, Sandwich, Kent, CT13 9NJ, UK.* E-mail: Robert.docherty@pfizer.com

Peter J. Dunn, *Pfizer Worldwide Research and Development, Sandwich Laboratories, Kent, CT13 9NJ, UK.* E-mail: peter.dunn@pfizer.com

Adam T. Gillmore, *Pfizer Ltd, Research Active Pharmaceutical Ingredients, Ramsgate Road, Sandwich, Kent CT13 9NJ, UK.* E-mail: adam.gillmore@pfizer.com

Simon Hamilton, *MSD Ltd (a subsidiary of Merck & Co., Inc), Hertford Road, Hoddesdon, Hertfordshire, EN11 9BU, UK.* E-mail: simon_hamilton@merck.com

Dave Laffan, *AstraZeneca PR&D, Silk Road Business Park, Macclesfield, SK10 2NA, UK.* Email: dave.laffan@astrazeneca.com

RSC Drug Discovery Series No. 9
Pharmaceutical Process Development: Current Chemical and Engineering Challenges
Edited by A. John Blacker and Mike T. Williams
© Royal Society of Chemistry 2011
Published by the Royal Society of Chemistry, www.rsc.org

Trevor Laird, *Scientific Update LLP, Maycroft Place, Mayfield, East Sussex, TN20 6EW, UK.* Email: Trevor@scientificupdate.co.uk

Suju P. Mathew, *Chemical Research and Development, Pfizer Global Research and Development, Sandwich, CT13 9NJ, UK.* E-mail: Suju73@yahoo.com

Ian F. McConvey, *AstraZeneca, Pharmaceutical Development, Charterway, Silk Road Business Park, Macclesfield, Cheshire, SK10 2NA, UK.* E-mail: Ian.McConvey@astrazeneca.com

Stephen McGhie, *Technical Shared Service, Global Manufacturing and Supply, GlaxoSmithKline, Shewalton Road, Irvine, KA11 5AP, UK.* E-mail: Stephen. X.McGhie@gsk.com

Mark B. Mitchell, *GlaxoSmithKline, Five Moore Drive, Research Triangle Park, NC 27709, USA.*

Mike J. Monteith, *GlaxoSmithKline, Five Moore Drive, Research Triangle Park, NC 27709, USA.* Email: mike.j.monteith@gsk.com

Paul Nancarrow, *School of Chemistry and Chemical Engineering, Queen's University Belfast, Stranmillis Road, Belfast, BT9 5AG, Northern Ireland.*

Kevin Roberts, *Institute of Particle Science and Engineering and Institute of Process Research and Development, School of Process, Environmental and Materials Engineering, University of Leeds, LS2 9JT, UK.* Email: K.J.Roberts@leeds.ac.uk

Yong Tao, *Pfizer Inc, Research Active Pharmaceutical Ingredients, Eastern Point Road, Groton, Connecticut 06340, USA.* E-mail: yong.tao@pfizer.com

Stefan Taylor, *Pharmaceutical Sciences, Pfizer Global R&D, Ramsgate Road, Sandwich, Kent, CT13 9NJ, UK.*

Nicholas M. Thomson, *Pfizer Ltd, Research Active Pharmaceutical Ingredients, Ramsgate Road, Sandwich, Kent CT13 9NJ, UK.* Email: Nick.Thomson@pfizer.com

Mike T. Williams, *133, London Rd., Deal, Kent, C14 9TY, UK.* Email: mike_t_williams2007@yahoo.co.uk

Stuart Young, *New Product Introduction Centre of Excellence, Global Manufacturing and Supply, GlaxoSmithKline, Temple Hill, Dartford, DA1 5AP, UK.*

CHAPTER 1
Introduction

A. JOHN BLACKER[a] AND MIKE T. WILLIAMS[b]

[a] University of Leeds, Institute of Process Research and Development, Leeds, LS2 9JT, UK; [b] 133 London Road, Deal, Kent, C14 9TY, UK

1.1 Process Research and Development in Context

Pharmaceutical process research and development (R&D) is a complex, challenging and exciting endeavour that crosses the boundaries between synthetic organic chemistry, process technology and chemical engineering. This book will explore the various aspects of process research and development for small-molecule manufacture that must be brought together to provide sufficient quantities of reliable and cost effective medicines, made with low environmental impact, to make the drug both a success for the company and a safe, affordable and sustainably produced medicine for society.

The pharmaceutical industry has grown inconceivably since early drugs such as aspirin and penicillin were discovered and developed. The growth was first fuelled by post-war increased healthcare requirements in North America, Western Europe and Japan. The undoubted impact of wide access to medicines has been the increase in life expectancy, which has doubled and continues to increase.[1] The growing populations within the economies of Asia and South America are not only starting to benefit from cheaper and more available medicines, but especially in the case of Asia are increasingly responsible for their production. Traditionally poor economies such as Africa are also starting to access cheaper, more widely distributed, medicines which may help alleviate suffering and improve mortality rates. More recently in Western economies there has been an increased demand for medicines to manage lifestyle in areas

RSC Drug Discovery Series No. 9
Pharmaceutical Process Development: Current Chemical and Engineering Challenges
Edited by A. John Blacker and Mike T. Williams
© Royal Society of Chemistry 2011
Published by the Royal Society of Chemistry, www.rsc.org

such as type II diabetes, anti-cholesterolaemics and infertility, and this is similarly expected to continue rising.[2]

The discovery of a drug candidate by medicinal chemistry is the first step in a long journey to the marketplace, and the vast majority of candidates fall by the wayside. The timescale of the overall drug discovery and development process, and the high attrition rate, are illustrated in Figures 1.1 and 1.2.[1,2] As the drug progresses through clinical trials, the demands for material increase. Not only is drug substance required for clinical trials, but also for analytical, stability,

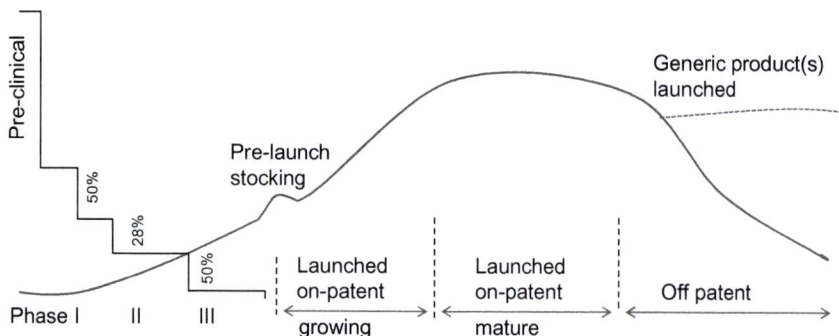

Figure 1.1   Material volume profile with respect to drug lifecycle, and approximate development failure rates.

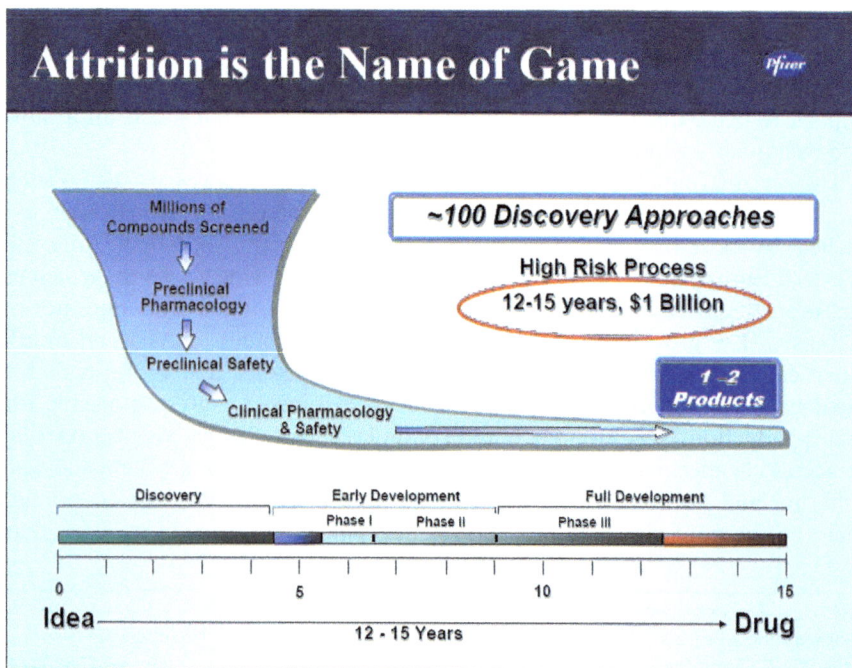

Figure 1.2   Attrition in the R&D process (reproduced with permission of Pfizer).

formulation and *in vitro* studies. Whilst the total amount of drug active required depends upon its activity, effect and physical properties, typical requirements are 1 kg at phase I, 50–100 kg at phase II and at phase III up to 1 tonne. As the drug progresses through the development pipeline, the project team estimates the likely commercial demand profile and ensures that it orders a sizeable contingency to avoid demand outstripping supply. Once launched, the successful drug sees growth in volume that for high-dose drugs (g d^{-1}) may exceed 1000s of tonnes per annum, or for low-dose drugs 100 kg per annum. The period approaching expiry of the main patent sees a significant change in the management of drug supply. Competition from generic producers affects the amount of drug manufactured, partly controlled by their interest in attacking the market and partly by the originating company's strategy. Over the past decade, only about 25 new chemical entities (NCEs) have been approved to enter the global market each year, and the cost of discovering and developing each NCE has been estimated to be in the range of \$500–2000 million, depending on the therapeutic area and developing company.[3]

As the drug development programme proceeds, the process R&D group will scale-up new synthetic processes from the laboratory, through to the pilot plant and ultimately into full-scale commercial manufacture, if the programme is successful. In the course of this endeavour, the process R&D group will serve three disparate customers:

- The development process requires escalating quantities of the active pharmaceutical ingredient (API) of the requisite purity for toxicological, clinical and other studies.
- Production requires a synthetic route and a robust manufacturing process to operate if the API is commercialised.
- Regulatory agencies require that the developed process is demonstrated to reproducibly produce the API of the approved high quality, when operated within defined and validated parameters.

1.2 Aims and Scope of the Book

The book is aimed at chemistry, engineering and pharmacy under- and post-graduate students and early to mid career professionals, but may also be of interest to those in allied disciplines such as biologists, medical and business people. To adequately cover the field, the book considers each aspect of process R&D more or less chronologically as it occurs during a project. For example, deciding what equipment will be required for manufacture requires good definition of the chemical process, which in turn necessitates a clearly defined route. The interdependency of each area must be recognised and the book will try to point these out within each chapter. Process R&D expertise resides largely within industry, which is why most of the chapters are authored by currently practicing experts. A number of academic institutions and funding organisations now recognise the need for strategic research and specific training requirements to support the pharmaceutical industry.[4]

This book thus differs from most others that have been published in the area as it is organised to try to walk the reader logically through key aspects of the pharmaceutical chemical R&D process, covering the essential aspects encountered in both early and late stage process development toward manufacture. In this way it is hoped that the reader will gain an appreciation of what is involved in working in this environment. Rather than relying on separate case history chapters, this text incorporates mini-case histories within the chapters to bring to life different aspects of the development process. The book aims to provide an overview of:

- How safe and scalable synthetic routes are designed, selected and developed.
- The importance of the chemical engineering, analytical and manufacturing interfaces.
- The importance of the green chemical perspective and solid form issues.

Every pharmaceutical company planning to launch, or be responsible for manufacturing, medicines must employ or have access to process development capability. The consequences of failing to develop manufacturing processes that give consistent, high-quality product can have serious adverse consequences for both the consumer and the producer. Many steps are put in place, by both the company developing the drug and independent regulatory bodies, to ensure this does not occur. These control measures provide the most important framework which the process R&D scientist must be aware of and operate within. However, this is only one aspect of the many varied responsibilities that these teams of skilled development professionals have.

Although this book aims to cover the breadth of process development activities as an advanced single volume text, it will not be able to provide the depth of coverage of a comprehensive handbook. Areas within product development that the book is unable to cover include formulation, packaging and distribution, or wider issues around generic drugs. Furthermore, whilst it is recognised that biological medicines such as vaccines, antibodies, peptides and nucleotide-based therapeutics are an important class of medicines that are being increasingly adopted to treat patients, small-molecule drugs remain the largest part of the market. Since the skills and infrastructure required to develop biomedicine processes are so different, it is beyond the scope of this book to discuss their development and manufacture. The reader interested in this area might refer to the book by Dutton and Scharer.[5]

1.3 Outline of Contents

The starting point for process development usually follows the identification of a bioactive molecule. To increase understanding of its effects, more material is required and commercial pressures often demand rapid scale-up to make kilogram quantities. To produce these quantities the company must be able to access, either internally or externally through contract or collaboration, a laboratory and people having the equipment and expertise. The chemical route is often inherited from the medicinal chemists and, whilst sub-optimal, is often used with the minimum number of changes to ensure safe but rapid delivery. Since the failure rate of early phase drugs is so high, process R&D effort is minimised to avoid wasted resources. Nevertheless, a host of factors need to be considered, including availability of materials, chemical safety and whether to accept chromatographic separations. Looking towards the need for larger quantities of the API, one of the first considerations is whether the route initially used is suitable. Often quite substantial changes are made which markedly improve the efficiency, cost, reliability and environmental impact of the route. Successful scale-up should then entail detailed pre-work to understand, for example, the rates of reactions, pre-equilibria, physical aspects such as multi-phasic mixing and reactor design (an activity involving both chemists and engineers). Even at this stage a broad idea of the type of equipment and methods of separation and purification should be part of the project team thinking. When scaling-up batch reactions beyond a laboratory scale of about 1 L, process safety aspects must be considered, as the ratio of surface area to volume and the ability to remove heat both decrease, and the ability to remove heat is limited. The consequent auto-heating of the reaction can lead to decomposition of any thermally unstable materials present in the reactor, runaway reactions and potentially to an explosion.

As shown in Figure 1.3, late phase clinical trials require substantial quantities of the API, which may require tonne volumes of starting materials in a carefully planned production campaign involving multiple batches.

Since the API is used in human clinical trials, the manufacture requires careful regulation and quality assurance to ensure that the product is fit-for-purpose with a traceable origin. A key advisor in this, and an essential part of the process R&D team at all stages of development and production, is the analyst, responsible not only for measuring the API purity, but also for understanding impurities that emanate from the process. If appropriate process analytical technology (PAT) systems are put in place, with the capability to analyse and interrogate the data, much can be learnt about the process from initial bulk campaigns that can assist in further scale-up. A high-purity API is usually obtained by controlled crystallisation processes in which the crystal lattice rejects impurities. Much science is involved in optimising such processes to ensure that a consistent product is made in high yield. Finally, as the drug moves towards approval, its manufacture must be planned. A dedicated plant is normally employed, and all the process R&D studies to date are used to define its layout and construction. The plant must not only produce the required quality of product to cover predicted future demands, but must also, for example, be designed to be safe, easily cleaned, low maintenance and cost efficient.

The book is organised in the following manner:

Chapter 2 provides some historical background to how pharmaceutical companies themselves emerged. This then leads to an examination of how the discipline of pharmaceutical process R&D has evolved, and the significant changes that have occurred in the past 20 years.

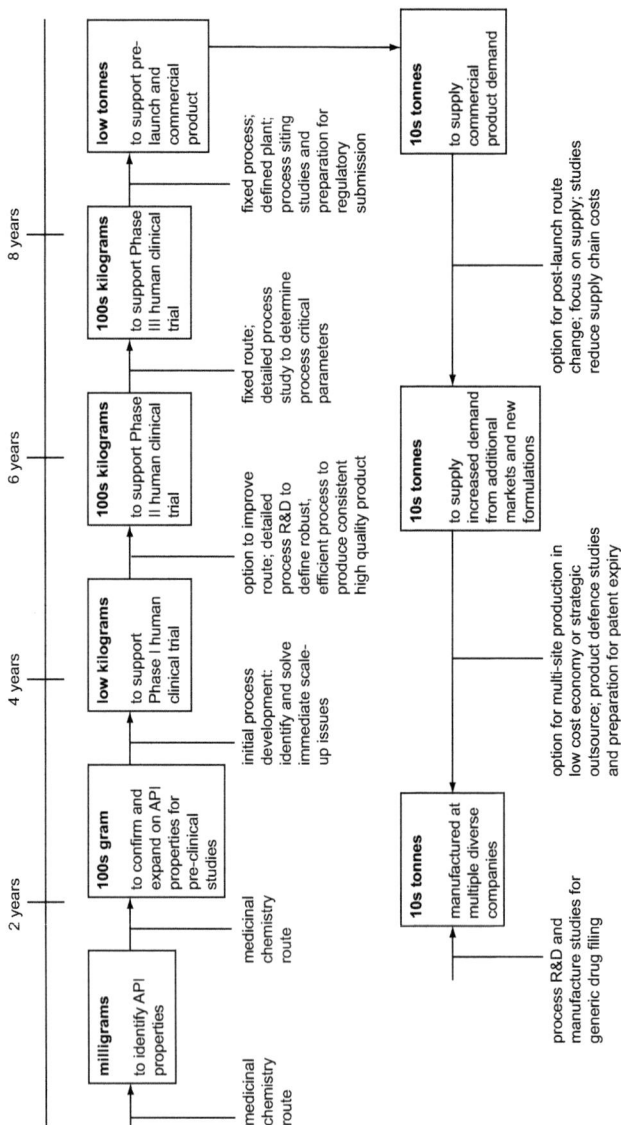

Figure 1.3 Typical volumes of drug required, approximate timescale across the drug lifecycle and some activities required to support the production.

Chapter 3 sets out what a synthetic or semi-synthetic API is, together with some of its typical physicochemical characteristics. An examination of some common features of small-molecule APIs leads into a discussion of how this impacts the design of drug syntheses, and the types of reactions that can be used on scale and frequently occur in API syntheses.

Chapter 4 covers the interface with medicinal chemistry, the high attrition rate in early development and the "fit-for-purpose" strategies that are often used to progress drug candidates through key early decision points with limited resources. The safety, reliability and efficiency criteria that guide decisions about expedient syntheses used to prepare multi-kilogram early development batches are discussed and exemplified.

Chapter 5 picks up the story once the drug candidate has successfully cleared early development hurdles, perhaps achieving clinical proof-of-concept (POC). The factors are discussed which affect the timing and intensity of the search for longer term synthetic routes. For APIs of only moderate complexity, a large number of potential routes will be possible. The criteria used to compare these candidate routes, and decide upon the one which will be developed for registration and production, are laid out with a range of examples.

Chapter 6 focuses on the importance of green chemistry within the pharmaceutical industry; the metrics used to assess the green chemical performance of routes and processes, and the importance of solvent selection, are also highlighted. Three case histories are then used to illustrate the progress that is being made in the "greening" of API syntheses, and the importance of biocatalysis.

Chapter 7 introduces the need for deeper understanding of chemical processes to ensure robust and reproducible operation. The use of different calorimetric techniques to determine reaction kinetics and test mechanistic models, particularly for catalysts, has over recent years been an important addition to the tools available for helping improve process performance.

Chapter 8 considers the crucial aspect of process safety. When up-scaling from the laboratory to plant the impact of process failure increases, and both chemical and operational hazards must be carefully evaluated. The identification and measurement of hazards will lead to adoption of a safe operating regime.

Chapter 9 discusses the key chemical engineering interface by considering the kinetic, thermodynamic and physical aspects of reactions that affect selection of reactor type, feed regime, types of mixing required and understanding of molecular interdependencies. The result is often improved yield, higher selectivities and more robust operation.

Chapter 10 introduces the importance of solvent selection, particularly in liquid–liquid extraction. The design and understanding of these systems is recognised to be crucial in sustainable processing due to its impact upon reaction efficiency, product separation, purification and operability.

Chapter 11 discusses some of the recent innovations and tools being adopted by the process chemist which assist in the gathering and interpretation of data.

Chapter 12 looks at the key interface between analytical and process development scientists, and the essential role analysts play in helping process chemists develop more efficient processes. The evolution of analytical equipment and techniques used to provide more rapid and detailed assessments of product purity, reaction kinetics and the levels of impurities is discussed.

Chapter 13 presents information required about the product in solid form. Detailed understanding of the crystallisation process is essential as this is the primary method for purification of the API, which must be produced to a high specification.

Chapter 14 discusses technical transfer: the end result of successful process R&D, but the beginning of a successful product launch. Many issues such as ensuring regulatory compliance, maintaining a robust process, ensuring supply and managing costs are discussed.

Chapter 15 looks at current trends in the pharmaceutical industry, how these have been impacting the conduct of process R&D, and what are likely to be the future developments and challenges.

1.4   Developed Processes in Exemplar Commercial Drugs

Process development is well illustrated through exemplars and each of the chapters describes aspects of process R&D with real case studies. To support this it is useful to describe pertinent examples of commercial pharmaceuticals: some recently launched ones that use the latest technologies; some older drugs that have fully established manufacturing processes. Figure 1.4 shows, row by row, selected analgesic/cardiovascular, antiviral and so-called lifestyle management

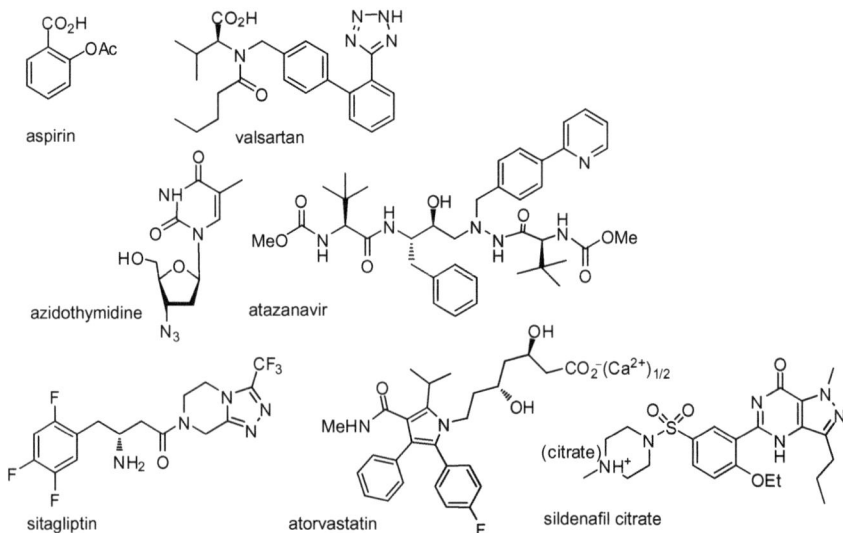

Figure 1.4   Structures of selected analgesic/cardiovascular (aspirin, valsartan), antiviral (azidothymidine, atazanavir) and so-called lifestyle management drugs (sitagliptin, atorvastatin, sildenafil citrate).

drugs, which are illustrative of the trend for increasing levels of molecular sophistication in management of the therapeutic area.

As the molecules increase in their complexity, so do the demands of process R&D. This particularly affects the cost of goods, which is becoming ever more important. The cost of the API as a fraction of the product selling price is generally between 10 and 25%. Formerly this was considered relatively unimportant, with the focus on material supply rather than manufacturing efficiency. However, in more recent times the absolute rather than relative cost of manufacture has been put under the spotlight, particularly for high-volume drugs or those sold into competitive markets. The result is an increased demand on cost control, whose impact is on the efficiency of the process and production. For example, most companies operate continuous manufacturing improvement programmes that examine the costs across the whole supply chain, including the outsourcing of registered starting materials to custom fine chemical suppliers increasingly operating in low-cost economies.

One further aspect of process R&D that requires discussion is patenting. Most important is the composition of matter patent covering the API molecule, because it can be clearly defined and is difficult to obviate. When speaking of drug patent expiry it is normally this patent being referred to. Product application patents that define the therapeutic area are also important, but require evidence of action and can be open to challenge. Besides these, a successful product defence strategy might include other composition patents around, for example, salt form, polymorph, formulation and delivery mechanism. Process patents can be useful to extend the originator's exclusivity after expiry of the main patent. The problems with process patents are: it is difficult to close off effective methods for making the pharmaceutical, since creative chemists will often find alternative routes and processes; by law they disclose the route and process, albeit often with minimal detail; they are difficult to enforce; they can be expensive if the company is maintaining a large portfolio; failure to patent opens the possibility of another applicant filing one as a block, encouraging public disclosure at the opportune time. The area of patenting is complicated and is not covered further here; the interested reader is directed to the book by Grubb and Thomsen.[6]

Aspirin was developed by Bayer at the end of the 19th century and may be one of the most widely ever used drugs, with therapeutic benefits still being identified. The drug is one of the simplest molecules; however, in its time the process R&D was demanding because of the need for large quantities of high-quality material. The Kolbe–Schmitt process of treating sodium phenolate with carbon dioxide is used to this day, with heating to cause rearrangement to sodium salicylate followed by acid precipitation and further purification by crystallisation or sublimation. Subsequent acetylation with acetic anhydride and controlled crystallisation from the acetic acid solution gives technical grade aspirin, which is further purified and formulated. The annual production of aspirin exceeds 20 000 tonnes.[7] Subsequent generations of analgesics, such as the "profens", opiates and COX-2 inhibitors, are more active, selective and molecularly complex, with attendant process development and manufacturing issues.

An example of increasing molecular complexity, and therefore more complex processing, is provided by cardiovascular drugs. Widely used thiazide diuretics, such as indapamide, are relatively simple sulfonamides that are both straightforward to manufacture and low cost to produce. The beta-blockers generally have a C_3, 1-amino-2,3-diol pharmacophore which is relatively straightforward to produce from glycidyl ethers; more recent versions are optically active but do not present any particular difficulty. The peptidomimetic angiotensin converting enzyme (ACE) inhibitors were introduced in the 1980s. The simpler ones are relatively straightforward to produce from amino acids,[8] whilst more recent ones such as fosinopril are more complex and require substantial processing. Other types of anti-hypertensive agent are calcium channel blockers such as the dihydropyridines and benzothiazepines; these typically employ 5–10 chemical stages to produce them. The sartans are angiotensin II receptor antagonists and are generally more complex and more challenging to produce than earlier generations. One example is valsartan (see Figure 1.4), the annual production volume of which may exceed 300 tonnes. Synthesis of the tetrazole moiety, common to most sartans, has exercised process R&D chemists as it requires the use of azide salts that are potentially explosive.[9a] After a number of iterations the reported process now employs tin azide in a process controlled by on-line analytical technology.[9b] The most recent generation of anti-hypertensives are renin inhibitors, exemplified by, first in class, aliskiren. This is a complex structure, which has required creative route design and substantial process development to give a manufacturing process able to generate the drug efficiently at the large volumes required and within the desired cost. Whilst the actual manufacturing route is not known, Scheme 1.1 shows a likely route based on publications.[10]

For many years, one of the few effective anti-virals was GSK's acyclovir, active against herpes simplex and other viruses. With the advent of the AIDS epidemic, GSK's knowledge of nucleoside analogues enabled them to rapidly develop and launch azidothymidine (AZT), which for a number of years was a first-line HIV therapy and gave them a strong market lead in the area. The process required thymidine produced by fermentation and an elegant series of steps to introduce the azido group stereospecifically. Subsequent nucleoside antivirals have become more selective but more complex and yet the processes to make them are still semi-synthetic, as this proves the most efficient means of production. As the mechanism of HIV infection was increasingly understood, the next generation of medicines, protease inhibitors, were invented, developed and brought to market. These peptide isosteres presented a number of process development challenges, including lengthy routes, difficult substitution patterns, multiple chiral centres and complex heterocycles; many routes have been reported and enormous effort expended to develop efficient, cost effective and robust processes.[12] One example is atazanavir; whilst the exact manufacture route is not known, Scheme 1.2 shows a likely route based on literature reports.[11] The results of these efforts are successful medicines that can be afforded and are now being introduced to needful populations in Asia and Africa.

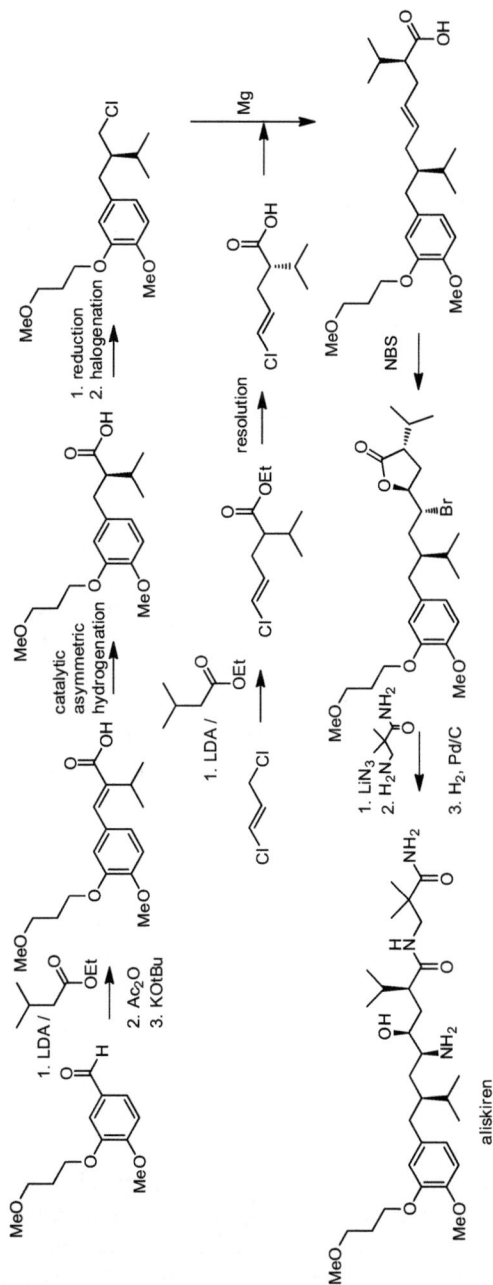

Scheme 1.1 Convergent synthetic route to the complex structure of aliskiren.

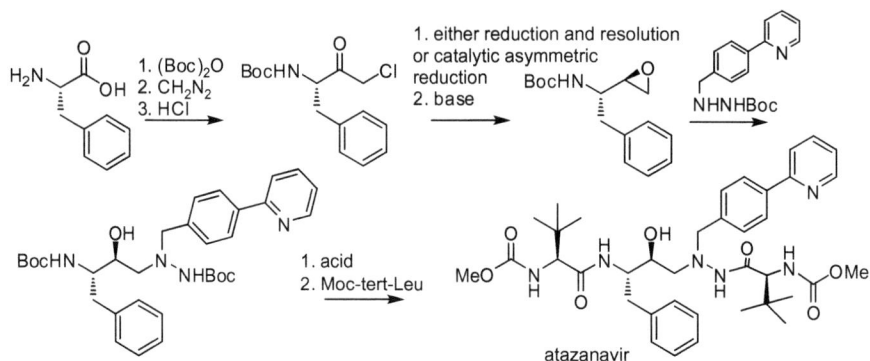

Scheme 1.2   Linear synthetic route to the complex structure of atazanavir.

Atorvastatin, a cholesterol lowering agent and the world's leading selling prescription drug, is a recent example of a high-volume drug for which there have been several generations of ever more efficient processes. Routes to the optically active 3,5-dihydroxyhexanoate pharmacophore have been highly innovative and much effort has been directed at coupling C_5 and C_2 compounds based on acetoacetate and acetate. Processes have required low-temperature reactions in some cases, as well as catalytic asymmetric ketone reductions (both chemo- and biocatalytic) that have stretched the process development community. A recent innovation in this area has been the use of continuous flow processes, enabling both selective reactions and high-intensity production.[13] A further line of process R&D has been the use of aldolase enzymes to carry out stereocontrolled synthesis of the 3,5-dihydroxyhexanoate starting from acetaldehyde in a three-component coupling. This has involved modification of both the enzyme and host organisms to enable processing of the functionalised, unnatural, materials such as a cyanomethyl side-chain to produce the corresponding lactol, (4R,6R)-6-(cyanomethyl)-4-hydroxytetrahydropyran-2-ol. Several further steps produce the required protected version of (3R,5R)-7-amino-3,5-dihydroxyhexanoate that can be coupled with a functionalised diketone to produce atorvastatin. The process improvements have resulted in a substantial decrease in the costs of production and the drug will soon become a generic medicine.

Two other so-called "lifestyle drugs" are worth discussing as they again illustrate the change that has occurred in the last decade towards much more efficient processes: these are sildenafil and sitagliptin. Process improvements in the formation and coupling of heterocycles have resulted in a much less wasteful Pfizer process for sildenafil, that has gone from an initial 1300 L waste per kg product to the vastly improved current commercial process generating 7 L waste per kg product.[14] Sitagliptin is a Merck drug effective for the treatment of type II diabetes. One of the processes challenged accepted wisdom in the type of substrate used in asymmetric hydrogenation. Chemists working on the project found that they were able to enantioselectively hydrogenate an enamine, as opposed to enamides that had always been used up to that point. This enabled the route to be shortened to three stages with two isolations, providing a very efficient process to the API.[15]

Whilst the above examples provide only a snapshot of activities in the process development and manufacture of drugs, they do show the type of issues that are faced by development teams, and the extraordinary level of innovation that is applied to enable high-quality, low-cost, low-waste processes that can be used for possibly decades of manufacture. It is hoped the reader will enjoy understanding what activities are carried out during process development and how these are linked together to produce the desired outcome of a safe and efficiently manufactured product.

References

1. J. A. DiMasi, *Clin. Pharmacol. Therap.*, 2001, **69**, 297.
2. C. G. Smith and J. Y. O'Donnell, *The Process of New Drug Discovery and Development*, Informa Healthcare USA, New York, 2006.
3. C. P. Adams and V. V. Brantner, *Health Econ.*, 2010, **19**, 130; J. A. DiMasi, R. W. Hansen and H. G. Grabowski, *J. Health Econ.*, 2003, **22**, 151–185; J. DiMasi, H. Grabowski and J. Vernon, *Drug Inf. J.*, 2004, **38**, 211–223.
4. www.iprd.ac.uk; http://www.nottingham.ac.uk/dice/; http://www.bath.ac.uk/csct/dtc/; http://engineering.mit.edu/research/labs_centers_programs/novartis.php; http://www.ices.a-star.edu.sg/ices/home.do (last accessed 12 February 2011).
5. R. Dutton and J. Scharer, *Advanced Technologies in Biopharmaceutical Processing*, Blackwell, Oxford, 2007.
6. P. Grubb and P. Thomsen, *Patents for Chemicals, Pharmaceuticals and Biotechnology, Fundamentals of Global Law Practice and Strategy*, Oxford University Press, Oxford, 5th edn, 2010.
7. *Kirk-Othmer Encyclopedia of Chemical Technology*, ed. M. R. Thomas, Wiley, New York, 2006, vol. 21, p. 1.
8. A. Geo and K. Mundinger (BASF), DE 4 407 985, 1993.
9. (a) G. Sedelmeier, *Synthesis of Tetrazoles on an Industrial Scale*, presented at the 26th SCI Process Development Symposium, 10 December 2008, Cambridge, UK; (b) J. Wiss, C. Fleury and U. Onken, *Org. Process Res. Dev.*, 2006, **10**, 349.
10. N. E. Mealy, J. Castanar, R. M. Castanar and J. S. Silvestre, *Drugs Future*, 2001, **26**, 1139; P. Herold and S. Stutz (Speedel Pharma), WO 02 02508, 2002; P. Herold, S. Stutz and F. Spindler, WO 02 02487, 2002 and WO 02 02500, 2002; A. J. Minnaard, B. L. Feringa, L. Lefort and J. G. De Vries, *Acc. Chem. Res.*, 2007, **40**, 1267–1277; S. Hanessian, S. Guesné and E. Chénard, *Org. Lett.*, 2010, **12**, 1816.
11. A. Faessler, G. Bold, H.-G. Capraro, M. Lang and S. C. Khanna (Novartis), WO 97 40029, 1997; Z. Xu, J. Singh, M. D. Schwinden,

B. Zheng, T. P. Kissick, B. Patel, M. J. Humora, F. Quiroz, L. Dong, D.-M. Hsieh, J. E. Heikes, M. Pudipeddi, M. D. Lindrud, S. K. Srivastava, D. R. Kronenthal and R. H. Mueller, *Org. Process. Res. Dev.*, 2002, **6**, 323.

12. K. Izawa and T. Onishi, *Chem. Rev.*, 2006, **106**, 2811.

13. L. Proctor, P. J. Dunn, J. M. Hawkins, A. S. Wells and M. T. Williams, in *Green Chemistry in the Pharmaceutical Industry*, ed. P. J. Dunn, A. S. Wells and M. T. Williams, Wiley-VCH, Weinheim, 2010, p. 221.

14. P. J. Dunn, S. Galvin and K. Hettenbach, *Green Chem.*, 2004, **6**, 43.

15. J. Balsells, Y. Hsiao, K. B. Hansen, F. Xu, N. Ikemoto, A. Klausen and J. D. Armstrong III, in *Green Chemistry in the Pharmaceutical Industry*, ed. P. J. Dunn, A. S. Wells and M. T. Williams, Wiley-VCH, Weinheim, 2010, p. 101.

CHAPTER 2

Process Research and Development in the Pharmaceutical Industry: Origins, Evolution and Progress

TREVOR LAIRD

Scientific Update LLP, Maycroft Place, Mayfield, East Sussex, TN20 6EW, UK

2.1 Historical Perspective

When I was researching this section on the history of process R&D in the pharmaceutical industry, I was surprised to find that the origins of the industry were much older than I thought, with companies such as Merck being founded in 1688 in Darmstadt, Germany, and Geigy (a predecessor of Ciba-Geigy and Novartis) and Allen and Hanbury's (a predecessor of Glaxo) beginning in the 18th century. In Europe, these companies started out making and selling herbal or plant medicines and only later moved into chemically manufactured drugs. Others, such as Hoechst, Bayer and, much later, ICI had their origins in the dyestuffs and chemicals industry, and came later to pharmaceuticals as an extension of these products. Already it had been found in the 1880s that some of these dyestuffs could be used as antipyretics and analgesics (for example, the drug kairin was launched by Hoechst in 1883). By 1888, Bayer had launched the synthetic drug phenacetin, made in three chemical stages (ethylation, reduction, acetylation) from p-nitrophenol (Figure 2.1). They also produced the drug sulfonal, made by the route shown in Scheme 2.1, but because of

RSC Drug Discovery Series No. 9
Pharmaceutical Process Development: Current Chemical and Engineering Challenges
Edited by A. John Blacker and Mike T. Williams
© Royal Society of Chemistry 2011
Published by the Royal Society of Chemistry, www.rsc.org

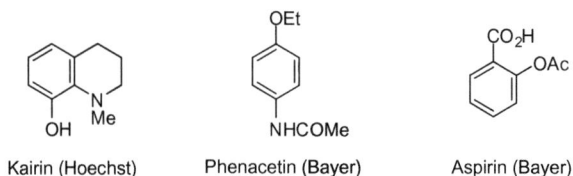

Figure 2.1   Kairin, phenacetin and aspirin.

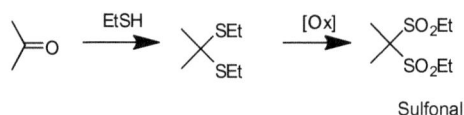

Scheme 2.1   Bayer synthesis of sulfonal.

complaints about the smelly thiol raw material, they had to move production to a more remote location. By 1899, Bayer were producing their major blockbuster aspirin, a drug which is still made in vast quantities ($\sim$40 000 tonnes per annum) using phenol as starting material.[1]

In Germany in the 1880s it was forbidden to patent pharmaceutical final products, so it was particularly important to patent the process for making it. The law changed in 1891, but by that time German chemists were already expert at process R&D and optimisation and this has continued to be the case.

In 1898, for example, Merck already had a central research laboratory in Germany, much of it concerned with introducing new products to the production plant, only handing the processes to production when they were fully developed. The emphasis on process chemistry was partly because Merck purchased the discovery products from outside sources, particularly universities, so the company was much more focused on getting the product into production quickly, and on process optimisation, rather than drug discovery. One of their major products was the barbiturate veronal (diethylbarbituric acid), which Merck estimated first-year demand in 1903 at 500 kg.[2] This quickly increased in 1904 to 5000 kg and to keep up with demand an improved process at lower cost was developed. By 1905, Merck had registered 65 patents on veronal and processes to make it. This was partly because of competition from Bayer, who also had the rights to veronal, and, it is said, had developed a better process to make it.[2]

In the US, most of the large pharmaceutical corporations began life as traditional medicine sellers. Lilly's best selling product around 1900, succus alterans, used as a blood purifier and also for syphilis, was derived from a traditional Creek Indian medicine. Only later did Lilly begin to work with universities and develop their own new products. It was Eli Lilly, the grandson of the founder of the company, who first took an interest in the manufacturing processes for pharmaceuticals, focusing on efficient manufacture, written instructions to process workers (rather than verbal commands), in-process controls and quality control of products, as well as meticulous record keeping. He even introduced some elements of automation and was well aware of the economies of scale. He was fascinated by efficiency in manufacture and had the bare bones of a standard operating procedure for change control.[3]

> *"No changes in details of manufacture or packaging shall be made except by written memoranda. No written memoranda seeking to effect changes in details of manufacture or packaging shall be authority for such changes unless it bears the following stamp."*
> Memo to Supervisors in 1914 composed by Eli Lilly.

Eli Lilly and his father created a new Department of Experimental Medicine and hired PhD scientists who maintained close contact with universities. The most successful collaboration was with Toronto University, leading to the development of insulin and a Nobel Prize for the Canadians. Lilly's contribution to the partnership was to develop methods of transferring from small-scale laboratory quantities to large-scale production, and this was a achieved in 1923, along with record profits. It was the profits from insulin that helped Lilly to create a new research laboratory in 1934 and paved the way for discoveries and processes for the manufacture of antibiotics in the 1940s. It was only later that Lilly embraced more chemical, as opposed to biochemical, pharmaceuticals.[3]

The first general purpose antibiotic used in modern medicine was prontosil, discovered by Gerhard Domagk in 1932 and launched by Bayer in 1935; Domagk won the Nobel Prize in 1939 for this discovery.[4] The formula (Figure 2.2) clearly shows the origins of prontosil in azo dye chemistry, but it is the presence of the sulfonamide moiety that was significant, prontosil being the first member of a class of antibacterial agents known as sulfa drugs. Later drugs did not need the azo group to function and the link between dyestuffs and pharmaceuticals has grown further apart ever since. Nevertheless, the process development aspects of dyestuffs intermediate chemistry, relying on synthetic routes to the basic building blocks of multifunctional aromatic and hetero-aromatic molecules, are still strongly represented in the starting materials used to make pharmaceuticals.

As synthetic chemistry developed, more complex pharmaceutical products became available on a large scale. For example, the adrenal corticosteroids

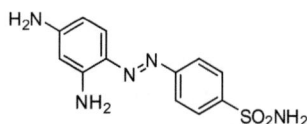

Prontosil

Figure 2.2 Structure of prontosil.

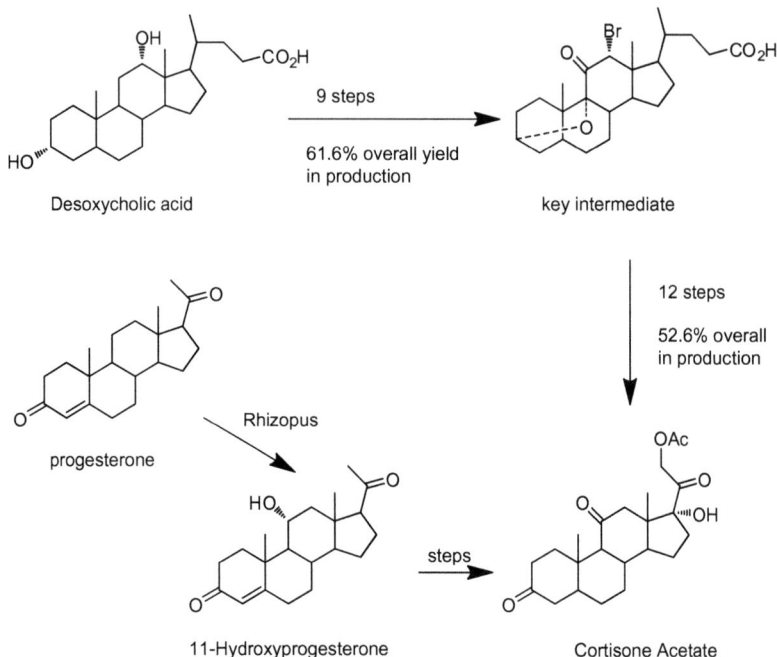

Scheme 2.2 Merck early synthesis of cortisone (and Upjohn shortened synthesis).

were recognised as early as 1927 and synthetic efforts towards cortisone soon followed. Most of the routes began with desoxycholic acid (DCA), easily obtained from cattle bile. A fascinating description of the development of a process to produce cortisone from DCA at Merck reveals the challenges of such a complex synthesis on scale in the early 1950s.[5] In those days, chemical development was carried out without the aid of NMR, mass spectrometry or modern chromatographic methods. Purity was judged by rigorously determined reproducible melting points, optical rotation and UV absorbance. Although thin layer chromatography (TLC) was discovered in 1938 in Ukraine, it was not until the mid-1950s that it began to be used widely in organic chemistry laboratories. The yields shown in a summary of the early process (Scheme 2.2) are therefore all the more remarkable. Subsequently, the development of microbial hydroxylation of steroids at Upjohn, which allowed progesterone, readily available from diosgenin, to be used as raw material, shortened the synthesis considerably.[5]

Many of the reagents and solvents used in the Merck process would not be allowed in large-scale processes today; oxidations were carried out with dichromates, brominations were done in benzene/methanol and chloroform was used for extractions. It was not until a couple of decades later, however, that the environmental impact of industrial processes would start to become an issue.

Most of the development work on cortisone at Merck was carried out on the production site, close to where manufacture was taking place.[5] This was so that any new improvements could be quickly introduced into the plant, in some cases without being piloted first, somewhat different from today's regulated environment. What is similar today is the close relationship between process development groups in industry and academic laboratories, enabling new reagents and methodologies to be quickly imported, improved upon and introduced into pilot plant or production.

Nowadays, most process R&D groups are located on research sites, to take advantage of the excellent spectroscopic and analytical facilities, and they may be thousands of kilometres away from where eventual production takes place. When I first joined Smith Kline and French (SK&F) in 1979, however, the process group was located on the manufacturing site in Tonbridge, UK, although even at that time there was no NMR or MS facility on the site. All drugs were analysed by accurate quantitative TLC, and even now this can be a better method of detecting new impurities compared to HPLC, where late eluters can occasionally go undetected.[6]

The increased complexity of new chemical entities, and the need to identify and characterise even small amounts of new impurities, has meant that modern process groups need to have access to the best instrumentation, and inevitably a location close to discovery chemistry is preferred. The physical separation, however, between process R&D and production can be a disadvantage, since inexperienced chemists may not realise what can cause problems on scale, and what delays processes in plant. Typical problems are physical issues such as emulsions during separations, possibly caused by adding aqueous to organic solutions, rather than the reverse operation, or particle size distribution problems caused by a poor crystallisation technique, resulting in long filtrations and poor washing-out of impurities in filters and centrifuges. There is nothing to beat the experience of a couple of weeks on the night shift in production to help focus attention on these issues and the knowledge gained can assist in developing more efficient and robust processes.

2.2 The Chemical Complexity of Modern Drugs; Implications for Process R&D

The pharmaceutical industry has always been involved with complex chemical products, ever since its roots in natural products and extracts. Fortunately, in the early days of the pharmaceutical industry, most of these complex materials could be produced by extraction and purification, without having to be synthesised from small-molecule building blocks. Even for the first semi-synthetic antibiotics, the penicillins and cephalosporins, the key heterocyclic nucleus was produced efficiently by fermentation and then a few chemical steps were added to convert the key building blocks, 6-aminopenicillanic acid or 7-aminocephalosporanic acid, to the desired new antibiotic. An exception, however, was the steroid family of drugs, which were sometimes manufactured by total synthesis (for example, at Roussel-Uclaf in France[7]).

In the 1960s, there was great interest in both academic circles and in industry in the prostaglandins (Figure 2.3) and many excellent synthetic routes were devised. Subsequently, companies such as Upjohn in the US and ICI[8] in the UK developed products based on prostaglandins and the task of the process chemist was to devise efficient manufacturing routes and processes. Admittedly, manufacture in prostaglandin chemistry usually meant a few kg a year, but even so this was a challenging target. It was, however, made much easier by the excellent academic work that had been published earlier; ICI's synthesis of racemic cloprostanol relied on a modified version of E. J. Corey's route to prostaglandin $F_{2\alpha}$. The challenges of this prostaglandin chemistry are described in the first book to be published on process development, by Stan Lee and Graham Robinson of ICI in 1995.[9]

In the 1980s, a milestone in synthetic pharmaceutical chemistry was the process chemistry and manufacture of imipenem by chemists at Merck. This complex and unstable molecule could not be made by semi-synthesis and an efficient total synthetic route needed to be designed and scaled up for manufacture. One of the key steps in the early development process was the use of a Mitsunobu reaction, only recently discovered in Japan, to invert the stereochemistry at the key position in the side chain. Another innovation was the use of diazo chemistry to effect a desired cyclisation.[10]

Chemical targets for the process chemist continue to become even more complex as natural products once again are shown to have useful properties, but are only available in microgram quantities (for example, from marine sources). The challenging task of the process chemist is highlighted by a series of papers on the synthesis and scale-up of routes to discodermolide (Figure 2.4)

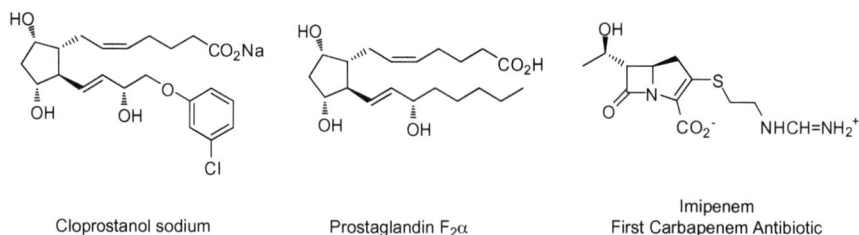

Cloprostanol sodium   Prostaglandin $F_{2\alpha}$ Imipenem
                                                                 First Carbapenem Antibiotic

Figure 2.3   Structure of marketed prostanoids and imipenem.

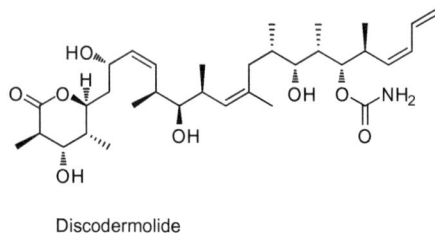

Discodermolide

Figure 2.4   Discodermolide.

at Novartis.[11] Once again, reliance on earlier work in the academic laboratories of Amos Smith in Pennsylvania and Ian Paterson in Cambridge, UK, enabled the process chemists to quickly synthesise building blocks in the laboratory and focus on the development of the later stages of this 30+ step synthesis, before going back later to re-optimise the early stages.

Process chemistry continues to rely heavily on synthetic methods introduced by academic professors, a trend dating back more than a century and emphasised by the speed in which the Wittig alkene synthesis was introduced in the 1950s into vitamin manufacture by BASF, less than two years after George Wittig's discovery of the reaction which bears his name. Today's chemists quickly adapt any new chemistry for their own needs and in the processes under development in the last decade we see lots of cross couplings (Suzuki, Sonogashira, Negishi) and aminations (Buchwald, Hartwig), as well as asymmetric reductions and alkene metatheses using catalysts developed in academic laboratories. The trend towards academic laboratories patenting these reactions and catalysts, however, has not been welcomed in industry, and there has been reluctance to pay license fees[12] for use of these catalysts. Many companies have designed their own catalysts/ligands to get around patents, rather than being held to ransom (from the point of view of being single sourced for a key catalyst/ligand/reagent) by a spin-off company from an academic institution.

In the 1960s, many drugs which contained asymmetric centres were marketed as racemates, and there were only a few cases of products being sold as the single enantiomer, where the undesired enantiomer had been shown to have toxic or otherwise undesirable effects. Most of these single enantiomer drugs were made *via* routes involving a resolution step somewhere in the synthesis, using methods which date back to Pasteur.[13] Nevertheless, with racemisation and recycling of the wrong enantiomer as well as the resolving agent, these processes could be made quite efficient. A good example of this is in the manufacture of (*S*)-naproxen (Scheme 2.3), where an efficient last step resolution/racemisation is key to the process efficiency.[14]

The advent of asymmetric synthesis in the 1960s to 1980s in academic laboratories spurred process chemists to use these discoveries in development and manufacture, and the more stringent regulatory requirements concerning single enantiomers meant that most drugs by the 1990s were developed in single enantiomer form. Many academic discoveries, particularly those from the laboratories of Sharpless, Katzuki and Jacobsen, were quickly translated into kilogram scale processes for the formation of key single enantiomer

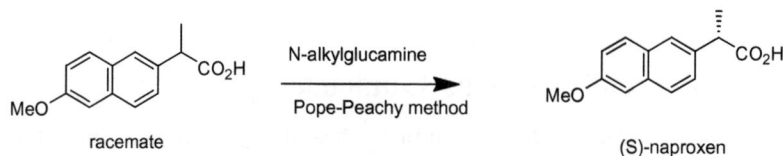

Scheme 2.3   Syntex naproxen resolution process.

epoxide intermediates, which could be further transformed into desired active pharmaceutical ingredients (APIs).[13] Asymmetric reductions, first practised on a manufacturing scale by Monsanto (for the manufacture of L-DOPA) in the 1970s and asymmetric isomerisations, by Takasago in the 1980s (menthol and other products), were built on discoveries of new ligands in the academic laboratories of Kagan (France) and Noyori (Japan). It was fitting therefore that the Nobel Prize should be awarded in 2001 for work in this area to Noyori, Sharpless and Knowles, and a shame (because of the rule that a maximum of three scientists can be awarded the prize) that Henri Kagan, who pioneered the development of chiral phosphorus ligands such as DIPAMP, used in the Monsanto process, was excluded. It was good, however, to see an industrial chemist, Bill Knowles, being rewarded for his work in this area.

One of the areas where attitudes have changed most over the last 20 years in process R&D is the acceptance of chromatography as a viable, and even environmentally attractive (if solvent recovery/reuse is efficient), method of manufacture, particularly for single enantiomer drugs and intermediates; often the "wrong" enantiomer can be racemised and the racemate reused as part of the process. Several single enantiomer drugs such as levetiracetam (UCB) and escitalopram (Lundbeck) are manufactured on a tonne scale from the racemate by chromatographic separation on an expensive chiral stationary phase, mostly using the technology known as simulated moving bed (SMB) or multi-column chromatography (MCC). The MCC technology operated by the French company Novasep, known as Varicol, is more flexible and even more productive for kilogram to tonne scale manufacture than SMB.[15]

The continuous nature of these processes, coupled with efficient racemisation/recycling of the unwanted enantiomer, highly efficient (99.97%) solvent recycling with minimal waste and, perhaps surprisingly, column lifetimes up to six years for dedicated processes, have ensured that these technologies have been the most cost efficient for the manufacture of certain drugs, even beating asymmetric synthetic methods and conventional resolutions.[15]

Nowadays, process chemists in the industry expect to be confronted with new molecules of increased complexity, predominantly having multi-functionalised aromatic or heterocyclic components as well as multiple stereogenic centres. These molecules need convergent synthetic routes where the building blocks either need to be made in-house, or their synthesis can be outsourced to a reputable supplier. Increasingly, these suppliers are from Asia, with China and India having vast numbers of companies dedicated to supplying kilograms of increasingly complex molecules to the pharmaceutical industry, as well as developing processes for their ultimate manufacture should the need arise.

2.3 The Trend Towards Outsourcing

Chemistry in the pharmaceutical industry has always been outsourced; it just depends where you start the synthesis. No-one these days would dream of starting with benzene as raw material for a drug when you could buy a

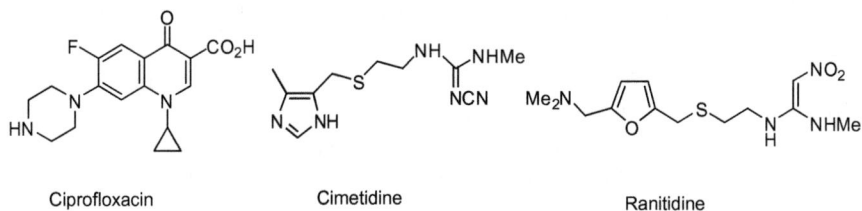

Ciprofloxacin   Cimetidine   Ranitidine

Figure 2.5 Ciprofloxacin, cimetidine and ranitidine.

disubstituted benzene as raw material. In the past, much of the molecular structure of a new drug was made by in-house manufacture, and most companies made their own drugs at in-house sites all over the world. An exception was Wyeth, who used a number of competent contract manufacturers in Europe and elsewhere to manufacture their APIs in the 1970s and 1980s. Other companies such as SK&F, Glaxo, Beechams and Wellcome (now, sadly, merged into one enormous conglomerate, GSK) would outsource manufacture of very early stage intermediates and manufacture the last few stages themselves, whereas those pharmaceutical companies that arose from the dyestuffs industry, such as ICI and Bayer, manufactured almost exclusively in-house. Thus for the drug ciprofloxacin (Figure 2.5), Bayer developed novel processes for the manufacture of the key raw material, cyclopropylamine, in-house.[1]

An important reason why companies used external contract manufacture in the 1980s was if the chemistry was potentially hazardous or smelly, or involved a technology that was unique. Thus companies such as Fine Organics in the UK built up skills in carbon disulfide chemistry to service part of the requirements for key intermediates for the blockbuster drugs cimetidine and ranitidine (Figure 2.5). In the 1970s and 1980s, SK&F outsourced an important sodium/liquid ammonia process to Lonza (Switzerland), involving reduction of an ester to an alcohol, before subsequently building their own facilities in Puerto Rico. This reduction step had previously been carried out with lithium aluminium hydride, consuming 95% of the world's supplies of this expensive reagent, but cheaper processes such as electrochemical reduction were piloted before finally settling on sodium/ammonia reduction as the best alternative. Even then, these processes were all eventually abandoned when the key alcohol intermediate became readily available by non-reductive processes involving hydroxymethylation of 4-methylimidazole.[16]

The rise of the small drug discovery companies, often misnamed "biotechs" but more appropriately termed emerging pharmaceutical companies, with no scale-up facilities at all, has spawned the emergence of new contractors with kilogram laboratory facilities who partner the emerging companies and deliver kilograms of high-quality API, as well as developing the processes for its manufacture. Larger pharmaceutical companies have examined this business model and have seen some benefits. The modern philosophy seems to be that a pharmaceutical company is responsible for discovery and development, but essentially all services, especially manufacture and possibly process R&D, can be outsourced. Resources can be purchased as and when needed, and costs can be saved by going to cheaper manufacturing areas, especially China and India.

In reality, it is not so simple as this and many process development chemists spend weeks of their time managing remote outsource projects when it would be quicker, possibly cheaper, but definitely more reliable, to do the work in-house. Not only that, but a lot of the valuable process information would be retained by the chemists in the pharmaceutical company, rather than in the minds of the chemists in the outsource company. This loss of information on how to develop and manufacture the company's own drugs is severely undervalued. At the time of writing, there seem to be some concerns resulting from poor experiences of some companies in Asia, with a variety of issues such as delays in deliveries, the appearance of new impurities in developmental drugs and poor physical form of APIs, so maybe there will be a backlash towards doing more critical work in-house, or externally in Europe or USA, in the future.

2.4   The Improvement in Process Safety Evaluation

Perhaps the most significant change to the process chemistry aspect of the pharmaceutical industry over the last 40 years has been the emphasis on safety testing processes from the point of view of thermal hazards and runaway reactions. A fundamental change occurred in the 1970s when, after several runaway reactions in the UK,[17] USA, Germany and elsewhere, mostly involving nitration chemistry or reduction of nitro compounds, a number of companies began investigating how these could be prevented in the future. More importantly, they began talking to each other and sharing information. It was mostly chemical engineers and institutions such as the Institute of Chemical Engineers who pushed this agenda in the UK by holding "Hazards" conferences, but safety legislation in many countries helped to accelerate the process of hazard evaluation. In the mid-1970s, in my time at ICI in Manchester, the company set up a central company-wide function for Thermal Hazard Evaluation headed by Norbert Gibson, who with his team developed many methods to assess the thermal hazards of processes and chemicals involved in these processes. His work was continued by the late Richard Rogers, who produced the classic text, with John Barton of the UK Health and Safety Executive (HSE), on *Chemical Reaction Hazards*.[18] Other companies such as Hoechst also went down the same track, and by 1976 already had a programme for safe running of reactions, founded by a Dr Schaefer (mentioned in another classic text[19] from Dr Theodor Grewer, formerly of Hoechst). The latter text, less widely known than Barton and Rogers, is much more appropriate for the organic chemist wishing to learn about thermal hazards, as is the earlier work by Yoshida on *Safety of Reactive Chemicals*.[20]

Also in the 1970s, companies such as Sandoz and Ciba-Geigy in Switzerland developed in-house calorimetry instrumentation and, rather than just use the instrumentation themselves, they made the decision to commercialise and market the products. The Contalab, made by Contraves from a Sandoz design, became available in the early 1980s as an automated reactor with calorimetry option, being able to follow heat release in a reaction with time (the instrument was subsequently sold to Mettler). Shortly afterwards, Mettler, then a Ciba-Geigy subsidiary, launched the RC1 reaction calorimeter, based on early prototypes at Ciba-Geigy. This useful piece of kit eventually became the workhorse of hazard evaluation of processes the world over, as well as being useful in the automation of process development.[16] What became clear from the early studies was it was not just sufficient to know the exothermicity of a reaction, but to know exactly *when* this heat was released, *how fast* and *whether* there was any delay or accumulation. A knowledge of the complexities of mechanistic organic chemistry, and the kinetics of each reaction step, helps to understand the data revealed by these calorimeters, and enables a true hazard assessment not possible before with manual systems such as Dewar calorimeters. As with all areas of process chemistry, detailed knowledge and understanding leads to better control, improved batch-to-batch consistency and therefore improved safety awareness.[16]

Dow in the US had also been heavily involved in assessing process hazards and had developed instrumentation which eventually became known as the Accelerated Rate Calorimeter, still being used today. The small-scale evaluation is particularly valuable when only grams of intermediates are available; a major advantage, compared to the RC1 and larger scale calorimeters, is that it also allows the exotherm to proceed in a "bomb" and measures data on the actual runaway, which is extremely useful in assessing a worst case scenario (time to maximum rate, pressure build-up, and other data). Other companies have designed and sold similar equipment and so these days most pharmaceutical companies have well-equipped hazard evaluation laboratories, and the standard procedures to go with them, to evaluate every stage of the process before scale-up. For the smaller company, it is possible to outsource this work to a specialist company (for example, Chilworth Technology or Hazard Evaluation Laboratories in the UK), or to ask the outsource provider to do the work in-house.

Despite the detailed knowledge built up in the 1970s and 1980s, runaways continue to occur, sometimes when using materials that are known to be problematic, or where there are already data from within the same company on an earlier runaway. Trevor Kletz, another pioneer in process safety,[21] terms this phenomenon "organisations have no memory", and emphasises the need for each generation of new recruits to industry to be comprehensively trained in safety, especially since most university education (particularly for chemists) is weak in this field. Only then will disasters be avoided.

Unfortunately, even today, many outsource companies do not have these facilities as standard, and because mostly they work at the early stages of development at a low kilogram scale, often with dilute inefficient processes,

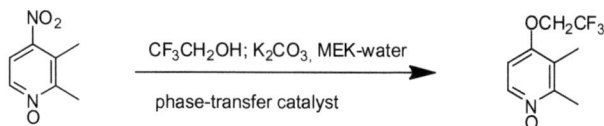

Scheme 2.4   Lansoprazole process.

very few problems such as runaways are seen. Occasionally, runaways are still being reported with nitro compounds; for example, a runaway occurred during the reaction shown in Scheme 2.4, to make an intermediate in the manufacture of the anti-ulcer drug lansoprazole at a Spanish generic company.[23]

There is, fortunately, modern comprehensive literature on the subject of thermal hazards in an excellent compendium entitled *Bretherick's Handbook of Reactive Chemical Hazards*, the first edition of which came out in the 1980s and the 7th edition (over 2600 pages) in 2006.[22] Les Bretherick, who worked at BP in the UK, did process chemists and engineers a tremendous service by compiling this first edition and Peter Urben, who has continued the compilation after the death of Dr Bretherick, is also to be commended for not only updating the compendium but also doing it with quirky humour. This database is now accessible on-line. The American Chemical Society (ACS) journal *Organic Process R&D* also publishes safety highlights in a special safety edition of the journal in the November/December issue each year, beginning in 2002.[24]

It is sad to see, however, that there are many chemists and engineers who, despite the availability of excellent literature, still remain ignorant of good practices in safety of scale-up and we still hear of unfortunate, but totally avoidable, deaths of employees working on processes which are highly dangerous. Many of these are surprisingly in the USA. Our knowledge of these incidents may be partly because US incidents are more widely publicised, with some investigations and findings freely available from the US Chemical Safety Board.[25] The reports from this body make excellent reading and their graphic videos can give out important training messages when shown to staff. In contrast, the UK government HSE does not disseminate the detailed findings of investigations so that all can learn from others' mistakes. UK scientists and engineers usually find out about incidents from what they read in the local or chemical press or on the internet. The internet, however, is a good source of information about runaway reactions and reveals that some manufacturers still have a lot to learn in terms of process safety. In the USA, over the last 20 years there have been 108 deaths associated with reactive chemical hazards, most of them avoidable, and there continue to be large numbers of incidents reported from all over the world.

When auditing a company for a potential supply of intermediates or APIs, therefore, it is vitally important to find out their policy and practices on thermal hazard evaluation, since if there is a runaway, this can affect security of supply.

Auditing for adherence to GMP is not sufficient; the audit needs to be carried out by a competent chemist or engineer, in addition to a QA or regulatory specialist. One US pharmaceutical company had three different APIs affected by three runaways/explosions at different factories in 2005, when key raw materials and intermediates suddenly became unavailable.

2.5 The Environment, Effluent Minimisation and Green Chemistry

Another major change over the last 30–40 years has been an increasing emphasis on the environmental aspects of process chemistry and on that much over-used word "green" chemistry. In the past, processes would be optimised to make them less costly and more efficient by increasing the yield at each stage and maximising the space-time yield (STY), thus minimising solvent and aqueous waste. Thus, in the past, environmental improvement was a by-product of cost reduction. Nowadays it is the other way around, with cost improvements often being a by-product of green chemistry changes. Bulk and fine chemicals have achieved a great deal in this area over the past 40 years and pharmaceuticals are only, in the last decade, beginning to catch up.[26]

One of the pioneers of the environmental movement in process chemistry was Roger Sheldon, formerly a process chemist with Shell and Andeno (now DSM) and subsequently a professor at Delft University of Technology in The Netherlands. It was his introduction of the *E* factor (kg waste per kg product; see Section 6.3.2) which emphasised that, compared to the bulk chemicals industry (*E* factor < 1–5), the fine chemicals sector (*E* factor 5–50+) and the pharmaceutical industry (*E* factor 25–100+) were dirty industries in terms of waste, in contrast to the normal clean image projected by pharmaceuticals. Of course, the *E* factor represents the actual amount of waste produced, defined as anything but the desired product, so naturally the pharmaceutical industry with its complex molecules and multi-step syntheses was going to come out badly in this calculation. However, it made process chemists and engineers in industry think about other reasons why the *E* factor was so high, and whether something could be done to change the attitudes in the industry.[27]

First of all was the attitude to catalysis. In the pharmaceutical industry, most processes used stoichiometric quantities, or even excesses, of reagents, and the by-products from these reagents, sometimes heavy metals, ended up in the effluent; often this waste needed expensive treatment. If these processes could be rendered catalytic, then not only could effluent be reduced, but manufacturing costs could be significantly lowered. Whilst this was not so important for drugs still under patent exclusivity, for the generic pharmaceutical industry cost reduction is paramount. Sheldon's experience at Andeno/DSM had been partly in the manufacture of generic APIs and intermediates, and he also retained important knowledge of catalysis in the manufacture of bulk chemicals from his days at Shell. This led him to promote the use of catalytic processes in fine chemicals and pharmaceuticals.[27] He introduced the term "atom efficiency" to relate the amount of product to the total amount of by-products produced in a chemical reaction; around the same time, Trost[28] introduced a similar term, "atom economy" (see Section 6.3.1), to assess synthetic chemistry in environmental terms.

A major factor in the environmental impact of a multi-stage pharmaceutical process is the amount of convergence in the synthetic route, as well as the number of synthetic steps and the yield. The more convergence, the lower the overall mass of the intermediates produced, and hence proportionately less effluent should be produced. While convergence has always been recognised as important for many reasons in process development and manufacture, the green chemistry movement has re-emphasised the significance for the environment. Fortunately, the advent of combinatorial chemistry in drug discovery means that modern drugs are usually made with convergent methods, and for reasons of economy, as well as the ease of outsourcing key intermediates, often classified as regulatory starting materials, this aspect of the industry has been well taken care of. However, the environmental impact of outsourcing is rarely assessed, and in practice the potential for pollution is effectively transferred to those countries who are adept at supplying building blocks and key intermediates, namely India and China. More needs to be done in these countries to bring environmental standards up to those of Europe, but it is good to see that some Indian generic companies are already adopting green technologies.[29]

Where the pharmaceutical industry still needs more focus is on the work-up and product isolation of intermediates and final products, particularly in the early phases of manufacture. Too often, particularly when dipolar aprotic solvents are so often used for reactions, the easy option in work-up is an aqueous quench followed by extraction with a second solvent. These processes generate huge amounts of waste which is not so easy to recover and reuse. Rather than use aqueous quenches, some companies have focused on direct crystallisations of intermediates and final product APIs from reaction media (sometimes called bottom-up or direct-drop processes),[30] in which water is avoided and the solvents (not now contaminated by water) can be better recycled. This may mean using a mixture of solvents for the reaction, with the ratio designed to maximise product recovery (Scheme 2.5).[31]

In the last decade the trend in academic green chemistry has been towards carrying out reactions in water,[32] but too often, in the experimental detail of papers, one finds that the product is isolated by solvent extraction, thus negating the environmental advantages. Nevertheless, organic chemists have

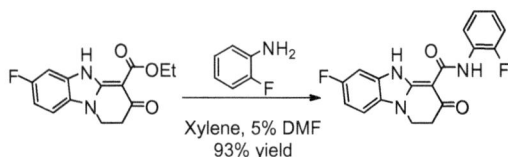

Xylene, 5% DMF
93% yield

Scheme 2.5   Use of mixed solvents to improve yield/productivity.

been surprised at the range of reactions that can be carried out in the presence of water, or, for even more environmental benefit, in the absence of solvent. Academic endeavours in this field can, however, fail to take account of the importance of mass transfer and the mixing sensitivity, which affects selectivity of many reactions on scale-up, and some procedures can be potentially dangerous to scale further, such as the preparation of Grignard reagents under highly concentrated conditions.[33]

The trend more recently in the last couple of decades, towards an increased emphasis on redevelopment of processes after launch of the API (sometimes called second-generation process development) – a subject often ignored by big pharma in the past until generic competition came along – has been a godsend for green chemistry. The opportunity to completely redesign a synthetic route to a key intermediate, or even an API, was frowned upon in many large companies in the 1980s because of the fear of introducing new impurities into APIs. This fear was irrational, since this was essentially what the generics industry did and the best generics companies always produced APIs of comparable quality with no new impurities over 0.1%.[34] However, now big pharma has embraced this idea with spectacular success, none more so than Pfizer.[35] One of the beneficiaries of this new trend has been biocatalysis (also touted by Sheldon in his early essays on green chemistry).[27] Of course, not all biological processes are environmentally friendly and many are run quite dilute. It is the advent of new enzymes or whole cells derived by directed evolution, DNA shuffling or similar processes that has enabled biocatalysis to provide efficient second-generation processes in such a spectacular way. The ability to design into biocatalysts some properties such as stability, or lack of substrate/product inhibition has allowed more efficient biocatalytic processes to be developed with excellent space-time yields. These methodologies are now being used in the manufacture of low-cost generics as well as in second-generation API processes, and are prominent in fine chemicals manufacture too.[36] There are now numerous examples of the use of ketoreductase (KRED) enzymes for asymmetric reduction of ketones, for example in the manufacture of atorvastatin[37] and montelukast intermediates[38] (Scheme 2.6).

The setting up of the Green Chemistry Institute Pharmaceutical Round Table[39] means that companies are sharing information and objectives and proposing targets for the future. Companies such as GSK, Lilly and Pfizer now have environmental targets for their processes in late development. We have indeed come a long way in the last 20 years – attitudes are changing at last.

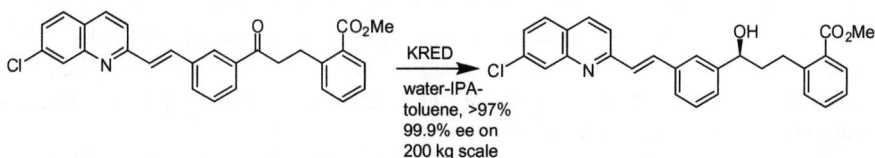

Scheme 2.6 Montelukast intermediate synthesis.

2.6 Significant Changes in Process R&D in the Last 20 Years

One of the most rewarding changes to take place, particularly in the last 20 years, has been the bringing together of organic chemistry with chemical engineering, not only in industry but also in academia. In previous decades these two important disciplines, particularly in UK universities, were in different faculties and contact was minimal. The advent of process chemistry has drawn these two disciplines together so that there are many cases of collaboration, usually on issues which affect process chemistry. It is therefore pleasing to see the new Institute of Process Research and Development (iPRD) at Leeds University,[40] set up by both chemists and engineers, gaining funding for new laboratories and prospering. Similar academic laboratories for process chemistry exist at the University of Zurich under the direction of Professor Jay Siegel.[41]

Organic process chemists now need to know much more than the organic chemistry they learned at university. Modern process chemistry in the last 20 years requires a knowledge of statistical methods of optimisation, often named "design of experiments" (DoE),[42] not only to enable true optimisation of reactions to be achieved, but also because regulatory authorities require processes to be validated, critical process parameters to be assessed and proven acceptable ranges to be established within which the process is robust. The change in emphasis of regulatory authorities from post-analysis approval of batches towards quality by design (QbD), in which increased process control and understanding are required, will only accelerate the trend towards use of DoE for fine tuning and quantitating the effect of parameter change on chemical reactions. What a pity, therefore, that this well-established and most useful interdisciplinary methodology, often taught in chemical engineering courses, does not usually appear on the syllabus of academic chemistry courses in the UK.

In the 1980s, practical organic chemistry was carried out in much the same equipment as in the 1880s, the major difference being a ground-glass joint in the flask instead of having to use connections *via* bored cork or rubber stoppers. Other disciplines (especially biochemistry and analytical chemistry) had moved into the modern era with lots of instrumentation, but organic chemistry had been left behind. In the 1990s, partly as a result of the expansion of combinatorial chemistry and the availability of automated equipment,[43] the capability of carrying out process experiments in parallel was developed. Since then, screening for catalysts, ligands and reaction conditions has been so much easier, and results have also been more comprehensive and meaningful. DoE experiments can now be conducted in small-scale vessels in parallel, saving much time and labour. We have come a long way in the 20 years since the early attempts at automation – many of these in-house work in pharmaceutical companies – were described.[16]

The expectation that organic chemists and chemical engineers will work together in teams, together with the increase in available instrumentation, has also enabled chemists to challenge the convention that small-scale processes should always be run batchwise (or semi-batch if dosing of reagents is involved). Small-scale continuous processes are now used in the fine chemicals industry for tonnage and even small kg amounts of intermediates, using either off-the-shelf or specially designed multi-purpose glass or Hastelloy reactors,[44] or even just plug-flow tubular reactors. Continuous processes are inherently safer than batch processes for hazardous materials and some companies have used this to advantage to manufacture and use reagents such as diazomethane, a reagent used in the synthesis of HIV drugs, routinely.[45] Whilst continuous processes are not ideal for all reactions, particularly those involving solids and viscous liquids, they could be more widely used. They may enable exothermic reactions to be carried out at more concentrated levels since efficient heat removal, which can affect selectivity of such processes, is likely to be much improved in continuous mode.[46]

Automation has also assisted in the screening of new chemical entities for improved salt forms[47] and new polymorphs.[48] It was the serendipitous discovery in the 1980s of a new and patentable polymorphic form of the anti-ulcer drug ranitidine, with improved properties, that enabled Glaxo to extend the patent lifetime of their blockbuster drug and delay introduction of generic equivalents. This patent is estimated to have generated billions of dollars in profit for GSK in the last two decades.[49] Since then the subject has been a hot topic in both the innovator and generic arms of the drug industry, with generic companies screening for new forms using unusual solvents and patenting many new forms of approved drugs. At the same time, both small and large pharmaceutical companies developing new drugs have become aware of the need to carry out early screening of polymorphs, using both in-house facilities or contract companies set up specially to deliver these services. Despite this attention to early-phase screening, companies still find surprises with new, more stable forms appearing late in the development. Often this can be in phase III, but occasionally cases are described where a new stable form appears after hundreds of batches have been manufactured. This is attributed to the influence of trace amounts of impurities which can inhibit or promote the nucleation or growth of particular forms.

The case of Abbott's ritonivir,[50] where the appearance of the new, less soluble form II caused issues with the formulation of the drug product, and subsequent temporary withdrawal from the market, is estimated to have cost the company millions, possibly billions, of dollars in lost revenue. There are now many well-documented cases where similar problems have occurred in late development[51] and I have heard of one where it occurred during the validation campaign. Generic companies have now built up expertise in making and developing stable formulations of metastable polymorphs to allow early entry on to the market, but manufacture of these less stable forms can be problematic.[52] This subject remains a fascinating area for academic and industrial research and readers are referred to special issues of *Organic Process Research & Development* and some excellent books for further examples.[48,49,53,54]

Academic chemical engineering groups have often focused their studies on the crystallisation steps of processes for the production of APIs, and in the past two decades significant advances have been made in the understanding of nucleation and crystal growth (see Section 13.3.5). From this knowledge, the optimal way to crystallise batches, with controlled cooling profiles and seeding at a specific point in the process, using specially prepared seeds with controlled surface characteristics, has emerged.[55,56] Chemists, however, often prefer to design crystallisations using an anti-solvent addition to a solution of the drug in a solvent since this can give optimal yields. Engineers have now developed more understanding of how the various process parameters affect this process and the physical properties of the product, particularly the particle size distribution and surface characteristics of the crystal.[57] There is, in the industry in the last decade, a much better understanding of how to ensure that the drug substance meets the needs of the pharmacist who has to formulate the API, and on the particle requirements for improved formulation.[58]

2.7   The Literature of Process Chemistry

Perhaps the most significant change for the process R&D community over the last 30 years has been the attitude of industry towards allowing publication of process-related papers and towards scientists giving presentations at conferences. As a result, there are now a large number of books on process chemistry, a prestigious journal and lots of conferences and courses covering all aspects of this wide-ranging subject. Apart from one or two articles in *Chemistry and Industry*, resulting from the highly successful Society of Chemical Industry (SCI) Process Development Conferences, begun in the early 1980s, and a few papers in *Journal of Organic Chemistry* and *Tetrahedron*, there was little process work published prior to 1990. One of the first comprehensive articles to set down the principles of process chemistry, development and scale-up appeared in the six-volume work *Comprehensive Medicinal Chemistry* in 1990.[16] Unfortunately this chapter is not discovered in a *SciFinder* search of the subject, so has not been widely read or cited. In it the author points the way forward by advocating the importance of chemical engineering issues, continuous processes, process hazards and safety, statistical methods of optimisation, automation, and crystallisation and polymorphism, as well as innovative organic chemistry. Today all these subjects are regarded as the realm of the process chemist but in the 1980s the focus was much more on plain organic chemistry.

The first small book to appear on process chemistry was that of Lee and Robinson (ICI in those days, now AstraZeneca) in the Oxford Primer Series in 1995,[9] but the best book on the subject, by Anderson,[59] did not appear until 2000. Ex-BMS scientist Neal Anderson's approach to process development is somewhat manufacturing orientated, and all the better for that. The focus is on producing processes that will scale well, on problem solving and particularly on process understanding and control. Since then, many new books have appeared covering this important subject[60–67] and more are planned (for example, a new edition of Anderson). There has even been a special issue of *Chemical Reviews* devoted to process chemistry.[68] One mammoth tome which deserves a mention is the single author work by the Indian chemist C. Someswara Rao[69] entitled *The Chemistry of Process Development in the Fine Chemical and Pharmaceutical Industry*, which at almost 1300 pages is like the bible of organic process chemistry. It is also extremely well referenced, with many of the references being to the journal *Organic Process Research and Development*, commonly known as OPRD, which began publication in 1997.[70]

The origins of how this publication came into being are interesting. I and a number of other chemists and engineers on the SCI Fine Chemicals committee, who had been responsible for the expansion of the SCI Process Development symposium from a one-day event in London to a residential three-day event in Cambridge, felt there was a need for such a process chemistry magazine or journal in the 1980s, particularly since SCI's magazine *Chemistry and Industry* was no longer interested in publishing papers from the symposium. However, SCI had an existing publication, *Journal of Chemical Technology and Biotechnology* (JCTB), which they felt could be used to publish process development work; the Fine Chemicals committee members, who rarely consulted JCTB, disagreed and the project was shelved. Subsequently, with my encouragement, Elsevier developed a similar idea for a magazine/journal devoted to process chemistry in the early 1990s, and even though the market survey looked good, the project was again shelved. Meanwhile, a group of US chemists who were involved with organising the annual one-day Mid West Pharmaceutical Process Chemistry Conference (MPPCC) had started to approach the American Chemical Society (ACS) about publishing a process chemistry journal. On one of my visits to the MPPCC in 1993, ACS was persuaded to undertake a feasibility study for a new journal. After strong advocacy from the then President of the ACS, Ned Heindel, the journal was launched in 1996 and I was installed as the Editor-in-Chief.

The journal provides the opportunity for industrial chemists and engineers to present their detailed work in a publication which is focused on process chemistry, rather than in an organic chemistry journal which does not wish to include discussions on important process issues such as impurity control, work-up and product isolation, scale-up, process safety or crystallisation and physical form issues. The presence of the journal has given the subject "academic respectability" and has helped promote the subject in universities. It has become a popular journal with university professors as well as in industry, and with medicinal chemists as well as process chemists and engineers.

The journal has also helped to create a process chemistry community which often meets at the increasing number of process events held at various parts of the world each year; for example, there is now a well established Japanese Process Chemistry Society which holds a very popular annual conference (in Japanese). The SCI Process Development Conference, really the starting point for the development of a process chemistry community in Europe, is still going strong and in its 28th year, while Scientific Update's twice a year OPRD conferences, which began in the late 1990s, have been held in the USA, Canada, Hong Kong and India, as well as in European venues. Even so, there is still room for smaller regional European events such as the Nordic Process Chemistry Forum. All these events provide a meeting for chemists and chemical engineers to get together to discuss process R&D and related issues and help build up the knowledge base for future generations.

2.8   The Future

For the large pharmaceutical companies, the main aim for the future must be to launch more drugs on the market and have less projects fail at the expensive phase III stage. If more drugs reach phase III, then this means process R&D departments will be much more involved in late-development activities, than as now, in quick (and dirty?) scale-up to make 1–2 kg by discovery-like synthetic routes. With the emphasis from regulatory authorities on QbD and process understanding, it will be an opportunity for chemists to become more involved in detailed development work (if the synthetic route is acceptable for manufacture) and technical transfer to manufacture or in second-generation process R&D (if the route needs changing) followed by transfer to manufacturing sites.

For those in emerging pharmaceutical companies, the challenges remain the same, but with limited resources, and to develop detailed process understanding (and retain that knowledge) using external resources is much more difficult. Choosing a good outsource partner is the key to success, and this means auditing those companies on the basis of scientific excellence, not just regulatory concerns; companies need to have good R&D proficiency with talented scientists and engineers, as well as good manufacturing capability.

For companies carrying out contract synthesis, process R&D and manufacture, the future looks particularly rosy, since large pharmaceutical companies (such as AstraZeneca, GSK, Merck and Pfizer) are reducing their scientific staff, and this inevitably means more outsourcing. Whether contract work will continue to be carried out in Europe and the USA or the trend towards moving work to India and China will further accelerate or, because of quality, delivery or communication problems, decelerate, remains to be seen. With costs increasing in Asia much faster than in the depressed Western world, then it will become progressively harder for Asian companies to compete. Asian companies who are not already proficient in DoE studies, parallel experimentation for true process optimisation and determination of proven acceptable ranges, chemical reaction engineering for scale-up studies, crystallisation/polymorphism control, *etc.*, will need to build up their skill levels to retain business; they cannot afford to assume that the good times will continue, without further investment.

For those postgraduates wishing to begin a career in process R&D, it is a difficult time, but jobs in the contract companies and emerging pharmaceutical companies should be available even if large pharma is not recruiting. Therein lies the hope for the future.

References

1. E. Verg, G. Plumpe and H. Schultheis, *Milestones; The Bayer Story, 1863–1988*, Bayer, Leverkusen, 1988.
2. C. Burhop, *Pharmaceutical Research in Wilhelmine Germany: The Case of E. Merck*, Max Planck, Bonn, 2008 (www.coll.mpg.de/pdf_dat2008_03online.pdf; last accessed 15 July 2010).
3. J. H. Madison, *Manufacturing Pharmaceuticals: Eli Lilly and Company, 1876–1948*, presented at the Business History Conference, 1989 (www.hnet.org/~business/bhcweb/publication/p0072-p0078.pdf; last accessed 15 July 2010).
4. For a review of antibiotics synthesis, see K. C. Nicolaou, J. S. Chen, D. J. Edmonds and A. A. Estrada, *Angew. Chem. Int. Ed.*, 2009, **48**, 660–719.
5. S. H. Pines, *Org. Process Res. Dev.*, 2004, **8**, 708–724.
6. T. Laird, *Org. Process Res. Dev.*, 2006, **10**, 167.
7. For a review of enantioselective synthesis of steroids, see A.-S. Chapelon, D. Morelada, R. Rodriguez, C. Ollivier and M. Santelli, *Tetrahedron*, 2007, **63**, 11511–11616.
8. ICI's product, cloprostanol sodium, was the active ingredient in Estrumate, a veterinary product for controlling oestrus in cattle.
9. S. Lee and G. Robinson, *Process Development: Fine Chemicals from Grams to Kilograms*, Oxford University Press, Oxford, 1995.
10. E. J. J. Grabowski, *Chirality*, 2005, **17**(S1), S249–S259.
11. S. J. Mickel, D. Niederer, R. Daeffler, A. Osmani, E. Kuesters, E. Schmid, K. Schaer, R. Gamboni, W. Chen, E. Loeser, F. R. Kinder Jr., K. Konigsberger, K. Prasad, T. M. Ramsay, O. Repic, R.-M. Wang, G. Florence, I. Lyothier and I. Paterson, *Org. Process Res. Dev.*, 2004, **8**, 122 (and previous papers in this series).
12. J. M. Hawkins and T. J. N. Watson, *Angew. Chem. Int. Ed.*, 2004, **43**, 3224–3227.
13. R. A. Sheldon, *Chirotechnology: Industrial Synthesis of Optically Active Compounds*, Dekker, New York, 1993.
14. P. J. Harrington and E. J. Lodewijk, *Org. Process Res. Dev.*, 1997, **1**, 72–76.
15. For a recent review, see *Green Chemistry in the Pharmaceutical Industry*, ed. P. J. Dunn, A. S. Wells and M. T. Williams, Wiley-VCH, Weinheim, 2010, ch. 12; for the example from GSK applied to the API radafaxine, see T. Grinter, *ibid.*, ch. 10.
16. T. Laird, *Development and Scale-up of Processes for the Manufacture of New Pharmaceuticals* in *Comprehensive Medicinal Chemistry*, ed. C. Hansch, Pergamon, Oxford, 1990, vol. 1, pp. 321–359.
17. For a survey of over 200 UK runaway reactions in the 1970s and 1980s, see P. F. Nolan and J. A. Barton, *J. Hazard. Mater.*, 1987, **14**, 233.
18. *Chemical Reaction Hazards*, ed. J. A. Barton and R. Rogers, Institution of Chemical Engineers, Rugby, 1993.

19. T. Grewer, *Thermal Hazards of Chemical Reactions*, Elsevier, Amsterdam, 1994.
20. T. Yoshida, *Safety of Reactive Chemicals*, Elsevier, Amsterdam, 1987.
21. T. Kletz, *Lessons from Disaster – How Organisations Have No Memory and Accidents Recur*, Institution of Chemical Engineers, Rugby, 1993, and other books by the same author; for a full list see http://en.wikipedia.org/wiki/Trevor_Kletz (last accessed 15 July 2010).
22. P. Urben, *Bretherick's Handbook of Reactive Chemical Hazards*, Academic Press/Elsevier, Amsterdam, 7th edn, 2006, vols. 1 & 2.
23. R. Barbas, M. Botija, H. Camps, A. Portell, R. Prohens and C. Puigjaner, *Org. Process Res. Dev.*, 2007, **11**, 1131–1134.
24. For the latest special safety issue, see *Org. Process Res. Dev.*, 2009, **13**, 1387–1430.
25. See www.csb.gov for some excellent teaching materials and videos, some of which are also available on YouTube.
26. *Green Chemistry in the Pharmaceutical Industry*, ed. P. J. Dunn, A. S. Wells and M. T. Williams, Wiley-VCH, Weinheim, 2010.
27. R. A. Sheldon, I. Arends and U. Hanefeld, *Green Chemistry and Catalysis*, Wiley-VCH, Weinheim, 2007.
28. B. M. Trost, *Science*, 1991, **254**, 1471–1477; for a recent review on atom economy, see A. Moores in *Green Catalysis*, ed. R. H. Crabtree, Wiley-VCH, Weinheim, 2009, vol. 1.
29. A. Bhattacharye and R. Bandichhor, in *Green Chemistry in the Pharmaceutical Industry*, ed. P. J. Dunn, A. S. Wells and M. T. Williams, Wiley-VCH, Weinheim, 2010, ch. 14.
30. C.-K. Chen and A. Singh, *Org. Process Res. Dev.*, 2001, **5**, 508–513.
31. J. H. Cohen, C. A. Maryanoff, S. M. Stefanick, K. L. Sorgi and F. J. Villani Jr., *Org. Process Res. Dev.*, 1999, **3**, 260–264.
32. C.-J. Li and T.-H. Chan, *Organic Reactions in Aqueous Media*, Wiley, New York, 1997; P. Grieco, *Organic Synthesis in Water*, Blackie, London, 1998.
33. E. J. Milton and M. L. Clarke, *Green Chem.*, 2010, **12**, 381–383.
34. W. Cabri and R. Di Fabio, *From Bench to Market, The Evolution of Chemical Synthesis*, Oxford University Press, Oxford, 2000.
35. J. L. Tucker, *Org. Process Res. Dev.*, 2006, **10**, 315–319.
36. A. Liese, K. Seelbach and C. Wandrey, *Industrial Biotransformations*, Wiley-VCH, Weinheim, 2nd edn, 2006.
37. S. K. Ma, J. Gruber, C. Davis, L. Newman, D. Gray, A. Wang, J. Grate, G. W. Huisman and R. A. Sheldon, *Green Chem.*, 2010, **12**, 81–86.
38. J. Liang, J. Lalonde, B. Borup, V. Mitchell, E. Mundorff, N. Trinh, D. A. Kockrekar, R. N. Cherat and G. G. Pai, *Org. Process Res. Dev.*, 2010, **14**, 193–198.
39. The ACS Green Chemistry Institute Pharmaceutical Round Table produces *Green Chemistry Highlights*, which are published in *Org. Process Res. Dev.* every six months.
40. www.iprd.leeds.ac.uk.

41. www.lpf.uch.ch.
42. R. Carlson and J. E. Carlson, *Design and Optimisation in Organic Synthesis*, Elsevier, Amsterdam, 2nd edn, 2005.
43. M. Harre, H. Neh, C. Schulz, U. Tilstam, T. Wessa and H. Weinmann, *Org. Process Res. Dev.*, 2001, **5**, 335–339.
44. D. M. Roberge, L. Ducry, N. Bieler, P. Cretton and B. Zimmermann, *Chem. Eng. Technol.*, 2005, **28**, 318–323.
45. L. Proctor and A. Warr, *Org. Process Res. Dev.*, 2002, **6**, 884.
46. T. L. LaPorte and C. Wang, *Curr. Opinion Drug Disc. Dev.*, 2007, **10**, 738–745.
47. P. H. Stahl and C. G. Wehrmuth, *Handbook of Pharmaceutical Salts, Properties, Selection and Use*, Wiley-VCH, Weinheim, 2002.
48. R. Hilfiker, *Polymorphism in the Pharmaceutical Industry*, Wiley-VCH, Weinheim, 2006.
49. J. Bernstein, *Polymorphism in Molecular Crystals*, Oxford University Press, Oxford, 2002, ch. 10.
50. S. R. Chemburkar, J. Bauer, K. Deming, H. Spiwek, K. Patel, J. Morris, R. Henry, S. Spanton, W. Dziki, W. Portere, J. Quick, P. Bauer, J. Donabauer, B. A. Narayanan, M. Soldani, D. Riley and K. McFarland, *Org. Process Res. Dev.*, 2000, **4**, 413–417.
51. S. Desikan, R. L. Parsons Jr., W. P. Davis, J. E. Ward, W. J. Marshall and P. Toma, *Org. Process Res. Dev.*, 2005, **9**, 933–942.
52. H. Chaudhary (Dr Reddy's, India), presented at a conference on Scale Up of Chemical Processes, Scientific Update, Portugal, 2006.
53. See *Org. Process Res. Dev.*, 2009, **13**, issue 6.
54. *Polymorphism in Pharmaceutical Solids*, ed. H. G. Brittain, Informa Healthcare, New York, 2009.
55. W. Beckmann, *Org. Process Res. Dev.*, 2000, **4**, 372–383.
56. H.-S. Tung, E. L. Paul, M. Midler and J. A. McCauley, *Crystallisation of Organic Compounds, An Industrial Perspective*, Wiley, New York, 2009.
57. T. M. Crowder, A. J. Hickey, M. D. Louey and N. Orr, *A Guide to Pharmaceutical Particulate Science*, Interpharm/CRC, Boca Raton, FL, 2003.
58. *Modern Pharmaceutics*, ed. A. Florence and J. Siepmann, Informa Healthcare, New York, 2009, vols. 1 & 2.
59. N. G. Anderson, *Practical Process Research and Development*, Academic Press, San Diego, 2000.
60. *Process Chemistry in the Pharmaceutical Industry*, ed. K. Gadamasetti, Dekker, New York, 1999; K. Gadamasetti and T. Braish, *Process Chemistry in the Pharmaceutical Industry, Vol. 2: Challenges in an Ever-Changing Climate*, CRC Press, Boca Raton, FL, 2008.
61. O. Repic, *Principles of Process Research and Chemical Development in the Pharmaceutical Industry*, Wiley-Interscience, New York, 1998.
62. *Chemical Process Research; The Art of Practical Organic Synthesis*, ed. A. F. Abdel-Magid and J. A. Ragan, ACS Symposium Series 870, American Chemical Society, Washington/Oxford University Press, Oxford, 2004.

63. J. H. Atherton and K. J. Carpenter, *Process Development: Physicochemical Concepts*, Oxford Science, Oxford, 1999.
64. L. K. Doraiswamy, *Organic Synthesis Engineering*, Oxford University Press, Oxford, 2001.
65. *Organometallics in Process Chemistry*, ed. R. D. Larsen, Springer, Heidelberg, 2004.
66. D. Walker, *The Management of Chemical Process Development in the Pharmaceutical Industry*, Wiley-Interscience, Hoboken, NJ, 2008.
67. *Drug Discovery and Development*, ed. M. S. Chorghade, Wiley-Interscience, Hoboken, NJ, 2006, vols. 1 & 2.
68. *Chem. Rev.*, 2006, **106**, 2581–3027.
69. C. Someswara Rao, *The Chemistry of Process Development in Fine Chemical and Pharmaceutical Industry*, Techbooks World, East Brunswick, NJ, 2nd edn, 2006.
70. www.pubs.acs.org/oprdfk (last accessed 21 September 2010).

CHAPTER 3

Active Pharmaceutical Ingredients: Structure and Impact on Synthesis

JOHN S. CAREY[a] AND DAVE LAFFAN[b]

[a] GlaxoSmithKline Pharmaceuticals, Old Powder Mills, near Leigh, Tonbridge, TN11 9AN, UK; [b] AstraZeneca PR&D, Silk Road Business Park, Macclesfield, SK10 2NA, UK

3.1 Introduction: What is the Active Pharmaceutical Ingredient?

Chemists working in the field of pharmaceutical process research and development are responsible for the synthesis of multi-kilogram quantities of an active pharmaceutical ingredient (API) to support clinical and toxicology evaluation studies and ultimately to devise a synthetic route suitable for commercial manufacture.

In the early stages of development the requirements of the API may be in the order of a few kilograms and this material could be produced in 20–50 L laboratory glassware. As development progresses, the requirements of the API may rise such that tens or even hundreds of kilograms are required; this could be produced in a multi-purpose pilot-plant facility where the scale of operation ranges from 100 to 2500 L. If the project is ultimately successful and the synthesis goes into routine manufacture, then the annual production may be in the order of tens to hundreds of tonnes and the scale of operation 8000–10 000 L in fixed

RSC Drug Discovery Series No. 9
Pharmaceutical Process Development: Current Chemical and Engineering Challenges
Edited by A. John Blacker and Mike T. Williams
© Royal Society of Chemistry 2011
Published by the Royal Society of Chemistry, www.rsc.org

equipment. However, the commercial scale of manufacture for some products may be as little as tens of kilograms. Many factors influence the amounts of drug substance required; the two biggest ones are the potency of the drug (the quantities decrease as the dose decreases) and the commercial success of the product.

Within the scope for discussion are fully synthetic molecules such as atorvastatin and esomeprazole (Scheme 3.1), which are two of the biggest selling pharmaceutical products and are synthesised from small organic molecules.[1–3]

Amoxicillin is a semi-synthetic agent prepared from penicillin G (Scheme 3.2), which is in turn produced by fermentation.[4] Paclitaxel was synthesised from 10-deacetylbaccatin (Scheme 3.2), which in turn was isolated from the needles of the European yew.[5] More recently, however, this method has been superseded by plant cell fermentation technology which produces paclitaxel directly.[6] Processes such as fermentation and natural product extraction fall outside the scope of this discussion.

Also outside the scope of this discussion are large biological molecules, known colloquially as "biopharmaceuticals", which encompass high molecular weight proteins and monoclonal antibodies and for which a New Biologicals Application (NBA) would be submitted to the regulatory agencies.

3.2 Physicochemical Considerations

The structure of the API that the process development chemist is required to synthesise is a function of the role the molecule needs to play in the human body and a facet of the challenges faced within drug discovery. Drug discovery is a multidimensional process in which a number of different components must be simultaneously optimised to converge on a viable drug candidate. These components include such criteria as potency against and selectivity for the biological target, bioavailability and adsorption, distribution, metabolism, excretion and toxicity (ADMET) considerations. The importance of modulating the physical properties of a potential drug candidate molecule to avoid being in highly unfavourable regions of property space is a key consideration in drug discovery programmes.

The majority of small-molecule APIs are intended for oral administration, as opposed to being injectables or administered topically or using an inhaler. The oral bioavailability of a drug is dependent upon solubility, permeability and transport characteristics across membrane barriers. These in turn are related to some of the physicochemical properties of the molecule. Through the analysis of chemical structures, four criteria were identified by Lipinski[7] and called the "Rule of Five". These "rules" have highlighted the potential importance of the physical properties of drug candidates. The "Rule of Five" states that poor absorption or permeability are more likely when (1) there are more than five H-bond donors (expressed as the sum of OHs and NHs); (2) the molecular weight is over 500; (3) the log P is over 5; and (4) there are more than 10 H-bond acceptors (expressed as

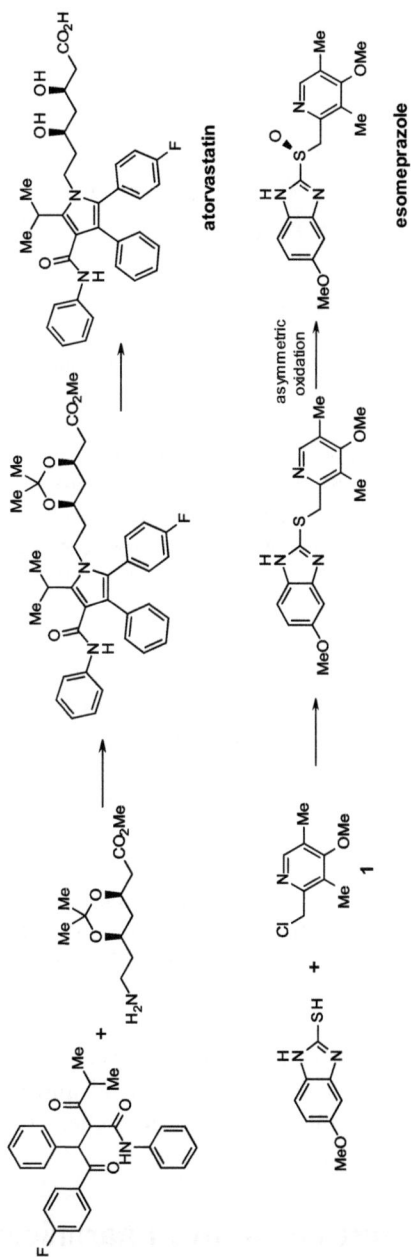

Scheme 3.1 Schematics for the syntheses of atorvastatin and esomeprazole.

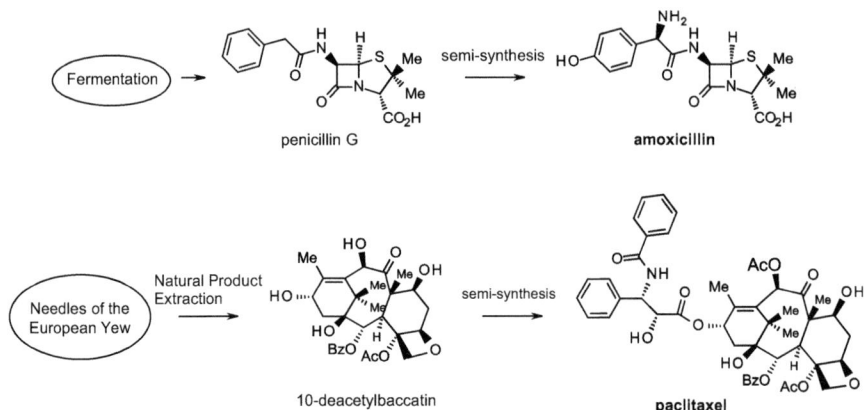

Scheme 3.2   Schematics for the semi-synthesis of amoxicillin and paclitaxel.

the sum of Ns and Os). Compound classes that are substrates for biological transporters, such as antibiotics, antifungal and vitamins, are exceptions to the rule. Approximately 90% of oral drugs obey at least two of these rules and 80% of marketed oral drugs obey all of them. These rules were augmented by Veber's criteria,[8] which added that a polar surface area of ≤ 140 Å^2 (or the sum of the hydrogen bond donors and acceptors ≤ 12) and the number of rotatable bonds of ≤ 10 are sufficient to predict oral bioavailability.

A high molecular weight may be beneficial for increased binding to the target, but has a negative effect on solubility and absorption.[9] Log P is the logarithm of the octanol/water partition coefficient of the molecule and is a measure of lipophilicity. A high log P means that the compound will be lipophilic and therefore have low aqueous solubility. Hydrogen bond donors and acceptors are required for substrate binding, so APIs must contain groups such as alcohols, amines, amides or sulfonamide, but their inclusion can make the molecules very polar and hence can decrease permeability. Rules for ADMET liabilities based upon molecular weight and log P have also been described[10] and, in general, increasing values in either parameter are generally detrimental to more than one ADMET parameter. The effect of aromatic rings within oral drug discovery programmes has recently come under scrutiny, with the rules of thumb that the fewer the aromatic rings contained in an oral drug candidate, the more developable is the candidate,[11] and also that increased levels of saturation increase the likelihood of higher solubility and lower melting point.[12]

3.3   Common Features of Active Pharmaceutical Ingredients

The APIs encountered by the process chemist are usually small molecules (< 550 MW), generally contain at least one N atom, an aromatic ring and approximately 50% are chiral. Nearly all of the APIs are solids, usually crystalline, and have the potential to form salts. The importance of the physical form for drugs is discussed in detail in Chapter 13.

If peptides and biologics are excluded from the list of top 200 branded and generic drugs (by value) in 2008,[1] the applicability of the Lipinski rules discussed in the previous section are clear. Within these limitations there is an extraordinary variety of structures. A look at the statin class of HMGCoA reductase inhibitors used for the treatment of cardiovascular disorders shows how structures can develop over time (Figure 3.1). The non-aromatic carbocyclic structures of the early statins, such as lovastatin (a natural product) and simvastatin, have been replaced by heteroaromatic structures in the more recent wholly synthetic compounds.[13]

The introduction of aromatic rings into candidate drugs is a common feature; indeed, from a recent survey of candidate drugs it was shown that there are on average two aromatic or heteroaromatic rings per candidate drug,[14] and a similar number in the top 100 selling drugs in 2008 (215 aromatic rings in 100 compounds).[1] Comparison of the structures from the *World Drug Index* and natural products from the *Dictionary of Natural Products* showed that aromatic rings were approximately twice as common in drugs (75.6% of drugs contained aromatic rings) as they were in natural products (37.9% contained aromatic rings).[15] These rings are frequently substituted, for example with F, Cl or Br, in order to block metabolic pathways and or to modify physicochemical characteristics. Understanding the construction and substitution behaviour for a wide range of aromatic systems is therefore essential for any process chemist in the pharmaceutical industry.

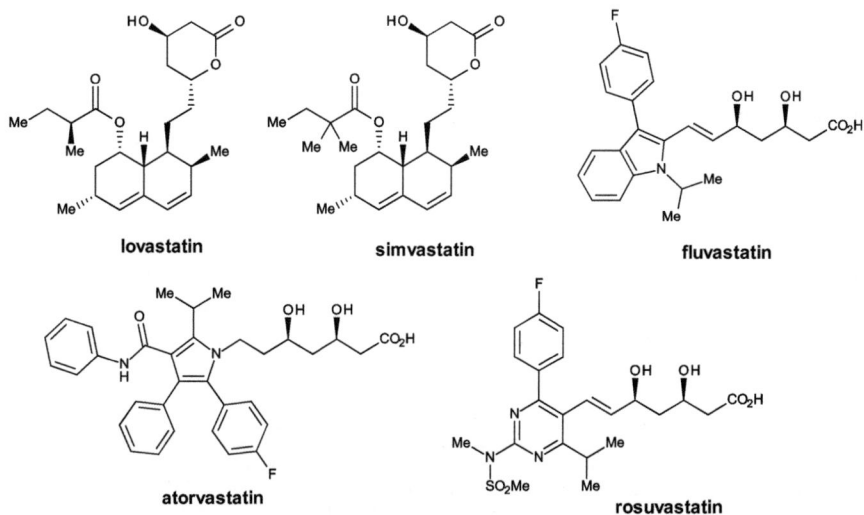

Figure 3.1 Molecules that belong to the statin class of cardiovascular agents.

The synthetic challenges that the process chemist faces are ultimately dictated by the molecules that medicinal chemists make. This is partly driven by design, but is also dictated by accessibility. Recent estimation of the number of drug-like organic compounds containing up to 13 atoms of C, N, O, S and Cl give a total of 970 million compounds.[16] Thirteen atoms is of course relatively small: only 5 of the 100 top-selling small-molecule drugs in 2008 had 13 atoms or less (excluding hydrogens) and the average was more than double this (>26 atoms). This total also ignores the compounds that contain other atoms, for example F or P, that are reasonably common in pharmaceuticals. Even with these restrictions this number is still considerably larger than all the compounds synthesised for screening and indicates that only a small portion of the available chemical space has been investigated.

There has been much debate about the presence of privileged structures in pharmaceuticals, an idea that was introduced in 1988 and referred to benzo-diazepines.[17] Since then, many structures have been proposed as privileged structures,[18] such as biphenyls,[19–20] benzopyrans,[19–21] benzofurans,[19,21] dihydro-pyridines,[19,21] purines[20,21] and indoles.[21,22] Although it is clear that certain sub-structures appear more frequently than others in drugs, it is not clear if this reflects their abundance in chemical libraries or because of the properties they confer.[18,19] For the process chemist it is important to recognise that developments in these areas will affect the candidate drugs put into development and the chemistry required to construct them.

For the purpose of this discussion the origin of these properties is irrelevant; it is merely important to recognise that these structures represent the nature of the synthetic challenge faced by modern process chemists.

3.4   The Synthetic Sequence

The synthesis of an API normally requires that multiple sequential chemical transformations are carried out to prepare the complex chemical targets. Process development chemists normally define a stage of the process as being from one isolated compound to the next and therefore the number of stages in the synthetic sequence is equal to the number of isolated compounds. A stage can be as short as one chemical transformation or even making a salt, or it can be more complex and involve the "telescoping" of multiple chemical trans-formations into a longer sequence. For example, in Scheme 3.3, stage 1 involves the combination of A with B to form A–B, which is then isolated; this is a stage that consists of a single chemical transformation. Stage 2 involves two chemical

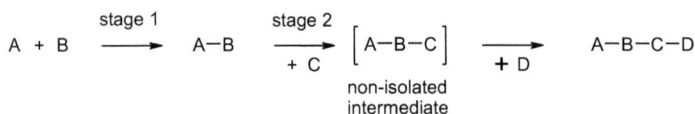

Scheme 3.3   Schematic showing non-telescoped and telescoped stages.

transformations, the first of which is the addition of C to give A–B–C, which is a non-isolated intermediate. This is then "telescoped" into the addition of D to give A–B–C–D, which is then isolated.

A survey has shown that, on average, the number of synthetic transformations within the synthetic sequence is eight, although this number can be as low as two and on occasions be greater than 20.[14] In natural product synthesis it is the norm to count the total number of steps in the synthetic sequence from commercially available, catalogue starting materials. For process development chemists, this definition of a starting material is not always appropriate because the majority of commercial compounds are not available on sufficient scale to be practical. The process development chemist must try and build a supply chain, looking to use contract manufacturers in the fine chemicals industry to transform readily available materials into the starting material for their synthesis; in practice these starting materials may well be more advanced than commercially available catalogue materials. The supply and definition of starting materials can be dynamic in nature; therefore good judgement must be used when using step counting to assess the relative merits of multiple synthetic sequences. A fundamental understanding of the genealogy of speciality chemicals and the capability of the fine chemical industry is beneficial for identifying a suitable starting material from which to build a lasting synthetic route. It is also worth noting that the term "regulatory starting material" in a pharmaceutical setting has implications and requires a deliberate data package to support its designation.

For the process development chemist the synthetic sequence has one primary role: to enable the assured supply of a high-quality API within a reasonable timeframe and for an acceptable cost. For the discovery chemist an ideal synthetic sequence is one whereby small quantities of related compounds can be prepared in an analogous fashion, ideally from common advanced intermediates. When the process development chemists become involved, the onus on the synthetic sequence moves from synthesising multiple compounds to the synthesis of a single target compound of specific interest. This change in emphasis may mean that a change in synthetic sequence is required or desirable. It may not be possible to immediately make the transition to the perfect long-term manufacturing route when a compound enters the development phase, but the process development chemist should be mindful of what the desirable features and attributes of a synthetic sequence are as the development process unfolds. Many of the desirable attributes that the synthetic sequence for an API should process are highlighted in Chapter 5 and in recent reviews.[23,24]

The synthetic sequence to be used must meet all the safety criteria set. Safety covers several facets: reaction/process safety, worker safety and, most importantly, patient safety. Before any chemical reaction or process can be scaled up it must be deemed safe to do so, as detailed in Chapter 8. If a reaction cannot be scaled up safely, then it should not be scaled up at all. The main types of issues are: (i) thermal run-away, (ii) gas evolution, (iii) potentially explosive, shock-sensitive materials, (iv) highly corrosive materials and (v) pyrophoric and highly flammable materials. A runaway reaction at large scale puts human lives and plant property at great risk, as some of these issues are exacerbated on scale-up as the surface area to volume ratio decreases.

As the scale of operation increases, so does the potential for exposure to substances harmful to health. An understanding of the toxicity of the chemical reagents and intermediates (for example, carcinogens or sensitizers) must be obtained in order to maintain a safe environment for operators. The toxicity of a chemical intermediate is an important factor to consider when deciding whether to isolate it or not. To minimise the potential exposure of a toxic intermediate it should be telescoped into the subsequent step rather than isolated. Where possible, hazardous chemicals should be replaced by less hazardous ones, for example the substitution of benzene by toluene. If this is not possible, then the quantity should be reduced. The final line of protection against chemical hazard involves the use of engineering controls and/or personal protective equipment.

The environmental challenge faced by process development chemists is not insignificant, and is covered in more detail in Chapter 6. The manufacture of an API produces relatively high levels of waste per kilogram of product when compared to other chemical industries.[25] There are two key environmental issues, namely environmental impact and sustainability, and these need to be addressed during development to ensure long-term manufacture is possible. The assessment of environmental improvement during process development is best achieved using a set of metrics.[26] Incorporation of green chemistry principles into synthetic route design makes sense, as very often the greenest process is also in the long run the most cost effective one.[23]

The ideal synthetic sequence needs to be economical to operate and meet all the cost of goods targets set by the business. The total cost is made up of two main components: (i) cost of the materials used in the processing (this includes the cost of the starting materials, the cost of solvents and the cost of reagents and catalysts required to carry out the transformations; these are the variable costs) and (ii) overhead and labour costs (these are the fixed costs, which involve the cost of the infrastructure, including maintenance and depreciation, employee costs, utilities, waste disposal and many other contributors). Many of these costs are directly linked to the concentration, time and efficiency of the process. Therefore the shorter the synthetic sequence, then the lower the cost is likely to be. When designing a synthetic sequence, the process development chemist should look to incorporate the most expensive raw materials or reagents or catalysts as late in the synthetic sequence as is possible.

The throughput of a process defines the amount of material (in kilograms or tonnes) that can be manufactured in unit time. The throughput of a synthetic sequence will affect both the cost and the lead time for delivery of the desired API. Therefore the shorter the synthetic sequence, then the higher the throughput is likely to be. An important goal of the process development chemist is to increase the throughput for the synthesis of their target API. One means of increasing throughput is to move from a linear synthetic sequence to a more convergent synthetic sequence. Scheme 3.4 shows two approaches to the metalloproteinase inhibitor **2**.[27] Both approaches use largely the same starting

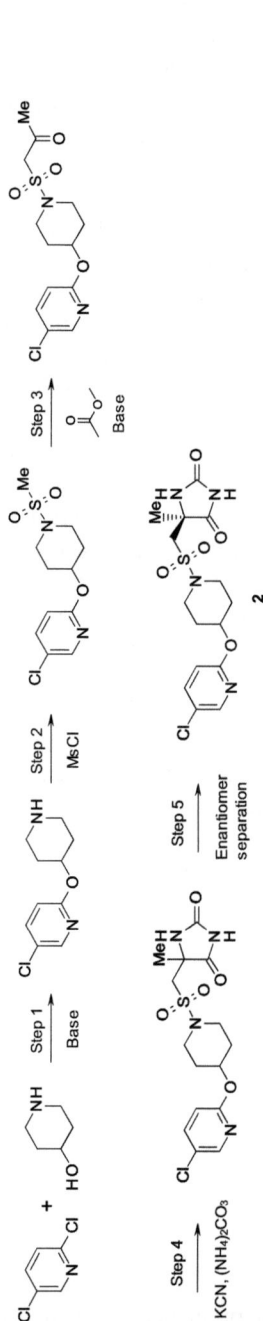

Scheme 3.4 Linear and convergent syntheses of metalloproteinase inhibitor **2**.

materials and both require the same number of synthetic transformations to reach the end product, but the convergent nature of approach 2 makes it much more efficient both in terms of yield and time required. Scheme 3.4 also demonstrates another important consideration of throughput and that is the position within the synthetic sequence of a low-yielding step, in this example the separation of enantiomers, where the maximum theoretical yield is 50%. It is much better to have the low-yielding step at the beginning of the sequence rather than at the end.

A strategy commonly used by process development chemists to improve throughput is to "telescope" multiple chemical transformations together into an extended stage. The gains in throughput arise through the time saved by avoiding the isolation of intermediate compounds. Crystallisation of a product followed by isolation by filtration and then drying of the wet cake is very often the bottleneck of a process in a pilot plant setting. The telescoping of chemical transformations is often required when the intermediates have poor isolation characteristics. Intermediates may be non-crystalline, lack chemical stability or be highly reactive/toxic, in which cases it may be preferable to react them *in situ*. The synthesis of the glycine antagonist 6 demonstrates this point (Scheme 3.5).[28] Stage 1 in the synthesis of 6 starts from 5-chloro-2-iodoaniline and involves a three-step telescoped sequence consisting of imine formation with ethyl glyoxylate, Mannich reaction using trimethyl(vinyloxy)silane and subsequent Wittig reaction. Both of the intermediates, imine 3 and aldehyde 4, are non-crystalline compounds which lack stability. The product 5, however, is a stable, highly crystalline compound that could be isolated in high purity by crystallisation. It should also be noted that the discovery route to the glycine antagonist 6 involved eight chemical transformations, while the route scaled up by the process development chemists contained only six.

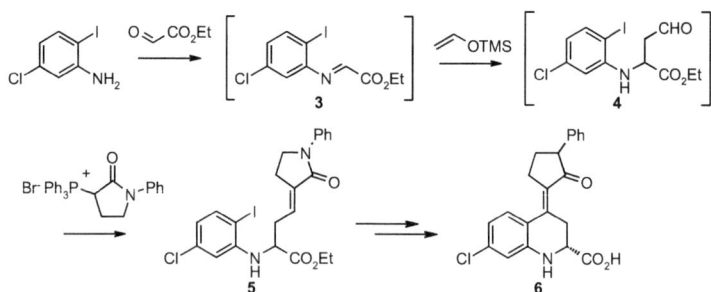

Scheme 3.5   Synthesis of glycine antagonist 6.

Many work-up and isolation procedures published by academic research groups involve removing the reaction solvent on a rotary evaporator, adding a non-water miscible solvent, washing with dilute acid, dilute base and then brine before drying over magnesium sulfate, filtering and then finally purification by chromatography. All of these activities are known as "unit operations" and if they all need to be performed that will negatively affect the throughput. The process development chemist will look closely at the work-up and isolation procedure and look to eliminate unnecessary solvent exchanges and washes, so as to minimise the number of unit operations. One approach that is commonly tried is the "direct drop" process, in which the selected reaction solvent dissolves the starting material, but not the product, so that the product can be isolated by filtration directly from the reaction mixture without needing any work-up procedure.[29]

An attribute that the synthetic sequence must exhibit is one of control. Control needs to be exerted over such matters as reaction yield, selectivity and quality. When planning to make a certain amount of an API, then having consistent yields will aid the scheduling and the purchase of starting materials. Secondly, making an API that fails quality requirements wastes time and resources. Finally, the regulatory agencies expect to see reproducible processes ahead of product approval. Analytical methods are only valid if the process performs consistently; this underpins the basis of volunteer/patient safety. Sufficient control points need to be built into the synthetic sequence to ensure that the final API meets all the quality criteria. Control points typically include isolated solid intermediates or extraction of acidic or basic compounds into an aqueous phase to remove non-ionisable impurities. Most purifications, either of intermediates or the API, are achieved by crystallisation. Therefore a synthetic sequence that incorporates crystalline intermediates is advantageous, as it should be possible to expel impurities when intermediates are isolated. Incorporating the principles of quality by design into process design is the cornerstone for ensuring product quality and process performance. Although maximising yield is important to increase throughput, it is not a driver for control. Process development chemists need to look for clean reactions, ones that form few if any impurities. It is much better to have one impurity at 10% rather than 10 impurities at 1% each. In both cases the losses to impurities are the same, but in the first scenario the chances of identifying the impurity and understanding its fate and effect are much higher. There is therefore a desire to have clean reactions, especially at the final stages of a synthetic sequence. The formation of impurities is unavoidable and the process development chemist should be looking to selectively form impurities that are easily removed during the work-up and isolation procedure, or are inert to the subsequent reaction conditions and therefore do not form new impurities. An example of this strategy is demonstrated during the synthesis of the GPIIb/IIIa receptor antagonist **10** (Scheme 3.6).[30] Enzyme-catalysed hydrolytic kinetic resolution of racemic ester **7** was achieved using *Candida antarctica* lipase B; this reaction was clean and consistently high yielding. Once the reaction was complete, the acid **8** was extracted into aqueous base (pH 7.5). Since the yield and purity of acid **8** was consistently high, it was acceptable to use the solution of acid **8** directly in the iodination reaction. This reaction used pyridine–iodine monochloride complex and was also carried out in aqueous base (pH 7.5). The amount of pyridine–iodine monochloride complex added needed to be controlled. Addition

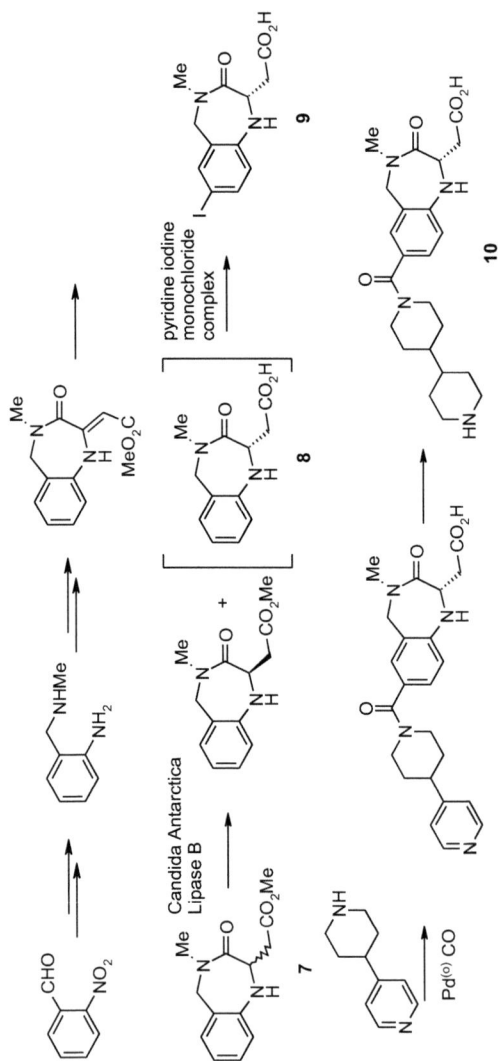

Scheme 3.6 Synthesis of GPIIb/IIIa antagonist **10**.

of insufficient reagent led to unreacted acid **8** being left over at the end of the reaction, while excess reagent led to the formation of two over-iodinated by-products. These impurities had the potential to react multiple times in the subsequent palladium-catalysed carbonylation reaction and hence form a multitude of new impurities. The control strategy that was adopted was to add sufficient pyridine–iodine monochloride complex such that 2% of the acid **8** was left unreacted; this ensured that no over-iodinated impurities were formed. The unreacted acid **8** was expelled when the product **9** was isolated by crystallisation.

The ideal synthetic sequence should be as short as possible and avoid the use of hazardous reaction conditions or reagents. Clean, high-yielding reactions that form few if any impurities are desirable especially towards the end of the synthetic sequence. This can generally be covered by the idea that pharmaceutical syntheses tend to perform simple chemistry on complex substrates. Techniques that increase throughput should be incorporated as well as devising a control strategy which ensures that the API produced meets all of its quality critical attributes.

3.5 Reactions Commonly Used

3.5.1 Introduction

A recent review of reactions used in the synthesis of APIs looked at the chemistry employed by AstraZeneca (AZ), Pfizer and GlaxoSmithKline (GSK) in the synthesis of their current (2006) development portfolios.[14] The data contained in that review will be used as the basis for this discussion (unless indicated otherwise).

In compiling this data the authors needed to select a starting point for each synthesis. The chemistry used to make the starting materials may be different to that used later in the drug substance synthesis. This may be particularly important with respect to adoption of asymmetric synthetic methods and in the synthesis of polysubstituted aromatics, where the data indicate that in both cases the complexity may be built into the starting material. The process chemist must develop a knowledge and appreciation of the chemistry used in preparing the starting materials in order to understand their quality, likely cost and availability.

The authors of the review used data from AZ, GSK and Pfizer development projects across all therapeutic areas from all the companies' locations. The data were sufficiently similar for each of the companies to indicate that there was little value in separating out by company and the data are therefore presented here together. As the data were for all compounds in development, it can be expected that, because of attrition, the data set contains many more early stage development compounds than late stage. Obviously, the types of chemistry will change during development, but it is impossible to tell from the data presented how.

3.5.2 Construction *versus* Modification

The reactions used in the synthesis of organic compounds can be broadly separated into those used for the construction of the molecule and those used

Table 3.1   Construction *versus* modification in API synthesis.

Construction[a]	Contribution (%)	Modification	Contribution (%)
Heteroatom alkylation and arylation	19	Deprotection	15
Acylation	12	Protection	6
C–C bond formation	11	Reduction	9
Aromatic heterocycle formation	5	Oxidation	4
		Functional group interconversion	10
		Functional group addition	3
		Resolution	3
Total construction	47	Total modification	50

[a]Miscellaneous accounted for 3% of the total and have been unassigned.

for modification (Table 3.1). The construction categories can be broken into acylation, aromatic heterocycle formation, C–C bond formation, heteroatom alkylation and arylation and some miscellaneous reactions. The modifying transformations are protection, deprotection, functional group addition (FGA), functional group interconversion (FGI), oxidation, reduction, resolution and some miscellaneous reactions.

Overall, approximately half of the reactions used are for construction and half for modification. It is part of the remit of the process chemist to increase the percentage of construction steps in a synthesis during development by minimising the use of protecting groups and redundant reduction/oxidation cycles.

3.5.3   Chirality

The requirement to produce chiral compounds as single enantiomers/ diastereoisomers for pharmaceutical application is well understood and publications over recent years have provided snapshots of asymmetric methods used on a process scale.[31] A typical target value for enantiomeric purity is 99.5%.

From the survey, 54% of the APIs were chiral with an average of two centres per chiral compound. Purchased chiral centres represented 55% of the total. This is driven by the desire to introduce chirality early in a synthesis, consistent with an assembly of smaller complex fragments and determined by their availability.

The predominant method of in-house generation of chirality is resolution (28% of the total and 62% of the in-house). Two-thirds of the resolutions were classical salt formations and the remainder were evenly distributed between dynamic kinetic, chromatographic and enzymatic methods. The availability of screening methods to develop classical resolutions, the increased understanding of crystallisations and the ease of scale-up continue to make this the preferred methodology for many chiral molecules.[32] This is exemplified by the synthesis of voriconazole (Scheme 3.7).[33]A diastereoselective Reformatsky reaction

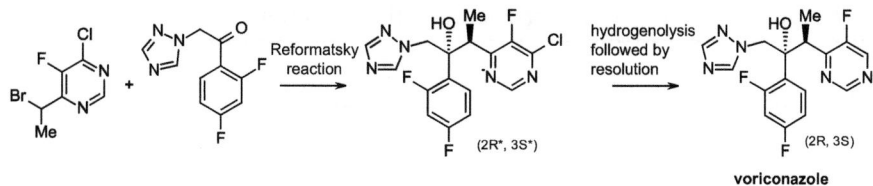

Scheme 3.7 Synthesis of voriconazole.

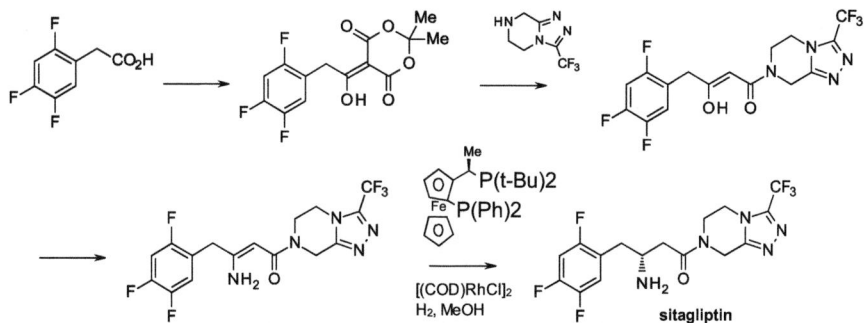

Scheme 3.8 Synthesis of sitaglipitin.

was employed to establish the relative stereochemistry, followed by camphorsulfonic acid resolution to obtain enantiomerically pure voriconazole.

Asymmetric synthesis only accounts for a smaller proportion, approximately 20% of the chiral centres generated, but it is noteworthy that the methods applied are catalytic in nature. Chirality may be expensive, which would suggest late incorporation into the synthesis is more desirable. In a recent example, the synthesis of sitagliptin (see Scheme 3.8, and also Chapter 6 for more details of this case history) was accomplished using catalytic asymmetric hydrogenation. In this case the chemistry was clean and the catalyst easily removed.[34] This synthesis is a rarity; usually when a chiral centre is generated later in the synthesis within a more complex substrate it seems that few methods exist that are sufficiently straightforward to be operated economically.

An important example of a catalytic asymmetric reaction is the sulfide to sulfoxide oxidation for esomeprazole (see Scheme 3.1). The use of a catalytic asymmetric oxidation of the sulfur for esomeprazole has resulted in a process that is significantly more efficient than the process used for the racemic oxidation in omeprazole.[3a]

3.5.4 Substituted Aromatic Starting Materials

The majority of APIs contain substituted aromatic sub-units. For reasons of control, substitution patterns are often set in the starting materials (see Section 3.5.13). The distribution of di-, tri- and tetrasubstituted benzenoid aromatics

Table 3.2 Distribution of benzenoid starting materials.

Substitution pattern	Number per 100 development compounds	Frequency (%)
1,2-C$_6$H$_4$	12.5	11
1,3-C$_6$H$_4$	13.3	12
1,4-C$_6$H$_4$	29.7	27
1,2,3-C$_6$H$_3$	10.2	9
1,2,4-C$_6$H$_3$	35.9	32
1,3,5-C$_6$H$_3$	3.9	4
1,2,3,4-C$_6$H$_2$	3.1	3
1,2,3,5-C$_6$H$_2$	1.6	1
1,2,4,5-C$_6$H$_2$	0.8	1
Total	111	100

purchased is shown in Table 3.2. While the most common of these have been prepared using standard electrophilic substitution reactions, a number have synthetically challenging substitution patterns and hence are expensive.

3.5.5 Heterocycle Occurrence and Formation

The dataset showed 88 heterocyclic aromatic rings per 100 APIs. Approximately half were made and half were purchased. The formation of saturated heterocycles is captured under the relevant bond-forming step (for example, acylation or alkylation). There is a strong preponderance of *N*-containing aromatic heterocycles (92% of the purchased and 98% of the synthesised contained a nitrogen). The most commonly occurring aromatic heterocyclic systems in this survey were pyridine, quinazoline, pyrazole, pyrimidine, 1,2,4-triazole and thiazole.

These data of course reflect the situation in development in 2006 and it is interesting to compare them with the heteroaromatics in marketed drugs. Using the top 100 selling branded drugs in 2008 as a comparison,[1] the six most common heterocycles are pyridine, pyrimidine, imidazole, thiazole, triazole and tetrazole and there are 66 heteroaromatics compared to 88 heteroaromatics per 100 compounds in the sample of development compounds from 2006. There are no quinazolines in the top 100 and only one in the top 200 selling branded drugs. This illustrates that heterocylic chemistry has always been an important area for the process chemist but the exact nature of the heterocylces has and probably will change over time.

Of the purchased heterocyclic starting materials, 73% were electron-deficient six-membered heterocycles and 25% were electron-rich five-membered heterocycles.

The maturity of the market for supply of substituted pyridines was such that in only 13% of these cases was it necessary to introduce a further ring substituent to attain the substitution pattern required in the target. Even though pyridine chemistry is well understood, some substitution patterns can still be difficult to prepare, for example the pyridine **1** (a starting material for esomeprazole, Scheme 3.9; see also Scheme 3.1).[35]

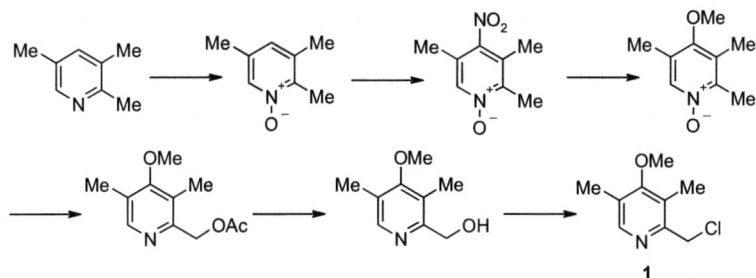

Scheme 3.9 Synthesis of the pyridine unit of esomeprazole.

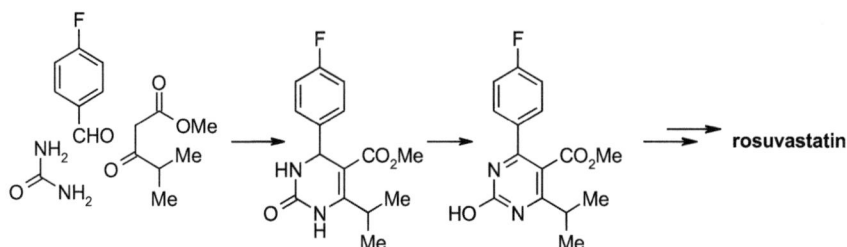

Scheme 3.10 Synthesis of rosuvastatin.

Mechanistically, most of the ring-forming reactions were classified as either condensations (44%) or cyclodehydrations (22%). In only 7% of cases were the heterocycles formed by a cycloaddition reaction. Scheme 3.10, exemplifying the formation of a pyrimidine for rosuvastatin, shows the general trend in the use of established condensation chemistry.[36]

3.5.6 Protections and Deprotections

Protections and deprotections account for 6% and 15%, respectively, of the total chemical transformations and therefore between them account for > 20% of all the transformations, almost two chemical transformations per molecule. The reason that deprotections significantly outnumber protections is that many of the purchased starting materials contain the protecting groups in place. For the 35% of syntheses achieved without the use of protecting groups, the average number of synthetic steps was 5.9, which is considerably shorter than the overall average of 8.1.

In contrast to many natural product syntheses in the literature, the process chemist is relatively rarely confronted by the need for selective deprotections. Protections and deprotections fall predominately into three groups: protection of the amino NH group, protection of a carboxylic acid and protection of a hydroxy group. Although there exist many different protecting groups[37] for protection and deprotection of multivarious functional groups, only a small number are routinely used in the preparation of APIs. The most common amino protecting group was Boc, followed by either benzyl or the Cbz group (that is, groups that can be removed by hydrogenolysis). In general, the carboxylic acid group is protected as a simple alkyl ester. A wider variety of protecting groups were used for the protection of hydroxy groups. The most common groups were benzyl ether, silicon-containing protecting groups or acetate esters. Compared to academic syntheses, the use of silicon protecting groups is small, because they reduce crystallinity in the intermediates and, except for TMS, they are expensive. High molecular weight protecting groups, such as trityl, are disfavoured as they are costly and reduce throughput.[24]

By-products from the deprotection must also be benign, especially during the final stage of a synthesis. Even a straightforward removal of a Boc group from a molecule containing aromatics/nitrogens can result in unacceptable levels of impurities resulting from reaction with the *t*-Bu cation.

By and large, protections and deprotections are designed out of a synthesis, for example the synthesis of GPIIb/IIIa antagonist **10** (Scheme 3.6). The development route contained no protecting groups whereas the original discovery route contained five protection/deprotection steps.[30,38]

3.5.7   Acylation

Acylation reactions, especially of nitrogen, are frequently used in the preparation of drug candidate molecules; the categories of acylation, and the occurences of each, are summarised in Table 3.3. Acylations comprised 12% of the reactions in the survey, and the dominance of *N*-acylations is to be expected from the frequency with which amides occur in drug molecules. Of the top-selling small-molecule drugs in 2008, 21% contained an amide and a further 9% a sulfonamide moiety.[1]

Within the amide formation category, the breakdown of *N*-acylation methods is tabulated in Table 3.4. Acid chlorides were by far the most common acylating agent. The use of mixed anhydrides provides an inexpensive and readily scaled process for *N*-acylation that is particularly valuable for cases prone to epimerisation. The moderately priced reagent carbonyldiimidazole (CDI) is gaining popularity, as reactions are readily scaled-up and worked-up.[39] New reagents such as propanephosphonic acid anhydride (propylphosphonic

Table 3.3   Categories of acylation reactions.

Acylation	*Frequency*
N-Acylation to amide	66%
N-Sulfonation to sulfonamide	9%
N-Acylation to urea	6%
Carbamate/carbonate formation	5%
Amidine formation	4%
O-Acylation to ester	4%
Other	5%

Table 3.4 Methods for *N*-acylation.

Method	Frequency (%)
Acid chloride	44
Coupling agent	25
Mixed anhydride	13
CDI	11
Other	7

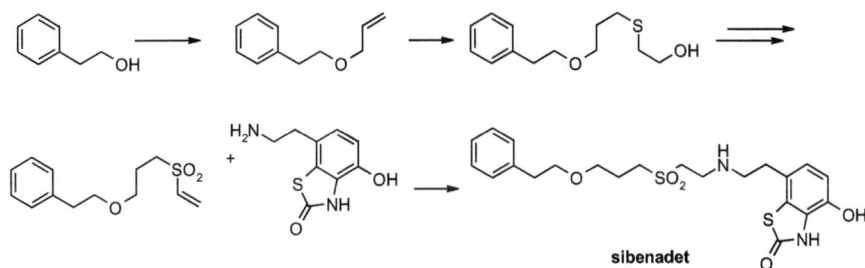

Scheme 3.11 Synthesis of sibenadet.

anhydride, T3P) are gaining usage.[40] Although amide formation with coupling reagents such as carbodiimides are frequently used for early development, for later development these reagents are usually developed out of processes as they are sensitisers and are relatively costly.

3.5.8 Heteroatom Alkylations and Arylations

This is the largest single class of reactions in the survey, making up 19% of the total. Typically about 90% of drugs are N-containing and an equal proportion are O-containing.[1] When synthesising drug candidates, it is not surprising that alkylation and arylation at nitrogen and oxygen emerge as major reactions in the construction of the molecules.

The synthesis of sibenadet hydrochloride (Scheme 3.11) shows a molecule constructed by *O*-, *N*- and *S*-alkylations.[41] A phase transfer alkylation of an alcohol is used to construct the O–C bond, a free radical addition to an alkene is used in the construction of the S–C bond and a conjugate addition of an amine to a vinyl sulfone is used in the formation of the N–C bond. It should be noted that the latter two addition reactions are highly atom efficient. The free radical thiol addition could be carried out neat on the laboratory scale, but had to be diluted with toluene for scale up, as it was a very exothermic reaction.

N-Substitution is typically achieved by one of three strategies: (i) direct reaction with alkyl-X or aryl-X, (ii) reductive alkylation using an aldehyde or ketone or (iii) acylation plus reduction of the carbonyl. S_N2 reactions are still widely used, although the stringent requirements to ensure that only vanishingly small levels of residual alkylating agents are present in drug candidates discourages such substitutions late in a synthesis.[42] The acylation/reduction strategy avoids the need to handle alkylating agents and would be more widely used if safer bulk amide reduction methods were developed.

Nucleophilic aromatic substitution has long been used in pharmaceutical syntheses for electron-deficient aromatics. However, the development of palladium-mediated aryl C–N bond-forming methodology, led by Buchwald and Hartwig,[43] is an example of where a technology advance has led to an increase in the use of a transformation in bulk syntheses (see Scheme 3.12 for the preparation of the 5-HT receptor antagonist **11**).[44] This reaction has also been adopted by medicinal chemists, which has in turn led to an increasing number of candidate drugs that require this type of reaction for their construction.

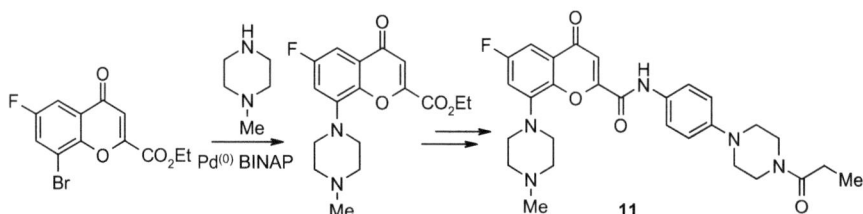

Scheme 3.12   Synthesis of the 5-HT receptor antagonist **11**.

Alkylations of phenols (23 per 100 syntheses) and alcohols (19 per 100 syntheses) to give ethers were encountered with moderate frequency, while alkylations at sulfur (12 per 100 syntheses) were less common. Examples of the Mitsunobu reaction for heteroatom substitution are found in early development, but this reaction is not often used on scale because of its poor atom economy and the thermal hazards associated with azodicarboxylates.

3.5.9   Oxidation Reactions

The use of oxidation reactions in the preparation of candidate drug molecules is low.

The most common is the oxidation of sulfide to sulfoxide or sulfone (for example, see Scheme 3.1 for the synthesis of esomeprazole).[3]

Adjustment of oxidation state from alcohol to carbonyl is rarely performed even though it is a common feature of many published target-orientated syntheses. The problems associated with many oxidation methods probably mean that syntheses requiring them are avoided if possible. The heavy metals used in many oxidation reactions can cause problems in removal and must only be present in trace amounts in the final product (<10 ppm). Many oxidising agents are high-energy species, giving rise to thermal hazards and a lack of chemoselectivity. Recently, more manufacturing friendly oxidation conditions have been developed, for example the TEMPO/bleach combination.[45]

3.5.10   Reduction Reactions

Reduction reactions are used much more frequently than oxidative transformations, totalling 9% of reactions (14% if reductive aminations and removal of

Table 3.5 Categories of reduction and reagents used.

Conversion	Heterogeneous Pd/Pt/Ni, H$_2$	Homogeneous catalyst, H$_2$	Hydride[a]	Borane	Other	Total
Imine/nitrile to amine	7		5	1	1	14
NO$_x$ to NH$_2$	16				5	21
Alkene to alkane	6	4	2			12
Alkyne to alkane	1					1
Amide to amine			3	3		6
Amide to aldehyde			2			2
Ester to alcohol			8	2		10
Ketone to alcohol			4	3		7
Aryl/heteroaryl to fully saturated	6					6
Other	5		5	1	4	15
Total	41	4	29	10	10	94

[a]LiAlH$_4$, NaBH$_4$, DIBALH.

benzyl-type protecting groups are included). The main reductive transformations are summarised in Table 3.5.

Catalytic hydrogenation over precious metal catalysts is the most frequently used technique (47%), followed by hydride (32%) and borane reductions (10%). Reductive amination is a frequently used technique to form C–N bonds and is usually carried out with sodium cyanoborohydride or sodium triacetoxyborohydride initially, but these reagents are often superseded by catalytic hydrogenation over an appropriate catalyst on industrial scale. Catalytic hydrogenation using hydrogen gas is the most atom efficient process. It is striking that none of the reductions of carboxylic acid derivatives employ catalytic hydrogenations. The global reduction of the paroxetine intermediate **12** (Scheme 3.13) using excess lithium aluminium hydride examplifies this type of approach.[46] Hydride and borane reagents are hazardous[47] and lead to both complex work-up procedures and high levels of waste.

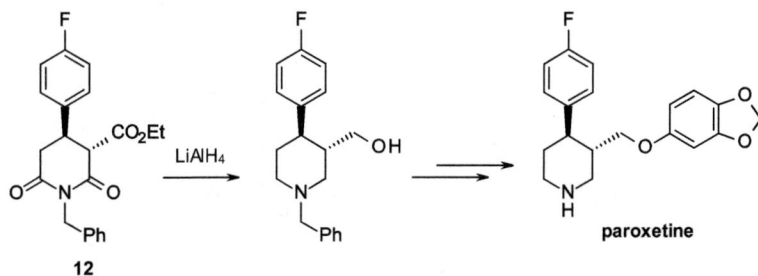

Scheme 3.13 Synthesis of paroxetine.

3.5.11   C–C Bond-forming Reactions

The main C–C bond-forming reactions used are shown in Table 3.6.

The transformations listed in the table are those that make up >10% of the C–C bond-forming reactions. This makes up about 59% of the total, which shows that much of this category does not fit into any significant trend.

The largest reaction type within the palladium reactions is the Suzuki reaction. Its popularity is derived from the easy accessibility of the two materials, convenient reaction conditions, broad functional group tolerance and easy removal of the inorganic by-product, although palladium removal can be problematic.[48] The boronate used in the Suzuki coupling is often formed using an organometallic derived from a halide, linking the increased use of the Suzuki reaction to the improved understanding and operation of organometallic chemistry. Organometallics used directly in C–C bond formation make up an important group of reactions themselves. The advances in understanding and operation of these reactions have made scale-up almost routine and can be carried out with confidence. The synthesis of the 5-HT$_{1D}$ receptor antagonist **13** (Scheme 3.14) shows the use of a palladium-catalysed Suzuki reaction.

Table 3.6   C–C Bond-forming reactions.

Reaction	Frequency (%)
Pd catalysis	22
Suzuki	(11)
Heck	(6)
Ester condensation	14
Organometallic	12
Aryl-metal	(7.5)
Directed lithiation	(2.5)
Grignard	(2)
Friedel–Crafts	10
Other	41

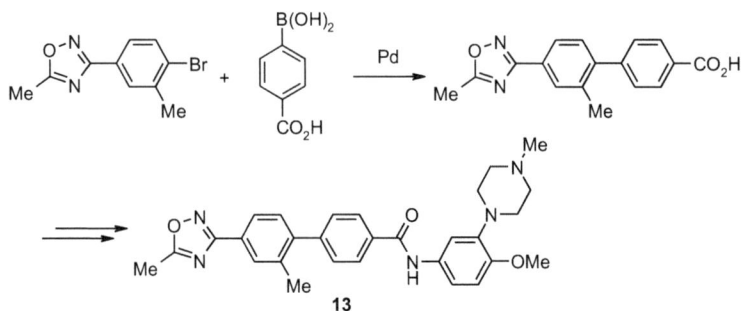

Scheme 3.14   Synthesis of the 5-HT$_{1D}$ receptor antagonist **13**.

The development route moved its position from the final to the penultimate step in the synthesis to control the levels of residual palladium in the final product.[24]

It is probably not surprising that ester condensation and Friedel–Crafts reactions still make up a large proportion of the C–C bond-forming reactions used. They are well-established reactions where the difficulties are well understood and the possible side reactions identified.

3.5.12 Functional Group Interconversions

Functional group interconversions (FGIs) account for 10% of the total reactions. The three largest categories in this class of reactions (acid to acid chloride, hydroxy to halide/sulfonate ester and amide to imidoyl chloride) are directly related to the main reactions used for molecular construction, acylation and heteroatom alkylation/arylation. The acid chloride figure is understated because where the acid chloride is used *in situ* it has been classified with the acylations.

3.5.13 Functional Group Additions

The main functional group additions (FGAs) are shown in Table 3.7 and form only 3.2% of the total. There is a general preference for purchasing starting materials that contain some functionality rather than introducing it. This is due to the problems of selectivity when introducing functional groups, making it generally more straightforward to control the quality of the starting materials than to control the selectivity of the FGA. For example, over-halogenation impurities are usually easier to remove from small molecules by distillation than from larger molecules by crystallisation.

Although some of the halogen will be retained in the final compound, the nitro group is rarely retained in APIs and sulfonations are usually followed by conversion to sulfonamides.

Fluorine is common in many drug substances but there were no examples of fluorination in the survey. This clearly understates the importance of fluorination to the process chemist. The difficulty of introducing it to any complex substrate means the process development is heavily dependent on specialist fine chemical manufacturers for the supply of starting materials. The importance of fluorination can be seen by considering that 16% of the top 100 small molecules from 2008 contain fluorine: there were 13 Ar–F, seven CF_3 and seven other aliphatic C–F groups.[1]

Table 3.7 Types of functional group additions.

Functional group addition	Frequency (%)
Halogenation (Cl, Br, I)	52
Nitration	36
Sulfonation	6
Other	6

3.6 Constraints on the Process Chemist

The process development chemist working in the pharmaceutical industry faces many constraints. Some of these constraints come from the equipment available, the regulatory environment and the business operating model.

Most pharmaceutical producers use multi-purpose batch reactors within their pilot plant facility for the manufacture of APIs during the research and development phase. By its very nature, multi-purpose equipment has to be adaptable for use and therefore may not be suitable for all situations. Typical multi-purpose batch plant will be glass lined and have an operating range from $-30\,^{\circ}C$ to $+150\,^{\circ}C$. Reaction temperatures outside of this range are possible but require more specialist equipment to be available, so the choice of reactors (size and stirring attributes) will be reduced. The equipment materials of construction can be limiting when highly corrosive reagents are planned to be used. The use of, or the liberation of, hydrogen gas, especially at elevated pressures, is another constraint upon equipment selection. Finally, the stirring and mixing characteristics of the reactors within a pilot plant suite may not be ideal for the all situations.

Equipment constraints also play a major role in the techniques used to purify intermediates or the API; hence the preference for crystallisations that can be performed in most multi-purpose reactors. It may be possible to purify lower molecular weight compounds by distillation or sublimation, although that usually requires specialised equipment to achieve the temperatures and vacuums required. Additionally, most distillations are performed in a continuous mode rather that batch mode to limit the time that the desired compound is held at the elevated temperature. Purification by column chromatography is a technique that is very commonly used in academic laboratories but is a technique that is designed out of most API synthetic sequences. It may be utilised for the rapid delivery of early development supplies where the quantities required are relatively small and the time pressures are acute, but longer term is avoided. In addition to the requirement to have specialist large-scale chromatography equipment, handling the large volumes of solvents that are needed can be prohibitive. In certain circumstances, such as the separation of a binary mixture (enantiomers), then certain continuous chromatography techniques can be employed, for example simulated moving bed (SMB) chromatography, but again specialist equipment is required.[49]

The production of APIs is a highly regulated business. Materials that have a significant negative environmental impact are often strictly controlled by regulations and may be subject to very demanding emission limits. Additionally, there may be plans in place for these emissions limits to be reduced at a future date and this needs to be taken into account. The use of chlorinated solvents such as dichloromethane is very common in an academic research setting. However, reducing the use of chlorinated solvents to meet environmental emissions regulations is one of many challenges facing pharmaceutical manufacturers.[50]

The API produced during pharmaceutical development is often dosed to volunteers in clinical studies and their safety needs to be protected.

The emphasis on quality for pharmaceuticals is most prominently manifested by the fact that not only do products have to meet strict specifications but also manufacturing processes and associated analytical methods must meet pre-set criteria. The concept of current good manufacturing practice (cGMP) must be adhered to. The API to be used in clinical studies needs to be produced according to cGMP and in accordance with the regulatory authorities. Impurities within the API should be controlled to acceptable low levels and in the case of heavy metal contaminants (such as palladium) or potential genotoxic impurities (for example, alkyl halides or alkyl sulfonates) these levels can be in the order of parts per million (ppm).

The business model operated by a pharmaceutical company can put constraints upon the process development chemist. The risky nature of innovative medicine discovery means that a great majority of drug candidates never reach the market place. The resource level assigned to a compound will therefore vary during the development lifecycle. During early development the process chemist frequently has to "make do" with a variant of the discovery synthesis, but once a project has shown a high likelihood of success then a degree of "catch up" will be required to provide both a robust route of manufacture and all the associated knowledge required. The strategies used to progress many drug candidates through key early decision points are discussed in detail in Chapter 4.

References

1. http://www.chem.cornell.edu/jn96/index.html (last accessed August 2010).
2. B. D. Roth, *Prog. Med. Chem.*, 2002, **40**, 1.
3. (a) H. Cotton, T. Elebring, M. Larsson, L. Li, H. Sorensen and S. von Unge, *Tetrahedron: Asymmetry*, 2000, **11**, 3819; (b) E. I. Carlsson, U. K. Junggren, H. S. Larsson and G. W. von Wittken Sundell, Pat. Appl. EP 074341, 2007.
4. A. Bruggink, E. C. Roos and E. de Vroom, *Org. Process Res. Dev.*, 1998, **2**, 128.
5. J.-N. Denis, A. E. Greene, D. Guenard, F. Gueritte-Voegelein, L. Mangatal and P. Potier, *J. Am. Chem. Soc.*, 1988, **110**, 5917.
6. P. G. Mountford, in *Green Chemistry in the Pharmaceutical Industry*, ed. P. J. Dunn, A. S. Wells and M.T. Williams, Wiley-VCH, Weinheim, 2010, p. 145.
7. C. A. Lipinski, F. Lombardo, B. W. Dominy and P. J. Feeney, *Adv. Drug Delivery Rev.*, 1997, **23**, 3.
8. D. F. Veber, S. R. Johnson, H.-Y. Cheng, B. R. Smith, K. W. Ward and K. D. Kopple, *J. Med. Chem.*, 2002, **45**, 2615.
9. M. A. Navia and P. R. Chaturvedi, *Drug Discovery Today*, 1996, **1**, 179.
10. M. P. Gleeson, *J. Med. Chem.*, 2008, **51**, 817.
11. T. J. Ritchie and S. F. Macdonald, *Drug Discovery Today*, 2009, **14**, 1011.
12. F. Lovering, J. Bikker and C. Humblet, *J. Med. Chem.*, 2009, **52**, 6752.
13. H. Stark, *Pharm. Unsere Zeit*, 2003, **32**, 464.

14. J. S. Carey, D. Laffan, C. Thomson and M. T. Williams, *Org. Biomol. Chem.*, 2006, **4**, 2337.
15. P. Ertl, S. Jelfs, J. Mühlbacher, A. Schuffenhauer and P. Selzer, *J. Med. Chem.*, 2006, **49**, 4568.
16. L. C. Blum and J.-L. Reymond, *J. Am. Chem. Soc.*, 2009, **131**, 8732.
17. B. E. Evans, K. E. Rittle, M. G. Bock, R. M. DiPardo, R. M. Freidinger, W. L. Whitter, G. F. Lundell, D. F. Veber, P. S. Anderson, R. S. L. Chang, V. J. Lotti, D. J. Cerino, T. B. Chen, P. J. Kling, K. A. Kunkel, J. P. Springer and J. Hirshfield, *J. Med. Chem.*, 1988, **31**, 2235.
18. J. Klekota and F. P. Roth, *Bioinformatics*, 2008, **24**, 2518.
19. D. A. Horton, G. T. Bourne and M. L. Smythe, *Chem. Rev.*, 2003, **103**, 893.
20. L. Costantino and D. Barlocco, *Curr. Med. Chem.*, 2006, **13**, 65.
21. R. W. DeSimone, K. S. Currie, S. A. Mitchell, J. W. Darrow and D. A. Pippin, *Comb. Chem. High Throughput Screening*, 2004, **7**, 473.
22. F. R. de Sa Alves, E. J. Barreiro and C. A. M. Fraga, *Mini-Rev. Med. Chem.*, 2009, **9**, 782.
23. T. Y. Zhang, *Chem. Rev.*, 2006, **106**, 2583.
24. M. Butters, D. Catterick, A. Craig, A. Curzons, D. Dale, A. Gillmore, S. P. Green, I. Marziano, J.-P. Sherlock and W. White, *Chem. Rev.*, 2006, **106**, 3002.
25. R. A. Sheldon, *Chemtech*, 1994, **24**, 38.
26. (a) D. J. C. Constable, A. D. Curzons and V. L. Cunningham, *Green Chem.*, 2002, **4**, 521; (b) D. J. C. Constable, A. D. Curzons, L. M. Freitas dos Santos, G. R. Geen, R. E. Hannah, J. D. Hayler, J. Kitteringham, M. A. McGuire, J. E. Richardson, P. Smith, R. L. Webb and M. Yu, *Green Chem.*, 2001, **3**, 7.
27. N. Barnwell, presented at the 27th SCI Process Development Symposium, 9–11 December 2009, Cambridge, UK.
28. A. Banks, G. F. Breen, D. Caine, J. S. Carey, C. Drake, M. A. Forth, A. Gladwin, S. Guelfi, J. F. Hayes, P. Maragni, D. O. Morgan, P. Oxley, A. Perboni, M. E. Popkin, F. Rawlinson and G. Roux, *Org. Process Res. Dev.*, 2009, **13**, 1130.
29. C.-K. Chen and A. K. Singh, *Org. Process Res. Dev.*, 2001, **5**, 508.
30. R. J. Atkins, A. Banks, R. K. Bellingham, G. F. Breen, J. S. Carey, S. K. Etridge, J. F. Hayes, N. Hussain, D. O. Morgan, P. Oxley, S. C. Passey, T. C. Walsgrove and A. S. Wells, *Org. Process Res. Dev.*, 2003, **7**, 663.
31. (a) J. D. Armstrong III, M. J. Martinelli and C. H. Senanayake, *Tetrahedron: Asymmetry*, 2003, **14**, 3425; (b) V. Farina, J. T. Reeves, C. H. Senanayake and J. J. Song, *Chem. Rev.*, 2006, **106**, 2734.
32. (a) U. C. Dyer, D. A. Henderson and M. B. Mitchell, *Org. Process Res. Dev.*, 1999, **3**, 161; (b) J. W. Nieuwenhuijzen, R. F. P. Grimbergen, C. Koopman, R. M. Kellogg, T. R. Vries, K. Pouwer, E. Van Echten, B. Kaptein, L. A. Hulshof and Q. B. Broxterman, *Angew. Chem. Int. Ed.*, 2002, **41**, 4281; (c) A. Borghese, V. Libert, T. Zhang and C. A. Alt, *Org. Process Res. Dev.*, 2004, **8**, 532.
33. M. Butters, J. Ebbs, S. P. Green, J. MacRae, M. C. Morland, C. W. Murtiashaw and A. J. Pettman, *Org. Process Res. Dev.*, 2001, **5**, 28.

34. K. B. Hansen, Y. Hsiao, F. Xu, N. Rivera, A. Clausen, M. Kubryk, S. Krska, T. Rosner, B. Simmons, J. Balsells, N. Ikemoto, Y. Sun, F. Spindler, C. Malan, E. J. J. Grabowski and J. D. Armstrong III, *J. Am. Chem. Soc.*, 2009, **131**, 8789.
35. (a) A. E. Brandstrom and B. R. Lamm, Eur. Pat. Appl. 103553, 1984; (b) S.-Y. Chou and S.-F. Chen, *Heterocycles*, 1997, **45**, 77.
36. (a) A. Matsushita, M. Oda, Y. Kawachi and J. Chika, WO Pat. Appl. 006439, 2003; (b) K. Yamamoto, Y. G. Chen and F. G. Buono, *Org. Lett.*, 2005, **7**, 4673.
37. T. W. Greene and P. G. M. Wuts, *Protecting Groups in Organic Synthesis*, Wiley-Interscience, New York, 3rd edn, 1999.
38. W. H. Miller, T. W. Ku, F. E. Ali, W. E. Bondinell, R. R. Calvo, L. D. Davis, K. F. Erhard, L. B. Hall, W. F. Huffman, R. M. Keenan, C. Kwon, K. A. Newlander, S. T. Ross, J. M. Samanen, D. T. Takata and C.-K. Yuan, *Tetrahedron Lett.*, 1995, **36**, 9433.
39. P. J. Dunn, W. Hoffmann, Y. Kang, J. C. Mitchell and M. J. Snowden, *Org. Process Res. Dev.*, 2005, **9**, 956.
40. H. Wissmann and H.-J. Kleiner, *Angew. Chem. Int. Ed.*, 1980, **19**, 133.
41. M. E. Giles, C. Thomson, S. C. Eyley, A. J. Cole, C. J. Goodwin, P. A. Hurved, A. J. G. Morlin, J. Tornos, S. Atkinson, C. Just, J. C. Dean, J. T. Singleton, A. J. Longton, I. Woodland, A. Teasdale, B. Gregertsen, H. Else, M. S. Athwal, S. Tatterton, J. M. Knott, N. Thompson and S. J. Smith, *Org. Process Res. Dev.*, 2004, **8**, 628.
42. *EMEA Guidelines on the Limits of Genotoxic Impurities*, CPMP/SWP/5199/02, EMEA/CHMP/QWP/251344/2006, European Medicines Agency, London, 2006.
43. S. Shekhar, P. Ryberg, J. F. Hartwig, J. S. Mathew, D. G. Blackmond, E. R. Strieter and S. L. Buchwald, *J. Am. Chem. Soc.*, 2006, **128**, 3584.
44. G. E. Robinson, O. R. Cunningham, M. Dekhane, J. C. McManus, A. O'Kearney-McMullan, A. M. Mirajkar, V. Mishra, A. K. Norton, B. Venugopalan and E. G. Williams, *Org. Process Res. Dev.*, 2004, **8**, 925.
45. M. M. Zhao, J. Li, E. Mano, Z. J. Song and D. M. Tschaen, *Org. Synth.*, 2005, **81**, 195.
46. (a) D. A. Greenhalgh and N. S. Simpkins, *Synlett*, 2002, **12**, 2074; (b) C. D. Gill, D. A. Greenhalgh and N. S. Simpkins, *Tetrahedron*, 2003, **59**, 9213.
47. *Bretherick's Handbook of Reactive Chemical Hazards*, ed. P. Urben, Butterworth-Heinemann, Oxford, 6th edn, 1999, vol. 2, p. 58.
48. (a) C. E. Garrett and K. Prasad, *Adv. Synth. Catal.*, 2004, **346**, 889; (b) J. T. Bien, G. C. Lane and M. R. Oberholzer, *Topics Organomet. Chem.*, 2004, **6**, 263.
49. E. Lang, E. Valery, O. Ludemann-Hombourger, W. Majewski and J. Blehaut, in *Green Chemistry in the Pharmaceutical Industry*, ed. P. J. Dunn, A. S. Wells and M. T. Williams, Wiley-VCH, Weinheim, 2010, p. 243.
50. D. J. C. Constable, P. J. Dunn, J. D. Hayler, R. Humphrey, J. L. Leazer, R. J. Linderman, K. Lorenz, J. Manley, B. A. Pearlman, A. Wells, A. Zaks and T. Y. Zhang, *Green Chem.*, 2007, **9**, 411.

CHAPTER 4

Rapid Early Development of Potential Drug Candidates

NICHOLAS M. THOMSON,[a] PIETER D. DE KONING,[a] ADAM T. GILLMORE[a] AND YONG TAO[b]

[a] Pfizer Ltd, Research Active Pharmaceutical Ingredients, Ramsgate Road, Sandwich, Kent, CT13 9NJ, UK; [b] Pfizer Inc, Research Active Pharmaceutical Ingredients, Eastern Point Road, Groton, CT 06340, USA

4.1 Introduction

Process chemistry has an extremely valuable role to play in the rapid development of potential small-molecule drug candidates. During the conversion of a small molecule into a medicine there are many hurdles to overcome, and attrition rates are high. Excellence in process chemistry, through rapid and fit-for-purpose design and enabling of synthetic routes, is essential during early development. Such efforts support the pharmaceutical industry in accelerating drug candidates through early toxicology and clinical milestones in order to identify potentially viable medicines. A fit-for-purpose approach also allows greater cost effectiveness, with resource intensive commercial route activities triggered only after key milestones are achieved.

Medicinal chemists do a wonderful job of balancing properties to identify a potential drug candidate, using a range of *in vivo* and *in vitro* models and studies. Physicochemical properties, such as lipophilicity, and pharmacokinetic properties, such as metabolism, are essential to drive predicted human exposure. Pharmacological properties, including potency, selectivity and toxicology, are essential to support predicted human toleration and efficacy. Finally, a target must also

RSC Drug Discovery Series No. 9
Pharmaceutical Process Development: Current Chemical and Engineering Challenges
Edited by A. John Blacker and Mike T. Williams
© Royal Society of Chemistry 2011
Published by the Royal Society of Chemistry, www.rsc.org

comply with pharmaceutical property requirements, such as stability, solubility and crystallinity. A medicinal chemist must balance all of this in order to trigger the development of a small molecule. It will then be several years before a commercially viable medicine can be brought to the market.

The first set of development hurdles involves assessment of animal safety in toxicology studies and human safety and toleration in first-in-human (FIH) healthy volunteer (phase 1) clinical studies (Figure 4.1). The phase 1 studies also define the acceptability of human pharmacokinetics. Phase 2a clinical studies offer the first opportunity to test for efficacy in patients, developing confidence that the molecule is exposed to the site of action, binds to the target and produces the desired pharmaceutical response. Completion of phase 2a clinical studies leads to a declaration of "proof-of-concept" (POC). Assuming the candidate also passes criteria relating to commercial viability (such as unmet medical need and market potential), then commercial development activities will likely begin in earnest.

During this early time horizon the imperative of the pharmaceutical industry is to drive at speed to key milestones in order to provide rapid feedback to ongoing research activities on follow-on programmes, ensure competitive advantage in reaching the market and identify potential attrition early. Attrition rates are routinely high across the pharmaceutical industry. Typical success rates are around 60% in the pre-clinical space, 70% in phase 1 and 40% in phase 2a. Hence any compound nominated for development (Figure 4.1) has a less than 1 in 5 chance of achieving a successful POC. The industry success rate, together with the drive for speed, inevitably provides a strong business rationale for an efficient and cost effective, fit-for-purpose, approach to early development. This is particularly true for the process chemistry community,

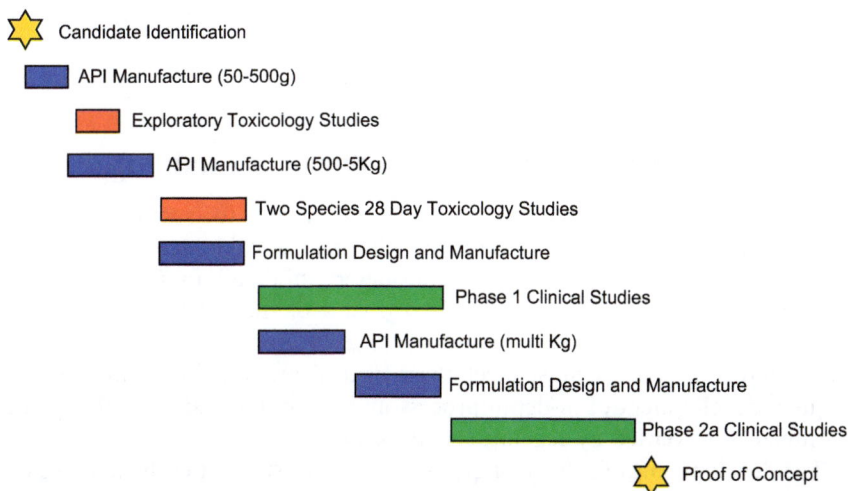

Figure 4.1 Typical candidate plan from candidate identification to clinical proof-of-concept (typically around four years from identification to POC).

who are typically on the critical path to early studies and require much investment when the risk of attrition is at its highest.

The drive for efficiency can provide a vibrant environment for the process chemist. As lead chemical series are translated into potential lead drug candidates by medicinal chemists, the process chemist must be on hand to track progress and influence and collaborate on synthetic route design. A synthesis that provides rapid access to desired analogues, and also offers a potential platform to provide multigram to multikilogram quantities, provides a win-win situation. Often the medicinal chemistry desire for a relatively linear route that offers opportunities for late stage analogue synthesis is contrasted against the process chemistry desire for a more efficient convergent approach. However, in early development a more linear route can offer significant advantage to the process chemistry community. The identification of an early core building block provides substrate for the process chemist to develop an appropriately robust process to key intermediates for early manufacture. Materials that are produced by the enabling efforts can provide key intermediates (or the technology to provide such materials through outsourcing activities) for analogue synthesis by medicinal chemists. Once a specific lead compound, the potential drug candidate, is identified and nominated, the process chemistry community can then turn their efforts to enabling the late stage process for rapid manufacture. However, in all aspects of process chemistry in early development the aim is not to develop the ultimate commercial route with in-depth process understanding and exquisite quality control. Rather it is to develop a fit-for-purpose process capable of supporting multikilogram synthesis with adequate control in a rapid timeframe. The opportunities for early engagement and partnership in synthetic design with the medicinal chemistry community, combined with a fit-for-purpose enabling approach to support early manufacture, can provide an exciting and vibrant environment for a strong synthetic process chemist, and ultimately a winning strategy for early development.

4.2   Criteria for Rapid Fit-for-purpose Enabling

The criteria for selection of a commercial route to support long-term manufacture and marketing of an active pharmaceutical ingredient (API) have been well documented;[1] they are considered in detail in Chapter 5. The SELECT mnemonic offers a valuable overview of the critical criteria in play during development of a commercial route (Figure 4.2). Any commercial manufacturing process must meet these critical criteria, in addition to subsequent development of in-depth process understanding and control of critical parameters, governed by regulatory authorities.[2]

The development of a fit-for-purpose process to support early advancement of a drug candidate to POC must also meet legislated safety, environmental, legal and control requirements, in addition to other key business efficiency requirements such as economic and throughput considerations. However, as

Criteria	Description
Safety	Safety of processes and human exposure to substances harmful to health
Environmental	Potential for wastage of natural resources and environmental exposure to harmful substances
Legal	Consideration of intellectual property rights and regulations that control use of reagents and intermediates
Economics	Meeting cost of goods targets for intended market and investment costs to support drug development requirements
Control	Control of quality, chemistry and physical parameters
Throughput	Time scale of manufacture in available plant and availability of raw materials

Figure 4.2   The SELECT criteria for the critical assessment of pharmaceutical processes and commercial route selection.[1]

the short-term manufacture of multikilogram quantities within a high attrition environment is a different proposition from long-term large-scale manufacture of tonne quantities for treatment of patients, the legal, business and regulatory hurdles are somewhat lesser. A useful set of criteria for early process development is offered in Figure 4.3. These criteria, combined with effective triggers for work, offer a framework for guiding a process chemist during early development and ensure a fit-for-purpose approach that drives speed and efficiency when effectively executed with contemporary enabling practices.

4.2.1   Safety

The primary requirement of an early development process is adequate safety, to ensure protection of operating staff, plant facilities and equipment, the environment and ultimately the patients (or volunteers) in a clinical study.

Many transformations that can be safely controlled and carried out on a laboratory scale can present a significant thermal hazard on a larger scale. Researchers at Johnson and Johnson[3] identified such an issue during the development of a potent and selective PPARα/δ dual agonist for the possible treatment of metabolic disorders such as diabetes, hyperlipidemia and atherosclerosis. The target compound **1** (Scheme 4.1) was prepared from the convergent coupling of the phenol **2** and thiadiazole **3**. The synthesis of the phenol **2** involved a potentially hazardous Baeyer–Villiger oxidation with *meta*-chloroperoxybenzoic acid (mCPBA) (Scheme 4.2). In order to circumvent these oxidative conditions, a literature synthesis[4] derived from methylhydroquinone (MHQ, **4**) (Scheme 4.3) was utilised to prepare pre-clinical supplies. Thus, MHQ was selectively protected, alkylated and then deprotected to afford the

Safety	Reaction Safety	Identification of significant thermal hazards or chemical incompatibilities through assessment and measurement.
	Toxicity and worker exposure safety	Identification of any highly potent, toxic, irritant or sensitizing reagents, solvents and intermediates that will require monitoring of the environment and people.
	Carcinogenicity and Genotoxicity	Identification of obvious genotoxic intermediates and raws.
	Operational Safety	Identification of highly hazardous chemicals or materials that cannot be safely handled in intended scale-up facility.
	Environmental	Identification of environmentally hazardous or banned chemicals, with consideration of the regulatory requirements in country where the work will be performed.
	Quality of API	Assessment of ability of route to provide material of appropriate quality (typically 98% purity for human use).
Reliability	Repeatable Yield	Identification of reactions with high variability in yield on previous runs.
	Reagent/solvent incompatibilities (quality)	Assessment of potential reagent/solvent incompatibilities that could cause impurities (e.g. amines with DCM or ketones)
	Mass and heat transfer sensitivity	Identification of reactions that are non reproducible based on mass or heat transfer limitations (e.g. where reaction performance is dependent on stirring rate, particle size or other parameter).
	Acceptable processing times and hold times	Identification of processes requiring very fast or very slow addition times, or lack of hold points due to stability or quality issues.
	Fits existing equipment	Identification of compatibility issues with scale-up equipment.
	Scalable isolation	Identification of isolation techniques that may not be amenable to scale-up (e.g. strips to dryness, lack of crystalline intermediates, foams, triturations).
	Processability at scale	Identification of processing issues such as poor physical properties (gumming, oiling), slow phase separations, poor crystallisations and slow filtrations.
Efficiency	Overall yield	Identification of overall yield based on average yields of historic runs (not the best) to ensure overall yield is >5%.
	Availability and cost of raw materials	Identify any raw materials that cannot be procured on a sufficient scale or are prohibitively costly.
	Chromatography	Identification of non scalable chromatographic separations. Chromatographic separations are increasingly becoming an important part of the tool kit for rapid early API manufacture.
	Throughput (volume)	Identification of high volume (high dilution) reactions and work-ups.
	Throughput (time)	Identification of long reaction times (>24 hours) or multiple and lengthy processing steps (e.g. multiple extractions).

Figure 4.3 Criteria useful for the critical assessment of pharmaceutical processes to support API manufacture in early development.

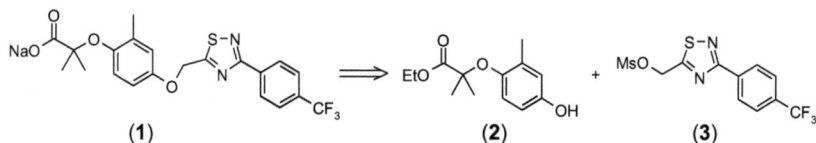

Scheme 4.1   Retrosynthetic analysis of thiadiazole **1**.

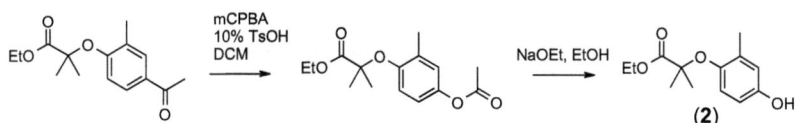

Scheme 4.2   Baeyer–Villiger oxidation route to phenol **2**.

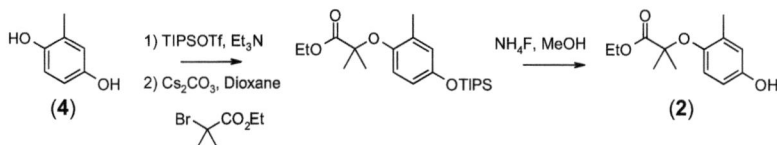

Scheme 4.3   Alternative MHQ route to phenol **2**.

desired phenol **2** in good yield and purity, following column chromatography. The phenol **2** was then reacted with the thiadiazole **3** under basic conditions, followed by caustic deprotection to afford the target compound **1**. Flash chromatography was utilised throughout the route to ensure appropriate purity, leading to a safe and fit-for-purpose synthesis.

In addition to thermal hazards, many reagents and reactions can present unacceptable toxicity and worker exposure safety issues. During the development of a bradykinin 1 antagonist **5** (Scheme 4.4) for pain and inflammation,[5] Merck researchers were faced with a hazardous Sandmeyer reaction, utilising copper cyanide under acidic pH conditions, leading to off-gassing of toxic and hazardous hydrogen cyanide. A key intermediate in the synthesis of **5** was the 1,2,4-oxadiazole **6**. The initial synthesis of **6** (Scheme 4.5) started from the inexpensive and readily available aniline **7**, which contains the desired halogen functionality with good regiochemical purity. The aniline **7** was converted into the nitrile **8** under Sandmeyer conditions by treatment with NOBF$_4$, to form a diazonium intermediate, which was converted into the nitrile **8** upon addition of CuCN and Cu(BF$_4$)$_2$. Addition of hydroxylamine and subsequent condensation with dimethylacetamide dimethyl acetal (DMADMA) furnished the desired 1,2,4-oxadiazole derivative **6**. In addition to the highly toxic Sandmeyer conditions, the route also proved to be relatively low yielding (20–30%), non-robust and difficult to control from a purity perspective.

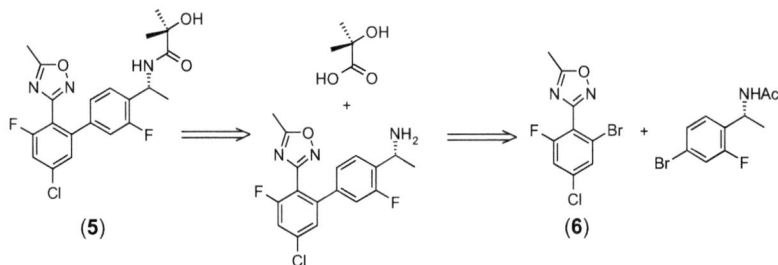

Scheme 4.4   Retrosynthesis of bradykinin 1 antagonist **5**.

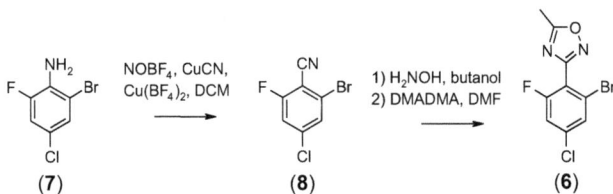

Scheme 4.5   Sandmeyer route to 1,2,4-oxadiazole **6**.

An alternative and less hazardous route involving regioselective halogen–metal exchange proved to be much more amenable to scale-up. Thus the aniline **7** (Scheme 4.6) was converted into the 1,2-dibromobenzene derivative **9** under modified Sandmeyer conditions, in good yield. The key halogen–metal exchange was then executed by treatment of **9** with iPrMgCl, followed by addition of DMF to afford the aldehyde **10**, with excellent 97:3 regioselectivity. Treatment of the aldehyde **10** with hydroxylamine provided a solid oxime **11**, which could be purified by crystallisation, leading to significant purging of the unwanted regio-isomer. The oxime was then converted into the desired 1,2,4-oxadiazole derivative **6** in two steps, *via* an amidoxime intermediate. With a safe and efficient route (45% yield over five steps) into the key intermediate **6**, the final API **5** was quickly prepared utilising an eight-step linear synthesis with an overall yield of 28%.

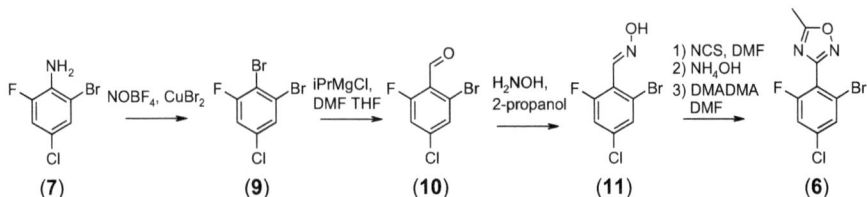

Scheme 4.6   Amidoxime route to 1,2,4-oxadiazole **6**.

In addition to the control of thermal hazards and toxicity and exposure issues, it is also essential that starting materials and reagents can be handled safely. This is exactly the situation that the team at Bristol Myers Squibb

(BMS) were confronted with when tasked with developing a synthesis of a tetrazole-based growth hormone secretagogue **14** (Scheme 4.7).[6] The medicinal chemistry route focused on the synthesis of a tetrazole core **13**, followed by appending the remaining side-chains (a relatively common strategy). One of the key reactions in the sequence used Mitsunobu-type reaction conditions to convert amide **12** to the tetrazole **13**, utilising diethyl azodicarboxylate (DEAD) and trimethylsilyl azide, reagents that are challenging to handle safely on scale. This, coupled with the highly exothermic nature of the actual transformation, meant that the reaction was not suitable for use and an alternative route to the tetrazole **13** was required.

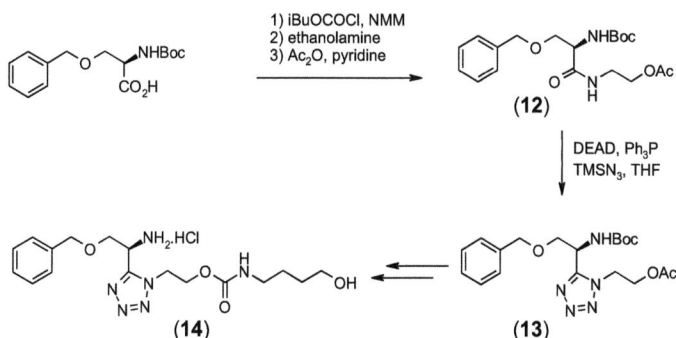

Scheme 4.7 Medicinal chemistry synthesis of tetrazole **14**.

Owing to the time constraints on the project, the team retained the same starting material and much of the original synthetic sequence, as is evident from Scheme 4.8. The key discovery was that they were able to prepare amidrazone **16** by reacting oxazoline **15** with excess hydrazine. This then allowed them to construct the tetrazole ring **17** *via* a diazotisation process. While this process was also highly exothermic, simply reversing the order of addition (adding amidrazone **16** and sodium nitrite to cold methanolic HCl) gave a readily controlled process that was scaled to 40 kg without incident, providing a fit-for-purpose route to the final target **14**.

As well as starting materials and intermediates, the minimisation, control or safe handling of by-products needs to be considered when scaling up. Chemists from Novartis had a need to monitor the presence of hazardous hydrazoic acid during the preparation of a tetrazole arising from formation and reaction of tri-*n*-butyltin azide with a nitrile (Scheme 4.9).[7] The hydrazoic acid is formed by reaction of tri-*n*-butyltin azide with trace amounts of hydrochloric acid.

Hydrazoic acid has a boiling point of 36 °C, forms an explosive gas mixture with air or nitrogen and is highly toxic. Controlling its presence to a very low level is therefore highly important. The Novartis scientists used a calibrated near-infrared (NIR) probe to monitor the concentration of hydrazoic acid in the reactor headspace of a 630 L vessel in the pilot plant. This allowed them to follow the concentration of hydrazoic acid in real time and respond accordingly if the level rose above a predetermined limit, for example by lowering the

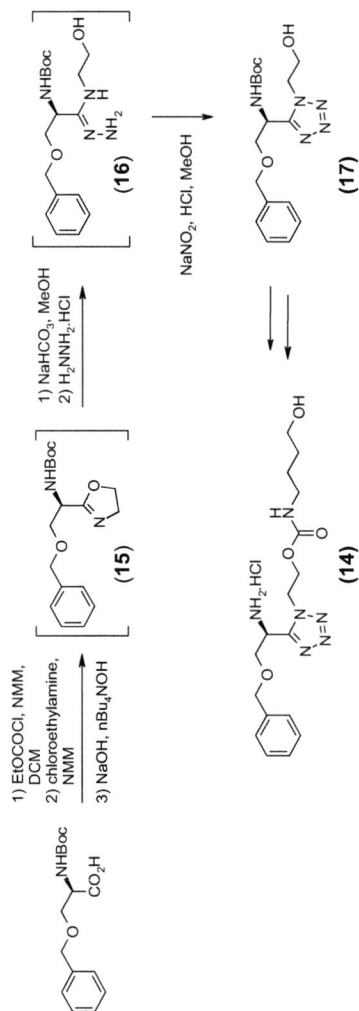

Scheme 4.8   Diazotisation route to tetrazole **14**.

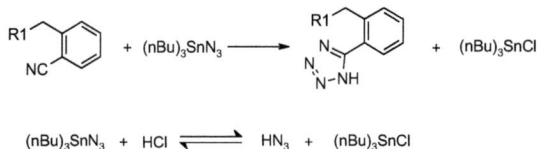

Scheme 4.9 Nitrile–azide cycloaddition route to tetrazoles.

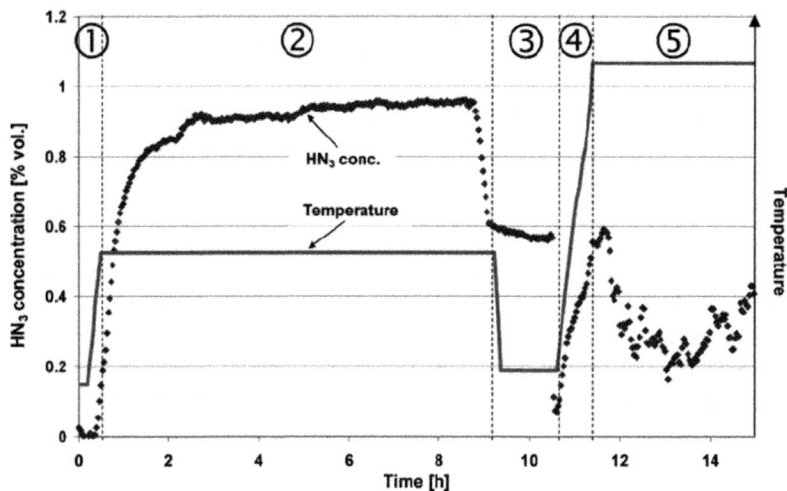

Figure 4.4 Concentration of hydrazoic acid as monitored by near-infrared spectro-scopy (reproduced from reference 7 with permission of the American Chemical Society).

reaction temperature. Figure 4.4 shows the NIR trace derived from the reaction, where tri-*n*-butyltin chloride and sodium azide are combined and the reaction is heated (1), the reaction temperature is held whilst tri-*n*-butyltin azide is formed (2), the reaction is cooled and a nitrile is added (3), the temperature is increased (4) and the reaction is held during formation of the tetrazole (5).

Many reagents and solvents can present a significant environmental impact, confronting the process chemist with a choice between changing the route or trying to control the environmental burden. While it is usually preferable to do the former, there are numerous examples where early phase projects have used the control option in order to maximise the speed of delivery.

For example, owing to a lack of alternative routes to a desired 1,3-disubstituted naphthalene core, a group at Merck decided to use a stoichiometric HgO-mediated selective decarboxylation reaction in their initial synthesis of a CCK(1R) receptor agonist (Scheme 4.10).[8] Whilst this pragmatic decision was feasible, given the relatively small scale of operations, it did impose additional burdens on the project team, as they were forced to monitor and control the mercury levels in all subsequent intermediates to ensure that the level in the API was acceptable and all the waste streams needed to be handled separately,

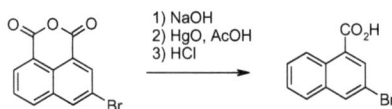

Scheme 4.10   Mercury-mediated regioselective decarboxylation.

incurring additional cost. Fortunately, treatment of a late-stage intermediate with a polystyrene-bound trimercaptotriazine (1,3,5-triazine-2,4,6-trithiol) resin reduced the level of mercury to an acceptable <3 ppm. The Merck solution provided an effective means to access early quantities of a potential drug candidate, but further development would certainly be required to support a future manufacturing process should the development programme be successful.

Conversely, researchers at Novartis found it unacceptable to utilise the environmentally hazardous solvent carbon tetrachloride in their synthesis of a novel and potent Flt3 kinase inhibitor.[9] A key intermediate **20** (Scheme 4.11) was prepared utilising a free-radical benzylic bromination in carbon tetrachloride to convert the substituted methylaniline **18** into the bromomethyl compound **19**. The enabled process (Scheme 4.12) involved only a two-step synthesis from commercially available benzonitrile **21**. Thus reduction of **21** with diisobutylaluminium hydride (DIBALH) cleanly afforded the aldehyde **22** in 98% yield. Subsequent reductive amination gave the desired intermediate **20** in 81% yield, providing a fit-for-purpose, more environmentally acceptable process that circumvented the troublesome radical-mediated step.

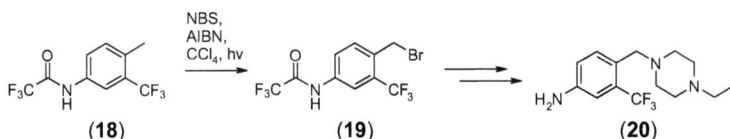

Scheme 4.11   Free-radical bromination of aniline **18**.

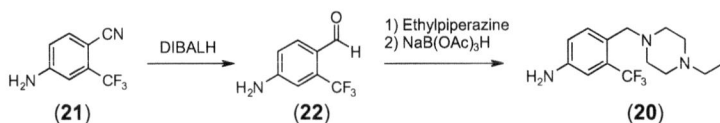

Scheme 4.12   Alternative nitrile reduction route to Flt3 kinase inhibitor **20**.

The preceding discussion has mainly focused on managing hazards posed by the use of specific reagents; however, the development chemist needs to remember that the material being prepared will ultimately be administered to human volunteers or patients. As a result, stringent purity standards are required and general guidance on acceptable levels of impurities,[10] metals[11] and residual solvents[12] is available. It is important to note that these levels are usually dependent on dose, duration of study, mode of administration, therapeutic area and other factors; therefore the final levels are set through discussions with toxicology experts.

In the case of reagents, intermediates or by-products that are recognised as having carcinogenic or genotoxic potential, more stringent controls are required. One commonly adopted approach is to control the level to the "threshold of toxicological concern" (TTC), defined as 1.5 µg d^{-1}. This recognises that complete elimination of the compound is not feasible and that, at this level, exposure to the compound will not pose a significant carcinogenic risk.[13] With this limit, a simple calculation allows an appropriate level for the impurity to be defined, and the analyst has to develop a method sensitive enough to detect the impurity at this level (usually in the low ppm range). It is important to consider all possible sources of potential mutagens or carcinogens, including those generated through side-reactions between reagents and solvents. Common examples of this include the generation of methyl methanesulfonate from the reaction between methanol and methanesulfonic acid,[14] and methyl or ethyl chloride from the reaction of hydrochloric acid with methanol or ethanol.[15]

4.2.2 Reliability

Any commercial process will be highly robust, with critical processing parameters well understood such that key quality attributes are controlled and the process is not being run close to the edge of failure. Operating to such standards requires significant time and effort which are not affordable to the early process chemist. Instead, critical criteria relating to robustness must be met to ensure successful scale-up without the burden of lengthy experimentation. A clear flag for early enabling is the identification of reactions with significant variability in yield, particularly if they reside towards the end of a synthesis.

Researchers at Abbott Laboratories enabled the original synthetic route of the H$_3$ antagonist **26** (Scheme 4.13),[16] to support delivery of multikilogram quantities for early development. The key step involved the coupling reaction of the aryl bromide **23** with pyridazin-3(2*H*)-one to form the arylpyridazinone **24**, and was found to be highly variable on scale when copper powder in pyridine was utilised to catalyse the reaction. A screen of metal catalysts led to optimisation of the copper-based conditions, with copper powder substituted by copper(I) chloride. A screen of ligands, bases and solvents identified the optimum conditions, which utilised catalytic copper(I) chloride in concert with an 8-hydroxyquinoline ligand (5 mol%) and potassium carbonate in DMF. These conditions were ultimately utilised to prepare **24** from **23** in around 85% assayed yield on a large scale, without any detectable *O*-arylation by-product. Of note, the alcohol **24** was not isolated owing to its high solubility in a range of solvents, but was instead converted to the desired API **26** *via* the crystalline tosylate **25**, which provided an excellent isolation and purification point.

Another important factor critical to early process enabling is sensitivity to heat transfer. Whilst chemical rates stay constant on scale-up, heat transfer rates between the bulk reaction medium and the vessel walls do not. This is because they are a result of the surface-area-to-volume ratio, and whilst the surface area

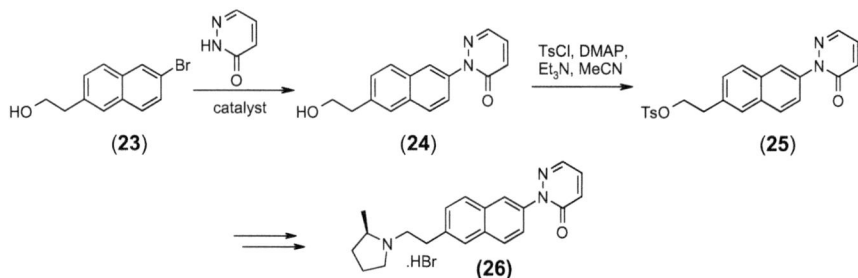

Scheme 4.13 Enabled route to H$_3$ antagonist **26**.

increases as a squared function of vessel radius, the volume increases as a cubed function. This means that as the vessel size increases, the surface-area-to-volume ratio decreases, resulting in poorer heat transfer between the reaction and vessel walls. As a result, heating and cooling become much slower.

A good example of heat transfer sensitivity is described by researchers at BMS, during their synthesis of a potassium channel activator drug candidate, BMS-180448 (Scheme 4.14).[17] The key reaction was a thermal Claisen rearrangement of 4-cyanophenyl 1,1-dimethylpropargyl ether {4-[(2-methylbut-3-yn-2-yl)oxy]benzonitrile, **27**} to 6-cyano-2,2-dimethylchromene (**28**). On a small scale this transformation proceeded well. However, during the course of laboratory scale-up an uncontrolled heating excursion occurred owing to a combination of the highly exothermic reaction and poorer cooling of the bulk reaction on the larger scale. Process safety testing (differential scanning calorimetry) revealed two broad exotherms with onset temperatures of 105 and 240 °C, respectively. Reaction kinetics data showed that the reaction occurred from 105 °C; however, a lack of sufficient cooling could push the reaction temperature above the secondary decomposition threshold. In one test, the temperature of the reaction rose from 180 to 445 °C within 33 seconds owing to self-heating. The BMS solution to this issue was to utilise a continuous plug flow reactor (PFR) rather than a standard batch mode vessel, as depicted in Figure 4.5. The cooling capacity of the reaction cell is greatly increased as the PFR technology provides a high surface-area-to-volume ratio, enabling effective management of highly exothermic processes. In addition, the size of the reaction at any one time is small, so if an exothermic runaway does occur the consequences are minimised, and the nature of the flow system means that flow rates and cooling can be adjusted in real time to re-establish control or stop the reaction. Thus the PFR provides a highly effective and rapid enabling solution to this heat transfer issue.

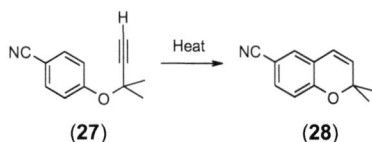

Scheme 4.14 Claisen rearrangement of propargyl ether **27**.

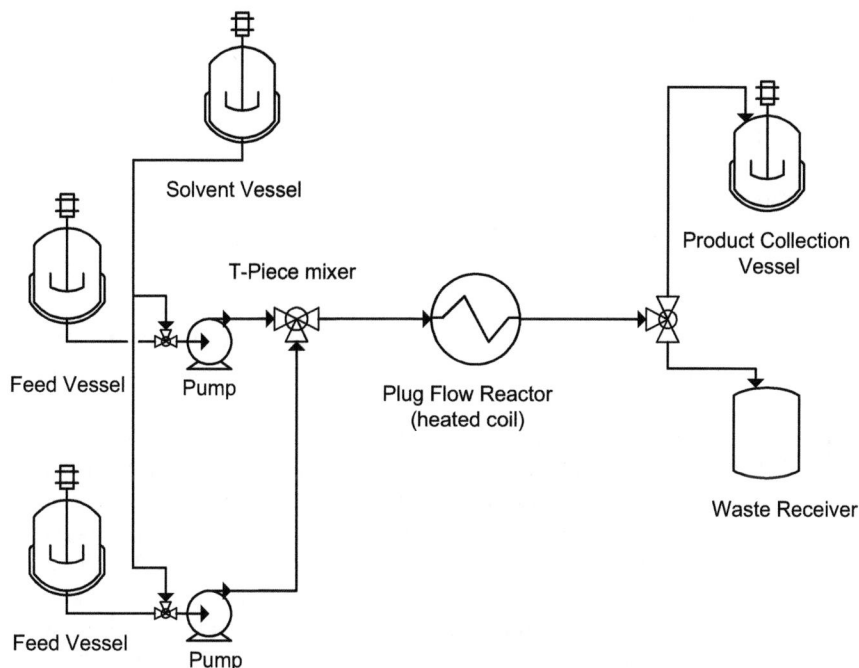

Figure 4.5 Schematic of a typical plug flow reactor system.

In addition to heat transfer rates, mass transfer rates are also generally poorer on scale for a given reactor system. Mass transfer describes mixing within or between phases, such as dissolution of solids or gases. Mass transfer rates will decrease on scale-up as the surface area for contact increases with the square of the radius, whilst the volumes of each phase increase as a cubic function. Moving from heterogeneous to homogeneous conditions is often an effective way of overcoming such issues. Alternatively, continuous processing or other mixing technologies can be an effective enabling approach.[18]

Chemists must enable their processes with an eye to available plant equipment in order to scale-up with reliability. Often an initial laboratory protocol cannot be duplicated on scale due to availability of large scale vessels or equipment. Such an issue was observed by Pfizer chemists during the development of the CCR1 antagonist **32** (Scheme 4.15) for the treatment of autoimmune disease and transplant rejection.[19]

The medicinal chemistry synthesis of **32** started from the lactone **29**, which was alkylated with prenyl bromide (1-bromo-3-methylbut-2-ene) to form **30**. The alkylation was highly exothermic and was run by addition of the prenyl bromide to a solution of the preformed lactone enolate under cryogenic conditions. The reaction was purposely stopped at around 70% completion to minimize formation of the bis-alkylation by-product **31**. Silica gel chromatography was required to separate **29**, **30** and **31**. The formation of dialkylated product and the presence of unreacted lactone are most likely due to proton

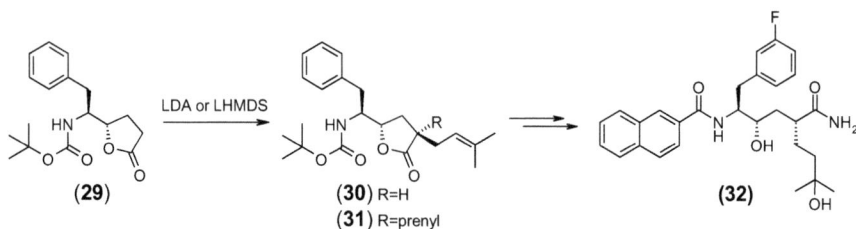

Scheme 4.15 Medicinal chemistry synthesis of CCR1 antagonist **32**.

exchange between the monoalkylated product **30** and the enolate of the starting material **29**, which occurs during the extended dosing time made necessary by poorer heat transfer on scale.

Initial enabling work focused on introduction of 1,3-dimethylimidazolidin-2-one (DMI) as a co-solvent to accelerate alkylation, and on assisting cryogenic temperature regulation by adding a pre-cooled prenyl bromide solution ($-78\,°C$) from a separate vessel. This allowed faster addition of the alkylating reagent to the lactone enolate solution whilst maintaining the temperature, significantly reducing the occurrence of proton exchange and subsequent bis-alkylation. These improved conditions gave the desired product **30** in typically 90–92% yield and >90% HPLC purity, with only 2–4% of the bis-alkylation product **31** and only 2% of starting material **29**. The crude product was directly taken to the next step after demonstrating that impurities at these levels could be purged in downstream steps, removing the need for chromatography.

This modification worked well at the kilogram scale, but the need for two cryogenic reactors proved to be a constraint to the existing Pfizer facility. Thus, to eliminate the need for the second cryogenic reactor, the process was modified further and the base (LHMDS) was dosed to a pre-cooled ($-78\,°C$) mixture of lactone **29**, prenyl bromide and N-methylpyrrolidin-2-one (NMP) in THF, giving an improved reaction profile (0.3–2% **29** and 1–2% **31**). This alkylation protocol worked well on a 25 kg scale, proved operationally convenient since only one cryogenic reactor was needed, and the reaction temperature was easily maintained by controlling the feed rate of the base. The crude product **30** could be directly carried forward to the next step as an oil.

Larger scale vessels and equipment can also have limitations in terms of processing parameters, *e.g.* high temperatures can be difficult to achieve. The 1,2,3,4,6,7,8,9-octahydroquinolizinium salt **34** (Scheme 4.16) is a key intermediate in the synthesis of the CCR5-antagonist MLN1251 **35**.[20] Whilst the team at Millennium Pharmaceuticals recognised that their initial process would not be suitable for large-scale production, they were unable to identify an alternative route within the time available. Therefore, in order to ensure delivery of material, they resorted to running the high-temperature cyclisation on a relatively small scale (~400 g), isolating the product by distillation at pot temperatures of up to 300 °C.

During the medicinal chemistry phase of a research programme, often little attention is paid to how a material is isolated. Frequently the product is

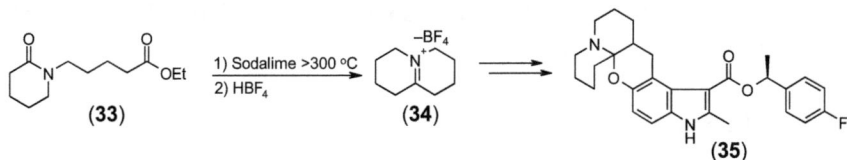

Scheme 4.16 High-temperature cyclisation of keto-amide **33**.

extracted into an organic solvent that is then concentrated to dryness and the residue purified by chromatography. As the scale of operation increases, this becomes more challenging and a significant proportion of process development is spent figuring out how best to isolate the material post-reaction and work-up. For example, during the synthesis of the H$_3$ antagonist **26** (Scheme 4.13), the alcohol **24** proved extremely challenging to purify *via* chromatography during the initial synthesis, whereas the introduction of the tosylate **25** provided a crystalline solid that was purified by crystallisation to afford the product in good yield (88%) and excellent purity (99% by HPLC), with the metal catalyst (Cu) level from the previous step controlled to <10 ppm.[16]

The isolation of the final API warrants specific discussion due to the importance of API properties on the formulation used in animal or human studies and the ability to achieve the desired exposure profile with acceptable stability and reproducibility. The selection and isolation of an appropriate solid form, such as a salt or free parent compound, and control of API polymorphism are key factors that must be managed during early development. As a compound moves through the development cycle, other factors such as crystal habit and particle size become increasingly important. These considerations are discussed at length in Chapter 13.

A critical skill of an early enabling process chemist is the identification of small-scale procedures that cannot be translated to a larger scale, for example common laboratory operations such as stripping to dryness, trituration of gummy solids and very slow filtrations. An excellent mechanism for identifying such issues is through the use of controlled laboratory reactors that mimic large-scale equipment.[21] Whilst they do not necessarily act as a precise engineering model for stirring efficiency or heat and mass transfer, problems such as extremely thick slurries that will not exit through the run-off valve, cleaning difficulties and encrustation on the heated jacket will highlight potential issues to the process chemist. Another benefit of the controlled laboratory reactor is that it can be upgraded to obtain more data than would be available from a conventional flask. Crude calorimetry data can be generated which will give an idea of potential thermal hazards, as well as an understanding of reaction rates and reagent stoichiometry.

GlaxoSmithKline (GSK) chemists came across an issue with slow filtration during the synthesis of a potent dual agonist of PPARα/γ for treatment of dyslipidemia.[22] Sticky agglomerates were formed during a process that required the presence of TiCl$_4$ as a water scavenger, leading to very slow filtration.

The issue was resolved by addition of sodium sulfate that led to an easily filtered, suspendable slurry.

4.2.3 Efficiency

An important factor to consider during the early stages of a project is the process efficiency, particularly any bottlenecks that could significantly hamper the speed of delivery or drive excessive cost. The overall yield of a process is a key consideration, particularly for linear routes. However, owing to the need for late-stage diversification in the medicinal chemistry process, linear routes are often provided as the starting point for process development. In these instances, the key points to focus on are low-yielding steps towards the end of the route, or specific inefficiencies that could cause a problem as the project progresses.

The initial medicinal chemistry route to the Pfizer A2a agonist **38** (Scheme 4.17), for the treatment of chronic obstructive pulmonary disease (COPD), was a linear sequence, starting from readily available guanosine **36**.[23] However, the final step was a low-yielding (53%) palladium-catalysed aminocarbonylation of iodide **37**. In addition to this low yield, the use of carbon monoxide would have been challenging and outsourcing the step to a specialist vendor (with attendant cost implications) was not considered a viable alternative. As a result, an alternative route was required, providing the team with the opportunity of designing a more efficient, convergent synthesis.

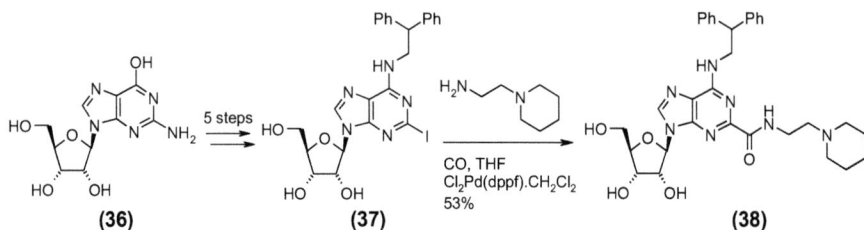

Scheme 4.17   Initial route to A2a agonist **38**.

The strategy adopted was to step-reorder the process (Scheme 4.18). Starting from 2,6-dichloropurine (**39**), a reasonably high-yielding route to the key heterocycle **40** was developed. In effect, the one-step aminocarbonylation reaction (53%) was replaced with a three-step cyanation, hydrolysis and amide formation sequence (47.6% overall), but because of the more convergent nature of the synthesis, the impact of this relatively low yield on the overall efficiency of the process was considerably less pronounced. This also removed the need for specialist CO technology, allowing the process to be operated within available plant equipment.

Whilst the primary driver for early phase projects is generally speed of delivery, the cost and availability of key raw materials is still an important consideration. During the development of iminosugar **42** (Scheme 4.19) in the laboratories of United Therapeutics,[24] a significant quantity of the key building block, lactone **41**, was required.

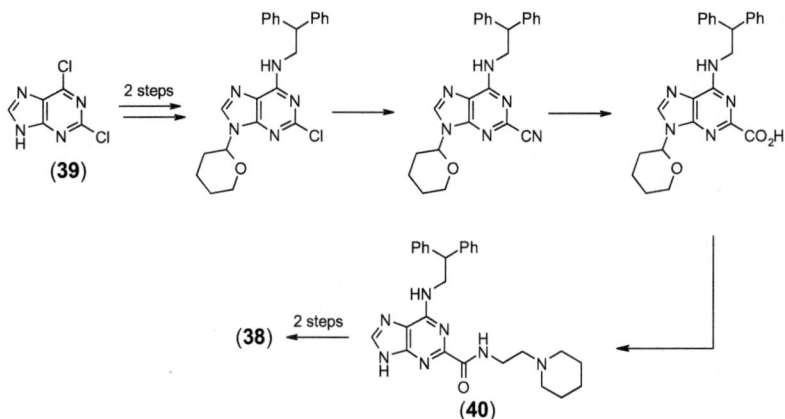

Scheme 4.18 Enabled route to A2a agonist **38**.

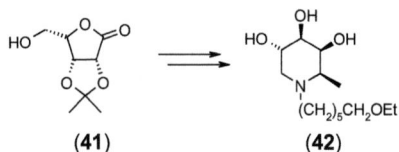

Scheme 4.19 Synthesis of iminosugar **42**.

Initially the material was prepared *via* a four-step sequence from D-gulono-lactone **43** (Scheme 4.20), which was very expensive and not readily available. The team therefore developed an alternative four-step sequence from readily available and relatively inexpensive D-ribose **44** (Scheme 4.21). The crux of this synthesis was inversion of the C-4 stereocentre by treatment of the mesylate **45** with potassium hydroxide, presumably *via* the intermediacy of an epoxide, to give the desired product in a relatively high-yielding cost-effective synthesis, from readily available raw materials.

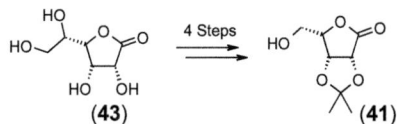

Scheme 4.20 Initial route to lactone **41**.

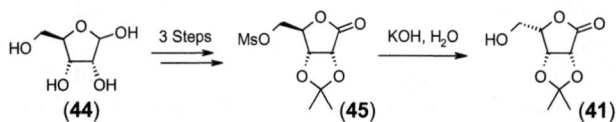

Scheme 4.21 Improved route to lactone **41**.

Historically, removing chromatographic purifications from a synthetic sequence has been a major focus of early process development, though the introduction of automated chromatographic systems at both laboratory and plant scale has meant that this is less of a concern and these separations are often retained, particularly in the early phases of process development.[25] However, there are still many instances where chromatographic procedures are replaced in early process development.

For example, during the initial development of a progesterone antagonist,[26] the Pfizer project team was faced with an inefficient and impractical chromatographic separation of the two regioisomeric pyrazoles **47a** and **48a** (Scheme 4.22). In the initial medicinal chemistry route, alkylation of the unsubstituted pyrazole **46** with chloromethyl methyl sulfide afforded a mixture of products **47a** and **48a** ($\sim$1:1) that was separated by chromatography.

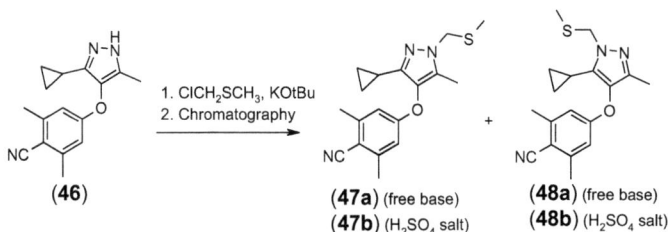

Scheme 4.22   Alkylation of pyrazole **46**.

In order to overcome this problem, an alternative isolation procedure was developed wherein the post-work-up product solution was treated with sulfuric acid, precipitating a mixture of the hydrogensulfate salts **47b** and **48b**. An additional reslurry of this initial mixture of salts in acetonitrile removed the unwanted regioisomer **48b**, affording high-purity **47b** in 37% overall yield on a multikilogram scale. An added advantage of this new process was that the malodorous residues derived from the chloromethyl methyl sulfide remained in solution and could be readily contained and disposed of.

One area in particular where chromatography has found extensive use, even through to commercial production, is in the separation of enantiomers by chiral chromatography. For example, during the development of the Pfizer farnesyl transferase inhibitor **51** (Scheme 4.23),[27] a key racemic intermediate **50** was prepared and then resolved by chiral simulated moving bed chromatography using a ChiralPak AD resin on up to 30 kg scale with excellent recovery of the desired enantiomer (42%).

Whilst the Pfizer researchers were able to manage this chromatography on scale, the synthesis of the racemic intermediate **50** suffered from another key efficiency concern: that of volumetric throughput (that is, how much solvent and aqueous waste is generated by the process). The main problem with the initial process was the addition of the methylimidazole Grignard reagent to the ketone **49**. The Grignard reagent was prepared *in situ* from the reaction of 5-bromo-*N*-methylimidazole with 1 M ethylmagnesium bromide. Four

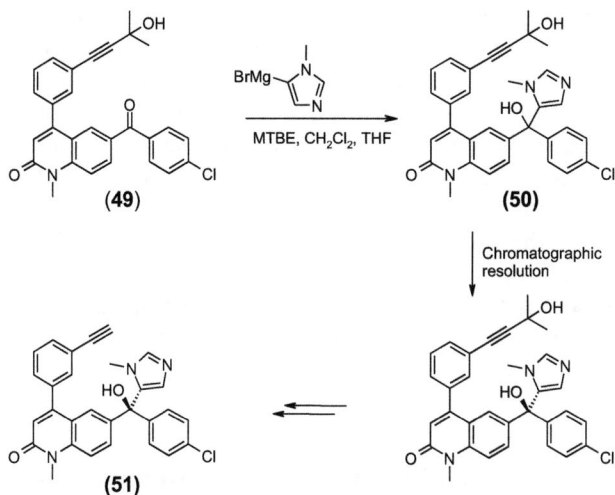

Scheme 4.23 Initial route to farnesyl transferase inhibitor **51**.

equivalents of the methylimidazole Grignard reagent were required in order to drive the reaction to completion. Reaction completion was also dependent on running the process at very high dilution in a complex mixture of solvents to ensure homogeneity of the reaction mixture. The high dilution severely hampered throughput and, despite considerable effort, the team was unable to improve this process.

Instead, an alternative synthesis of **50** was developed where the key to success was a change in protecting group strategy. The alkyne protecting group was changed from tertiary alcohol **49** to the trimethylsilyl (TMS) protected alkyne **52**, as outlined in Scheme 4.24. The Pfizer researchers found

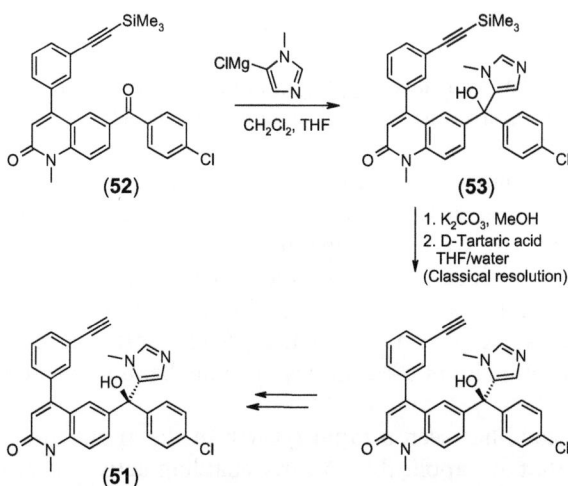

Scheme 4.24 Improved route to farnesyl transferase inhibitor **51**.

that the TMS-alkyne **52** was converted cleanly to the tertiary alcohol **53** by treatment with only two equivalents of the methylimidazole Grignard reagent under more concentrated conditions. Under the revised conditions the methylimidazole Grignard was prepared *in situ* from the reaction of 2 M ethylmagnesium chloride rather than 1 M ethylmagnesium bromide. Under these more concentrated conditions the reaction progressed to completion, even as a thin heterogeneous slurry. This reduced the total organic solvent requirement for the step by 38% and aqueous waste by 94%. In addition, deprotection of the TMS-alkyne was significantly easier, simplifying the downstream processing. In spite of the excellent throughput of the chiral chromatography, a more efficient diastereomeric salt resolution with D-tartaric acid was developed.

4.3   Technology for Rapid Fit-for-purpose Enabling

The use of technology to accelerate early process enabling efforts with small quantities of material has increased greatly in recent years, including (i) screening technologies, (ii) reactor control and data logging, (iii) processing technologies and (iv) analytical technologies.

Screening technologies offer accelerated entry into improved reaction conditions. Screening may be carried out with manual loading and sampling of a parallel reactor array (typically up to 12 reaction tubes of 0.5–10 mL scale) or can be fully automated with solid and liquid handling, full HPLC integration and sampling at multiple time points, often carried out on 96 reactions at once (<1 mL scale). Screening is usually used to find alternatives to toxic or thermally hazardous reagents, more suitable conditions in terms of processing qualities (for example, direct drop processes)[28] or to solve chemo- and regioselectivity issues. However, it may also be used to gather information on physical properties such as solubility, impurity purging and stability.

Reaction control and data logging capabilities are highly sophisticated on large-scale computer-controlled pilot plant vessels. Hence, a similar level of sophistication is required on smaller scale technologies that aim to mimic the larger scale equipment. Available laboratory systems are capable of controlling stirrer speeds, reagent addition rates and temperatures whilst logging data associated with these unit operations, leading to greater confidence of scale-up. The difference in reaction and jacket temperature $(T_r–T_j)$ is particularly useful as it gives an indication of the heat output of a reaction that can be used (with some caveats) to make process safety assessments and to understand the kinetics of a reaction which may be fundamental to predicting scalability.[21]

The last decade has seen a rapid growth in the application of processing technologies that are applicable to early enabling activities. A wide range of continuous processing or flow chemistry technologies are available to execute reactions, including microreactors,[29] static mixers[30] and continuous stirred tank reactors,[31] with the benefit that lengthy enabling effort can often be circumvented. Whilst it may be possible to run a reaction in a continuous fashion, consideration should also be given to the work-up. For early development candidates, a batch mode work-up is common place, although the use of tools such as a centrifugal extractor or a continuous crystalliser[32] may be considered. Another class of processing technology that is making a significant impact in the pharmaceutical sector is that of controlled crystallisation methods, which allow greater control of solid forms. Some examples of this type of technology are high-shear wet milling, sonocrystallisation and impinging jet crystallisation.[18]

Process analytical technology (PAT) describes a range of techniques that are used to provide analytical information about a reaction or process in real time. The most common technique is online or *in situ* IR spectroscopy, which can be further subdivided into near-IR (NIR) and mid-IR techniques.[33] In general, IR spectroscopy is a good tool to monitor the changes in chemical bonding in solution, such as the hydrogenation of double bonds, reactions at carbonyls and deprotonation or metallation of intermediates. Since it requires no sample preparation or work-up, a much better picture of the reaction status is obtained, reactive intermediates can be monitored directly and the real-time nature of the analysis allows reactions to be stopped at critical points prior to over-reacting. As well as following the appearance and disappearance of starting materials, intermediates and products in a qualitative sense, reaction systems can be calibrated to allow quantitative measures of reaction completion or impurity level, thereby allowing the technique to be used as a reaction completion monitoring tool. In addition to this, online IR can be used to give a much better understanding of the reaction system, helping the process chemist to direct problem solving efforts at the key issues which may not be apparent from offline analysis by HPLC, TLC or NMR. Limitations of the technique are that it will only analyse species in solution, aqueous solutions are generally unsuitable since the O–H stretching swamps most of the spectral region of interest, and the sensitivity of the technique is in the region of 10%, meaning that low-level impurities cannot be tracked using this method.

4.4 Conclusion

In order to accelerate potential drug candidates through the development lifecycle, rapid fit-for-purpose enabling of API processes is an essential requirement in the pharmaceutical industry. Through the use of effective criteria to trigger enabling activities, combined with rapid technologies to execute enabling packages, process chemists can play a critical role in delivering material rapidly to support toxicology and clinical studies. Once confidence in the drug candidate is achieved through a successful POC clinical study, the role of the process chemist then changes, as described in subsequent chapters.

References

1. M. Butters, D. Catterick, A. Craig, A. Curzons, D. Dale, A. Gillmore, S. P. Green, I. Marziano, J. P. Sherlock and W. White, *Chem. Rev.*, 2006, **106**, 3002.
2. See http://www.ich.org (last accessed 31 January 2011).
3. M. Reuman, Z. Hu, G. H. Kuo, X. Li, R. K. Russell, L. Shen, S. Youells and Y. Zhang, *Org. Process Res. Dev.*, 2007, **11**, 1010.
4. F. Mazzini, E. Alpi, P. Salvadori and T. Netscher, *Eur. J. Org. Chem.*, 2003, **15**, 2840.
5. K. Menzel, F. Machrouhi, M. Bodenstein, A. Alorati, C. Cowden, A. W. Gibson, B. Bishop, N. Ikemoto, T. D. Nelson, M. H. Kress and D. E. Frantz, *Org. Process Res. Dev.*, 2009, **13**, 519.
6. A. H. Davulcu, D. D. McLeod, J. Li, K. Katipally, A. Littke, W. Doubleday, Z. Xu, C. W. McConlogue, C. J. Lai, M. Gleeson, M. Schwinden and R. L. Parsons, *J. Org. Chem.*, 2009, **74**, 4068.
7. J. Wiss, C. Fleury and U. Onken, *Org. Process Res. Dev.*, 2006, **10**, 349.
8. J. T. Kuethe, K. G. Childers, G. R. Humphrey, M. Journet and Z. Peng, *Org. Process Res. Dev.*, 2008, **12**, 1201.
9. W.-C. Shieh, J. McKenna, J. A. Sclafani, S. Xue, M. Girgis, J. Vivelo, B. Radetich and K. Prasad, *Org. Process Res. Dev.*, 2008, **12**, 1146.
10. International Conference on Harmonisation (ICH), *Guideline Q3A(R2): Impurities in New Drug Substances*, ICH, Geneva, October 2006.
11. *Guideline on the Specification Limits for Residues of Metal Catalysts or Metal Reagents*, European Medicines Agency, London, February 2008; http://www.ema.europa.eu/pdfs/human/swp/444600enfin.pdf (last accessed 31 January 2011).
12. International Conference on Harmonisation (ICH), *Guideline Q3C(R4): Impurities Guideline for Residual Solvents*, ICH, Geneva, February 2009.
13. D. A. Pierson, B. A. Olsen, D. K. Robbins, K. M. DeVries and D. L. Varie, *Org. Process Res. Dev.*, 2009, **13**, 285.
14. A. Teasdale, S. C. Eyley, E. Delaney, K. Jacq, K. Taylor-Worth, A. Lipczynski, V. Reif, D. P. Elder, K. L. Facchine, S. Golec, R. Schulte Oestrich, P. Sandra and F. David, *Org. Process Res. Dev.*, 2009, **13**, 429.
15. Q. Yang, B. P. Haney, A. Vaux, D. A. Riley, L. Heidrich, P. He, P. Mason, A. Tehim, L. E. Fisher, H. Maag and N. G. Anderson, *Org. Process Res. Dev.*, 2009, **13**, 786.
16. Y.-M. Pu, Y.-Y. Ku, T. Grieme, L. A. Black, A. V. Bhatia and M. Cowart, *Org. Process Res. Dev.*, 2007, **11**, 1004.
17. R. J. Bogaert-Alvarez, P. Demena, G. Kodersha, R. E. Polomski, N. Soundararajan and S. S. Y. Wang, *Org. Process Res. Dev.*, 2001, **5**, 636.
18. N. G. Anderson, *Org. Process Res. Dev.*, 2001, **5**, 613.
19. B. Li, B. Andresen, M. F. Brown, R. A. Buzon, C. K.-F. Chiu, M. Couturier, E. Dias, F. J. Urban, V. J. Jasys, J. C. Kath, W. Kissel, T. Le, Z. J. Li, J. Negri, C. S. Poss, J. Tucker, D. Whritenour and K. Zandi, *Org. Process Res. Dev.*, 2005, **9**, 466.

20. M. Rönn, Q. McCubbin, S. Winter, M. K. Veige, N. Grimster, T. Alorati and L. Plamondon, *Org. Process Res. Dev.*, 2007, **11**, 241.
21. C. Bernlind and C. Urbaniczky, *Org. Process Res. Dev.*, 2009, **13**, 1059.
22. L. M. Oh, H. Wang, S. C. Shilcrat, R. E. Herrmann, D. B. Patience, P. G. Spoors and J. Sisko, *Org. Process Res. Dev.*, 2007, **11**, 1032.
23. S. Challenger, Y. Dessi, D. E. Fox, L. C. Hesmondhalgh, P. Pascal, A. J. Pettman and J. D. Smith, *Org. Process Res. Dev.*, 2008, **12**, 575.
24. H. Batra, R. M. Moriarty, R. Penmasta, V. Sharma, G. Stanciuc, J. P. Staszewski, S. M. Tuladhar, D. A. Walsh, S. Datla and S. Krishnaswamy, *Org. Process Res. Dev.*, 2006, **10**, 484.
25. C. J. Welch, D. W. Henderson, D. M. Tschaen and R. A. Miller, *Org. Process Res. Dev.*, 2009, **13**, 621.
26. P. A. Bradley, P. D. de Koning, P. S. Johnson, Y. C. Lecouturier, D. J. McManus, A. Robin and T. J. Underwood, *Org. Process Res. Dev.*, 2009, **13**, 848.
27. B. M. Anderson, M. Couturier, B. Cronin, M. D'Occhio, M. D. Ewing, M. Guinn, J. M. Hawkins, V. J. Jasys, S. D. LaGreca, J. P. Lyssikatos, G. Moraski, K. Ng, J. W. Raggon, A. M. Stewart, D. L. Tickner, J. L. Tucker, F. J. Urban, E. Vazquez and L. Wei, *Org. Process Res. Dev.*, 2004, **8**, 643.
28. C.-K. Chen and A. K. Singh, *Org. Process Res. Dev.*, 2001, **5**, 508.
29. (a) C. Wiles and P. Watts, *Eur. J. Org. Chem.*, 2008, **10**, 1655; (b) A. Palmieri, S. V. Ley, K. Hammond, A. Polyzos and I. R. Baxendale, *Tetrahedron Lett.*, 2009, **50**, 3287.
30. C. Brechtelsbauer and F. Ricard, *Org. Process Res. Dev.*, 2001, **5**, 646.
31. J. G. Van Alsten, M. L. Jorgensen and D. J. am Ende, *Org. Process Res. Dev.*, 2009, **13**, 629–633.
32. (a) J. G. Van Alsten, L. M. Reeder, C. L. Stanchina and D. J. Knoechel, *Org. Process Res. Dev.*, 2008, **12**, 989; (b) S. Lawton, G. Steele, P. Shering, L. Zhao, I. Laird and X.-W. Ni, *Org. Process Res. Dev.*, 2009, **13**, 1357.
33. K. De Smet, J. van Dun, B. Stokbroekx, T. Spittaels, C. Schroyen, P. Van Broeck, J. Lambrechts, D. Van Cleuvenbergen, G. Smout, J. Dubois, A. Horvath, J. Verbraeken and J. Cuypers, *Org. Process Res. Dev.*, 2005, **9**, 344.

CHAPTER 5
Route Design and Selection

MIKE BUTTERS

Pharmaceutical Development, AstraZeneca, Avlon Works, Severn Road, Bristol, BS10 7ZE, UK

5.1 Introduction

Process chemists in the pharmaceutical industry are responsible for designing an efficient synthetic route for the manufacture of an active pharmaceutical ingredient (API). During the course of a development programme, synthetic routes can be changed a number of times for a variety of reasons. In the modern pharmaceutical industry, process chemistry departments address this challenge in a highly organized and coordinated approach, often involving large numbers of chemists working towards a common goal. Given the increasing financial and time pressures on the industry, process chemists need to continually improve their capability for the design of new synthetic routes. The skills and knowledge required for the design of synthetic routes are multi-faceted and this chapter will discuss the various components of this important activity.

In common with most R&D activities, the design of a synthetic route involves a plan–do–review cycle, which is discussed in Sections 5.2–5.5. This cycle is made up of a number of individual steps (see Figure 5.1 below) and the process chemist will often repeat these steps in an iterative fashion before reaching a final decision on route selection.

Working through Figure 5.1, it can be seen that there are two alternative "recycle" options between the review and plan phases. Where experimental data do not support an idea, the chemist can choose to invest more effort to find a solution to the problem (recycle A). Alternatively, an idea can be eliminated from further investigation and alternative ideas can be created or prioritised

RSC Drug Discovery Series No. 9
Pharmaceutical Process Development: Current Chemical and Engineering Challenges
Edited by A. John Blacker and Mike T. Williams
© Royal Society of Chemistry 2011
Published by the Royal Society of Chemistry, www.rsc.org

1. Define the scope and timeframe of the project 2. Evaluate the current route 3. Exhaustive generation of ideas for synthesis 4. Sort and group into an 'Ideas Map' 5. Evaluation and prioritisation of the ideas 6. Generate work plan	PLAN
7. Agree responsibilities in the team 8. Source materials for the work 9. Complete phase of experimental work	DO
10. Review knowledge from experimental work 11. Eliminate and/or generate new ideas	REVIEW
12. Compare options and select route	SELECT

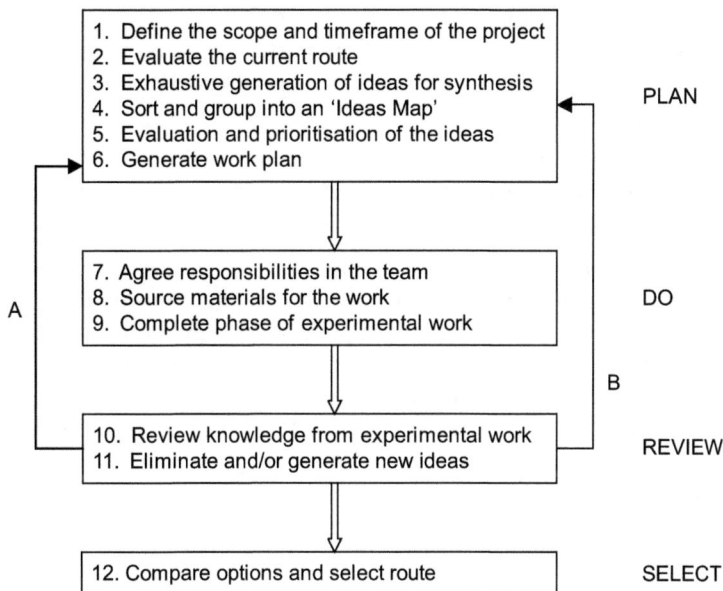

Figure 5.1 Plan–do–review process for route design and selection.

(recycle B). The degree of success in working through this plan–do–review cycle depends upon four key capabilities, which are discussed in Sections 5.2–5.5 below.

5.2 Responding to the Needs of the Drug Development Programme

The process chemist must understand the needs of the drug development programme before planning a route design programme. Route design is normally driven by (i) difficulties in supplying the API for the development programme and/or (ii) providing an attractive process to support regulatory submission and competitive commercial supply. If resources are available, it is more efficient to introduce the "commercial route" at the start of the drug development project lifecycle. For each drug development programme there are several factors which might influence the project management of route design:

(i) Project priority in the portfolio
High priority projects are strong candidates for early introduction of a commercial route. For lower priority projects, limited supply of resources may narrow the scope of the route design exercise.

(ii) Ability to meet bulk demands (quantity and timescale)
The technical difficulty of a target drug candidate can often take precedence over other factors and justify effort on route design to circumvent difficulties in supply.

(iii) Risk investment strategy
Route design activities should be planned to take account of GO/NO GO decision points associated with clinical or toxicological read-outs in the drug development project plan.

(iv) Structural similarity to other drug candidates
There is a benefit from investing in a route design programme that may serve more than one API in the discovery or clinical phase.

(v) Regulatory risks to a development candidate or commercial product
Changing the route of synthesis can result in a different impurity profile that may then need regulatory qualification. It is preferable to avoid repeat toxicological studies and instead schedule qualification of new-route material to fit in with existing (planned) studies. For a commercial product the benefit around cost savings should be weighed against the regulatory impact on the supply chain, especially where there is a change to registered good manufacturing practice (GMP) stages.

(vi) Cost of development
Complex drug candidates can be very expensive to manufacture on multi-kilogram scale and the introduction of an economical synthesis is an important contribution to reducing R&D expenditure.

(vii) Commercial product profile
The predicted volume at peak sales and the target cost of the API will vary according to factors like therapeutic area, dose and selling price. These relationships will be discussed in Section 5.3.4 and have a significant influence on setting design criteria for the synthetic route.

All of the above factors can influence the scope, timescale and design criteria of a route design programme.

5.3   Criteria for Route Evaluation and Selection

In the pharmaceutical industry, the process chemist is faced with a large number of constraints together with an opportunity to create significant economic value. Because of these factors, the process chemist will typically investigate a large number of potential synthetic routes to a single API target and strive to identify the most efficient option. Evaluation and comparison of different route options can be made using a set of criteria[1] (SELECT) which include safety, environmental, legal, economics, control and throughput considerations. Each of these criteria will be discussed below in the context of route design.

5.3.1   Safety

Safety is the most important of the SELECT criteria. If a route cannot be scaled up safely, then it should not be scaled up at all. The criterion of safety can be further subdivided into (i) thermal and reactive hazards and (ii) toxic hazards. A more detailed explanation of process safety is provided in Chapter 8. This section will therefore focus on representative case histories.

The potential for explosion is a threat that should be eliminated as soon as possible. 3-Nitropyrazole (**1**) (Scheme 5.1) was a starting material in the medicinal chemistry synthesis[2] of ICI-162,846 (**2**). The heat of decomposition for **1** was found to be >2800 J g^{-1} (a potential explosive) and therefore an alternative starting material was required.

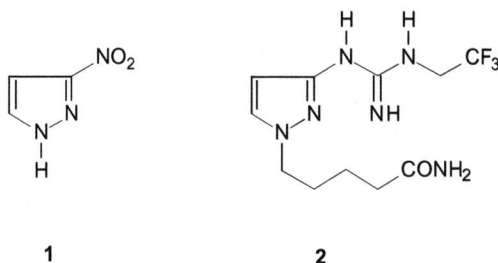

1   **2**

Scheme 5.1   Nitropyrazole (**1**) intermediate in the synthesis of ICI-162,846 (**2**) (reproduced with permission from Elseiver).

The Zeneca Pharmaceuticals drug candidate ZD-2079 (**6**) (Scheme 5.2) (a beta-3 agonist) entered development in 1991, intended for the treatment of non-insulin-dependent diabetes. The medicinal chemistry route to **6** presented a

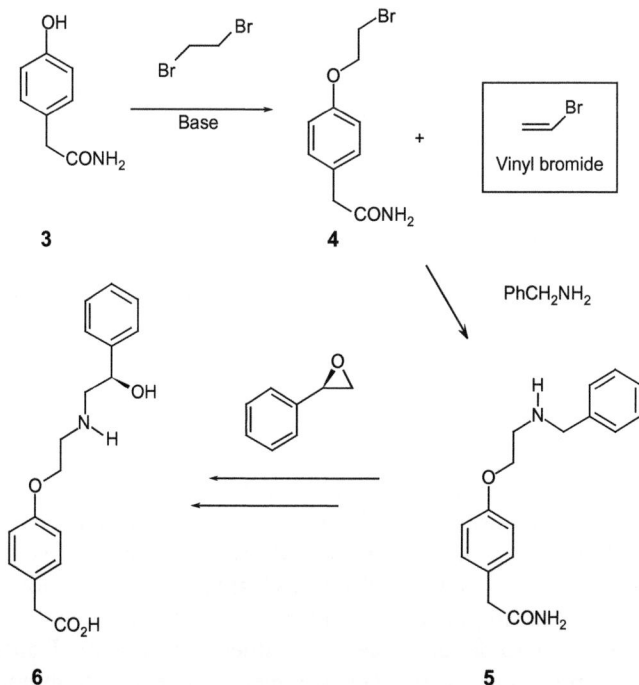

6   **5**

Scheme 5.2   Medicinal synthesis of ZD-2079 (reproduced with permission from American Chemical Society).

number of challenges for scale-up.[3] The generation of toxic vinyl bromide gas in step 1 was unavoidable, given that the reaction of dibromoethane with the phenolic starting material 3 required base. Vinyl bromide[4] is known to have genotoxic properties and cannot be readily removed by scrubbing on a plant scale. An alternative strategy for providing the two-carbon unit in step 1 involved ethanolamine derivatives (Scheme 5.3). *N*-Benzyloxathiazolidine *S*-oxide (9) was prepared by reaction of *N*-benzylethanolamine (8) with thionyl chloride. This cyclic derivative of ethanolamine provided activation of the oxygen toward nucleophilic attack while preventing intramolecular attack by nitrogen. This approach also circumvented the issue of controlling mono- *versus* di-substitution, as dibromoethane can form a diether with phenol 3. Reaction of 9 with the sodium salt of (4-hydroxyphenyl)acetamide (7) provided amine 5 in 64% yield (by comparison, the route in Scheme 5.2 gave only a 9% overall yield to 5).

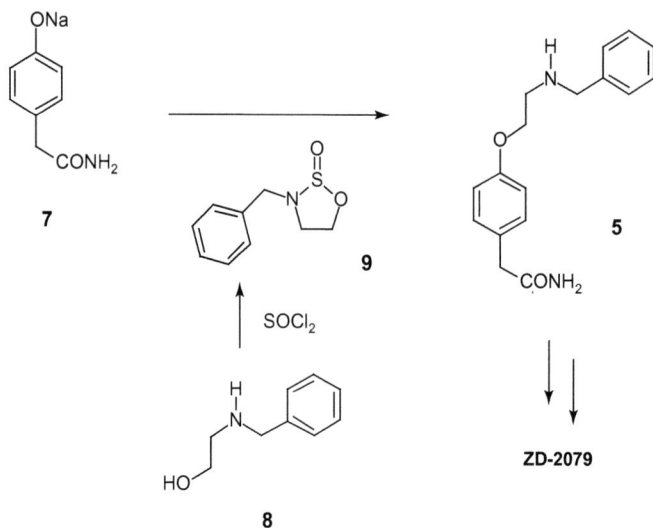

Scheme 5.3   Process R&D synthesis of ZD-2079 (reproduced with permission from American Chemical Society).

5.3.2   Environmental

Development of a process that is totally sustainable and has a low environmental impact is not always possible owing to the current limits of science and technology. However, process improvements can often be achieved and a fuller description of this activity is provided in Chapter 6. Understanding environmental issues and identifying opportunities in early development will maximize the chemist's ability to design "greener" synthetic approaches. Route design is an important opportunity to avoid environmental issues and to minimise waste.

ICI-162,846 (2) (Scheme 5.4) is a histamine H_2 blocker that was under development for the treatment of ulcers and gastric disorders. The medicinal chemistry route to 2 involved mercury oxide assisted conversion of a thiourea fragment **11** to give the key guanidine intermediate **12**.[2] Although it may be possible to purify the API to remove low levels of mercury, the environmental consequences of mercury-containing waste are serious.[5] A number of mercury compounds are known to accumulate in aquatic ecosystems, and high levels can be found in certain fish. It was very clear, therefore, that development quantities of **2** could not be manufactured using the mercury methodology, and the search for a new route (Scheme 5.4) was initiated. Treatment of the aminopyrazole intermediate **10** with a cyanamide reagent provided a more convergent synthesis of **12** and was used for the preparation of 100 kg quantities of **2**.

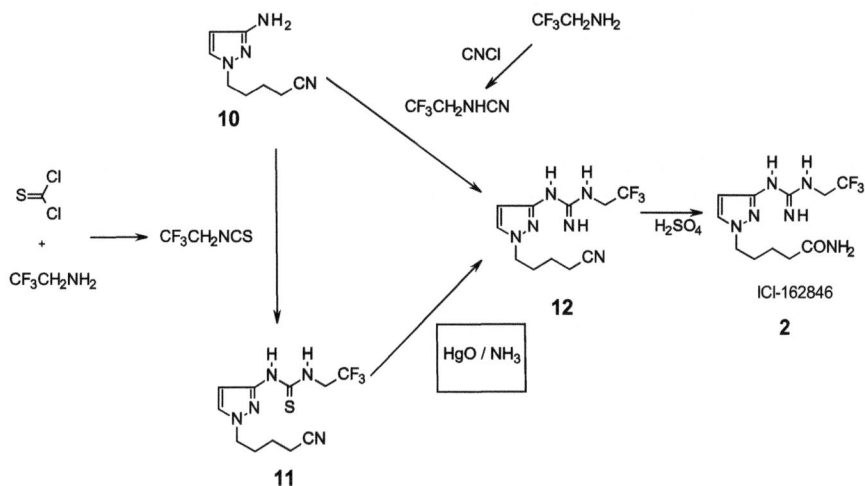

Scheme 5.4 Synthetic routes to ICI-162846 (reproduced with permission from Elsevier).

The medicinal chemistry synthesis (route A) of the glycoprotein IIb/IIIa antagonist lotrafiban (**15**) started from an aryl Grignard reagent, and the chiral centre was introduced using L-aspartic acid. This synthesis[6,7] involved 11 linear steps in an overall yield of 9% (Scheme 5.5) and the level of waste in this route would have a significant environmental impact.

Route B (Scheme 5.6) was quickly developed to support early clinical requirements and involved a one-pot procedure converting 2-nitrobenzyl alcohol (**16**) to intermediate **17**. Enzymatic resolution of **18** using an immobilized form of *Candida antarctica* lipase B gave the desired (*S*) stereochemistry. While this route was successfully scaled up to give kilogram quantities of **15**, it involved a wasteful late-stage resolution and low-yielding preparation of mono *N*-Cbz-4,4′-bipiperidine. Later in development, these issues were addressed through the introduction of route C (Scheme 5.7). Thus, enzymatic resolution of the simple benzodiazepine **19** proved advantageous in that the unwanted (*R*)-enantiomer **20** could be recycled. Furthermore, aminocarbonylation of **22** using 4-(piperidin-4-yl)pyridine gave a more efficient introduction of the bipiperidine unit. Including the recycle of the (*R*)-enantiomer **20**, this route provided **15** in a 29% overall yield, compared with 17% for route B.

13

(i) TFA, anisole
(ii) BOC-4,4-bipiperidine

(iii) aq NaOH, MeOH
(iv) aq HCl, Dioxane

14

15
lotrafiban

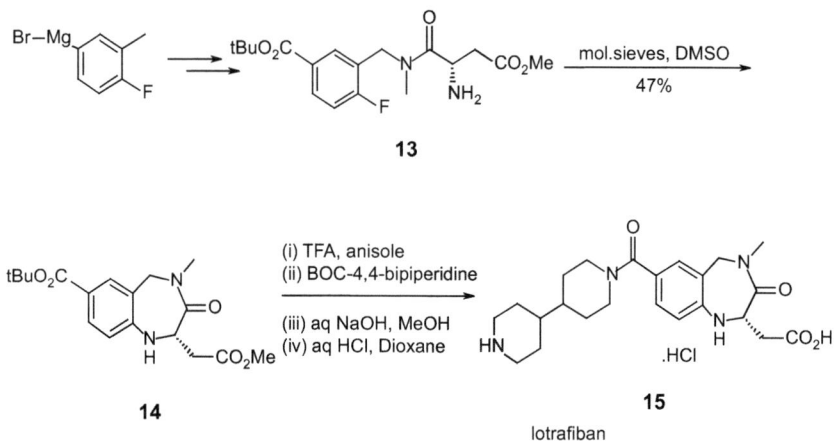

Scheme 5.5 Route A synthesis of lotrafiban (reproduced with permission from American Chemical Society).

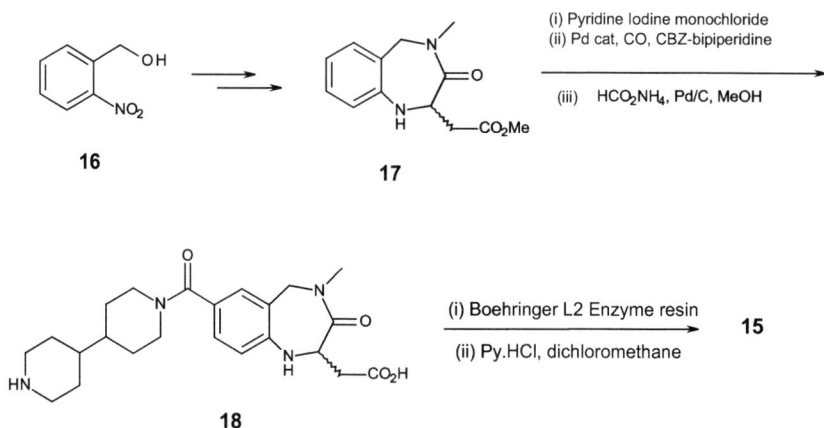

16

17

(i) Pyridine Iodine monochloride
(ii) Pd cat, CO, CBZ-bipiperidine

(iii)   HCO$_2$NH$_4$, Pd/C, MeOH

18

(i) Boehringer L2 Enzyme resin

(ii) Py.HCl, dichloromethane

15

Scheme 5.6 Route B synthesis of lotrafiban (reproduced with permission from American Chemical Society).

E-Factors and reaction mass efficiency (RME) are useful metrics that can be utilised to measure the environmental impact of a process, as discussed in Sections 6.3.2 and 6.3.3. A summary of these data for routes A, B, and C is provided in Table 5.1. Manufacture using route A would generate a significant amount of waste (>1.4 tonnes of waste per kg of API). The introduction of route C, however, provides a 5.4-fold improvement in waste reduction (reducing waste by >1.1 tonnes per kg of API).

5.3.3   Legal

It is important that the development and commercialisation of a pharmaceutical product can be performed without breaking laws or infringing valid intellectual

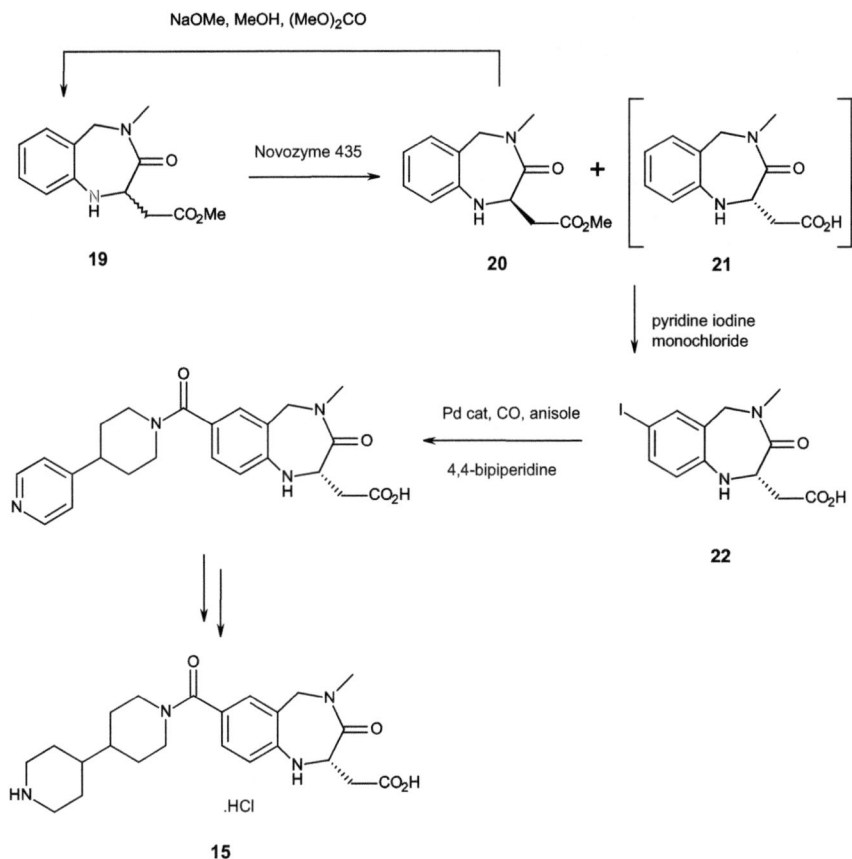

Scheme 5.7 Route C synthesis of lotrafiban (reproduced with permission from American Chemical Society).

Table 5.1 Comparison of environmental metrics for routes to lotrafiban.

Route	Yield (%)	E-Factor	RME (%)
A	9	1429	1.3
B	17	1173	2.6
C	29	262	7.6

property (IP). Legal issues can arise at any point in development and can justify a change in synthetic route or process irrespective of other potential issues. Types of legal issues fall into two major categories: (1) regulated substances [(a) use of controlled or banned substances;[8] (b) using unacceptable quantities of COMAH[9](control of major accidents and hazards, EU legislation) listed chemicals; (c) transportation of certain hazardous materials; (d) use of materials with third-party restriction[10] (for example, data from the notification of new substances NONS)]; and (2) patent infringement[11] [use of materials, technology or processes that potentially infringe current and valid third-party intellectual property]. In developing a commercially viable route to an API, there is a basic "freedom to operate" consideration. There may be little point developing a route to an API that uses an intermediate or process that is claimed in a valid third-party patent. Such a patent might exclude the process chemist from using the process or intermediate for commercial manufacture. These situations are generally dealt with on a case-by-case basis, and it is best to seek advice from a patent attorney (see Figure 5.2).

Acyclovir (**23**) (Scheme 5.8) was discovered, developed and marketed by GSK as an antiviral drug. In 1996, compound **23** accounted for 40% (about $1.25 billion) of the total antiviral agents market. Not surprisingly, therefore, this product engendered intensive competition among industrial research groups, which led to many challenges by generic companies to the patented Burroughs-Wellcome (BW) processes. An important example[12] is that of the Recordati Company, who successfully identified an arguably patent-free synthetic route to **23** by noting a small omission in the BW process patent. BW did not claim R = H (on a literal interpretation of the claims) in the definition of a key intermediate **24**, and the Recordati Company subsequently exploited this. In the

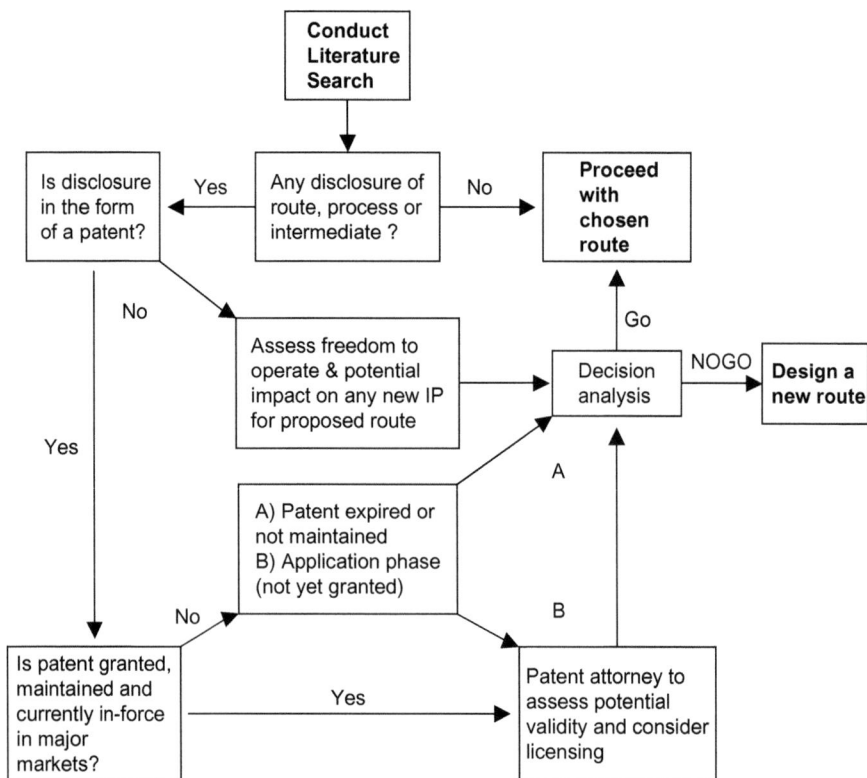

Figure 5.2   Decision process for patent protection of synthetic routes (reproduced with permission from American Chemical Society).

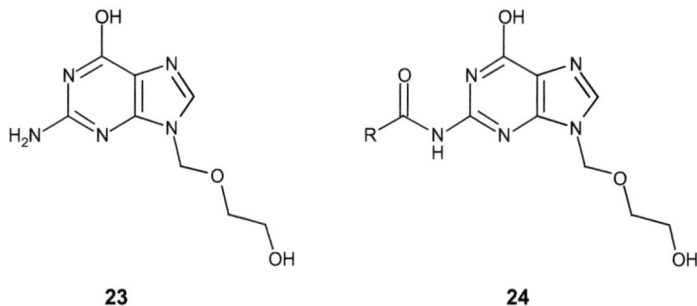

Scheme 5.8 Acyclovir and its precursor.

USA version it can be noted in the compound claims that R can indeed be H, therefore more clearly covering the *N*-formyl derivative. However, in the Italian/English version there is no explicit claim for a compound where R = H, which on a literal interpretation of the claims arguably allowed the Recordati Company to define an industrial process using the formyl as a 2-amino protecting group. However, the possibility in the USA and certain European countries of asserting patent infringement in respect of a technically "equivalent" process falling just outside the literal scope of the claims should be noted.

Astra developed remoxipride (**30**) (Scheme 5.9) for the treatment of schizophrenia and launched it in the early 1990s. The medicinal chemistry route[13] to **30** involved a classical resolution of racemic pyrrolidinamine **25**. This racemate was used in the commercial synthesis of sulpiride and readily available in bulk quantities. The resolution of **25** (route 1) was patented[14] by a third party, and various claims in the patent blocked commercialisation of this method. For example, a "composition of matter" type claim for the tartrate salt of enantiomerically pure pyrrolidinamine **26** was covered in the scope of this patent. It was therefore necessary to identify an alternative non-infringing route for long-term manufacture of **30**. A commercial process was established in which (*S*)-proline (**27**) (natural enantiomer) was converted to the amide **28**, alkylated with ethyl bromide to give **29** and reduced with Red-Al to provide **26** as the free base (route 2, Scheme 5.9). This route has the advantage that the starting material **27** is available as a single enantiomer compared with the resolution process, which has a 50% maximum theoretical yield.

5.3.4 Economics

The key variables determining the economic viability of a new drug are: (1) cost of goods (CoG), (2) product selling price, (3) marketing costs for the product and, in some cases, (4) product and/or technology licensing costs (see above section on legal issues). The term CoG is used to describe the total costs involved in the manufacture of a drug product (this includes API manufacture, formulation and packaging) expressed as a percentage of the selling price of the

Route 1

(i) L-(+)-tartaric acid

(ii) filter (R)-(Y)-tartrate salt

(iii) D-(-)-tartaric acid

(iv) collect (S)-(Y)-tartrate salt

25   **26**

Route 2

27   **28**

26   **29**

Remoxipride   **30**

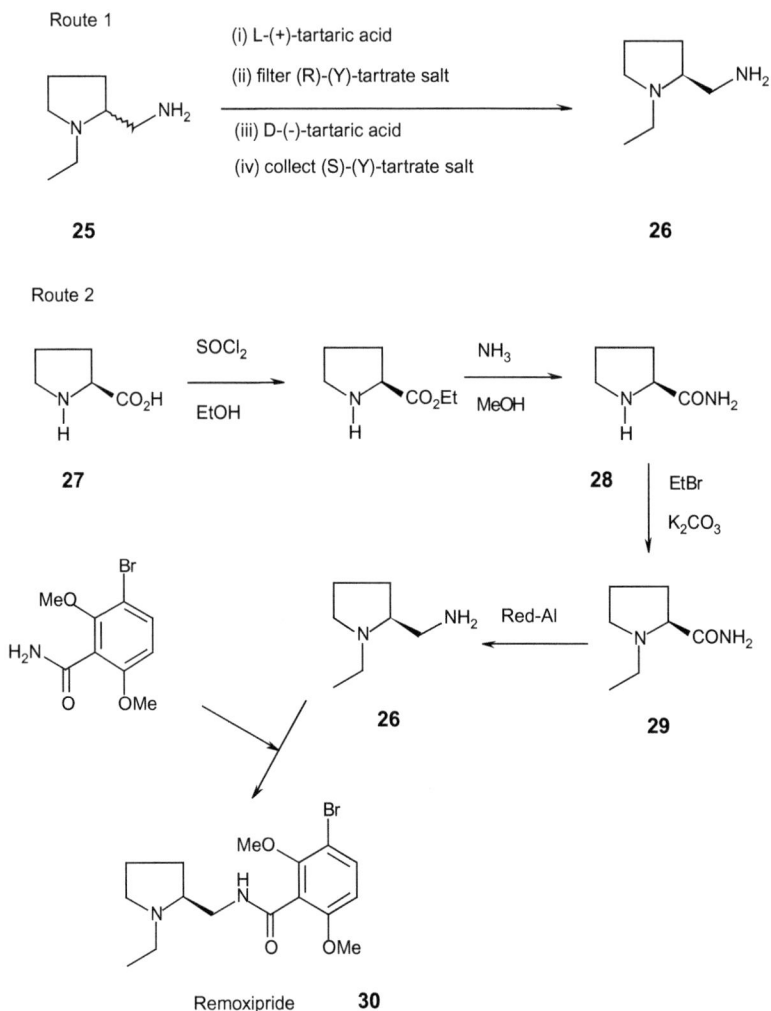

Scheme 5.9   Routes to remoxipride (reproduced with permission from Wiley-VCH Verlag GmbH & Co).

drug. It can be calculated using eqn (5.1), and an illustrated calculation is provided in Figure 5.3.

$$\%\text{CoG} = \frac{[(\text{Daily dose} \times \text{Cost of API}) + \text{Cost of formulation}]}{\text{Selling price}} \times 100 \quad (5.1)$$

API manufacturing costs are therefore a subset of CoG. The main types of economic issues associated with the manufacture of an API are: (1) failing to meet the CoG target, (2) unacceptable investment costs during development and (3) licensing costs for third-party IP.

Daily Selling Price = £2/day
Daily Dose = 0.2g/day
Bulk Drug Cost = £1,500/kg
Formulation Cost = £0.1/daily dose

Calculation: $\dfrac{(0.2\times1.5) + 0.1}{2}\times100$ = 20% Cost of Goods

Figure 5.3 Illustrated example of cost of goods calculation (reproduced with permission from American Chemical Society).

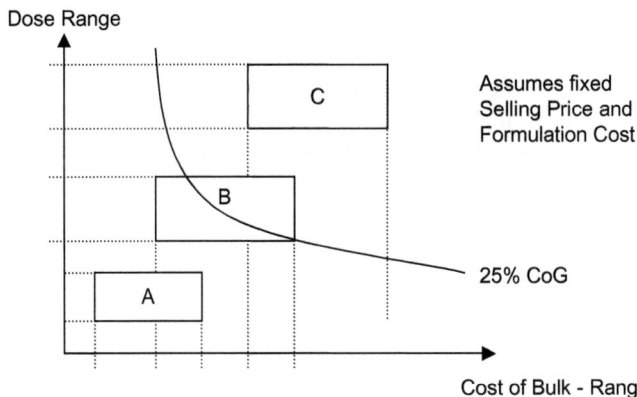

Figure 5.4 Modelling the economic viability of a drug product (reproduced with permission from American Chemical Society).

Predicting the capability of a process to meet the CoG target for the future market is a critical activity. An analysis published[15] by the Office of Technology Assessment suggests that CoG as a proportion of total product sales is, on average, 25%. Thus, on average, 75% of product sales contribute to marketing costs, research and development, operating costs and profit margin. Assuming that the target CoG is going to be 25%, the next step is to predict likely ranges for each variable in eqn (5.1) and assess the probability of meeting the target. In early drug development, however, it is helpful to model the economic viability by considering pricing scenarios and using expected ranges for daily dose and for API costs. The chart in Figure 5.4 can be constructed using eqn (5.1) and assuming a fixed selling price and a fixed formulation cost.

In Figure 5.4, the box labelled A represents a low-risk cost analysis; the expected dose range and API cost range place the box underneath the target CoG curve (25%) and suggest it should be easy to devise an economic process. Box C, on the other hand, illustrates a high-risk position where confidence in reaching an acceptable combination of dose and API cost is very low. The majority of early development projects fall into category A or B, where there is a low to medium risk of economic failure. It is expected, therefore, that the development chemist will be successful in establishing a cost-effective process.

ZD-3638 (**38**) (Scheme 5.10) is an atypical antipsychotic agent for the treatment of schizophrenia which was developed by Zeneca from 1993 to 1997. Early clinical evaluation predicted that the daily dose requirement was likely to be somewhere in the range 10–50 mg. An economic assessment of the drug candidate indicated that there was only a medium probability of meeting a 25% CoG target when using the early development route[16] (route 1, Scheme 5.10) with raw material costs of £1475 kg^{-1}. The most significant contributor to raw material costs was aldehyde **31** (£910 kg^{-1} of **38**), which was processed through five chemical steps to make **38**. In an addition reaction the (2-fluoropyridin-3-yl)lithium **35** (from LDA) was reacted with **34** to give **36** in a modest 65% yield. A series of alternative routes was evaluated which introduced the expensive aldehyde **31** at a later point in the sequence. In route 2 (Scheme 5.10), **31** was introduced in the final step. In addition, improved yields of the addition reaction between **35** and **39** (85%) were achieved using LiTMP. Raw material costs for route 2 were much lower (£789 kg^{-1}) and the contribution from **31** was only £294 kg^{-1} of **38**. This change of synthetic route significantly improved the probability of meeting the CoG target.

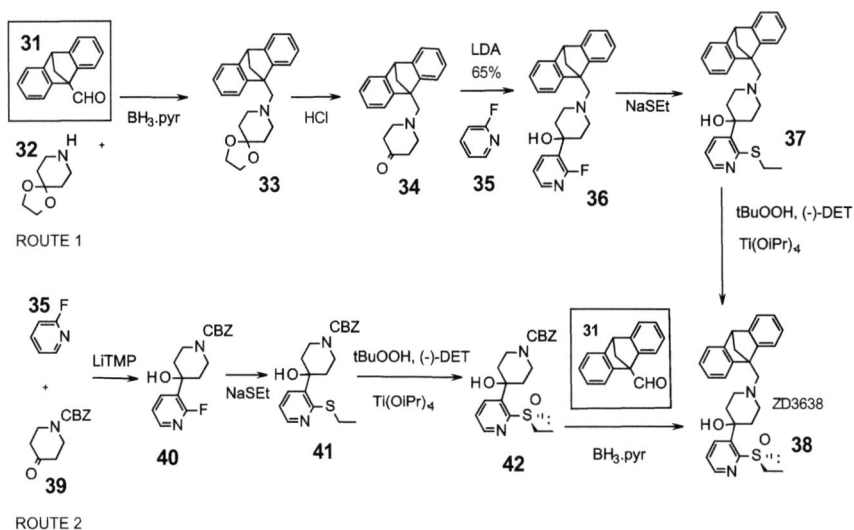

Scheme 5.10   Routes to ZD-3638 (reproduced with permission from American Chemical Society).

Candoxatril (**47**) (Scheme 5.11) is an orally active prodrug of candoxatrilat, a potent atrial natriuretic factor (ANF) potentiator indicated in the treatment of hypertension and congestive heart failure that was developed by Pfizer. While any new drug needs to demonstrate significant advantages over existing therapies, in the hypertension market it is also important that new treatments are economically viable due to the number of cheap, effective remedies already available. Candoxatril (**47**) contains a chiral centre and therefore an efficient

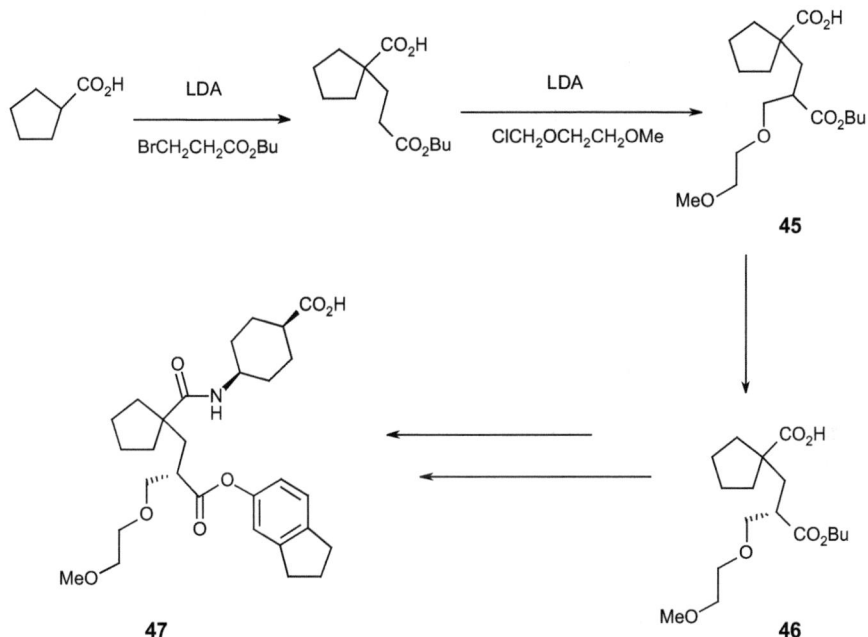

Scheme 5.11 Synthesis of candoxatril *via* resolution (reproduced with permission from Elseiver).

synthesis of the single enantiomer was required. The first development route[17] employed an expensive classical resolution of racemic glutarate **45** with (+)-pseudoephedrine to furnish the chiral glutarate **46** in 13% yield over four steps (Scheme 5.11). The new route[18] (Scheme 5.12) used the inexpensive starting material *tert*-butyl acrylate (**48**) in a Baylis–Hillman reaction to install the methoxyethyl side-chain and give acrylate **49** in a single step. An iodosulfonation–dehydroiodination sequence gave the tosyl acrylate **50**, which then underwent an addition–elimination step to produce the geometrically pure alkene **51** ready for hydrogenation investigations. A ruthenium/BINAP-based catalyst was employed in the hydrogenation to complete the new route to **46** in 33% overall yield, an increase of 2.5-fold over the medicinal chemistry route. An analysis of the relative costs of the two routes showed that the asymmetric hydrogenation route was approximately three times cheaper than the resolution route. This brought the cost of the API below the percentage CoG target.

5.3.5 Control

To conduct clinical trials in the USA it is necessary to manufacture an API according to FDA published guidelines[19] and in compliance with ICH guidelines (Q3, Q6 and Q7).[20] Equivalent guidelines are in place for the EU[21] and the rest of the world. These guidelines serve to protect patient safety during the clinical trials, and this is achieved through setting appropriate quality criteria[22]

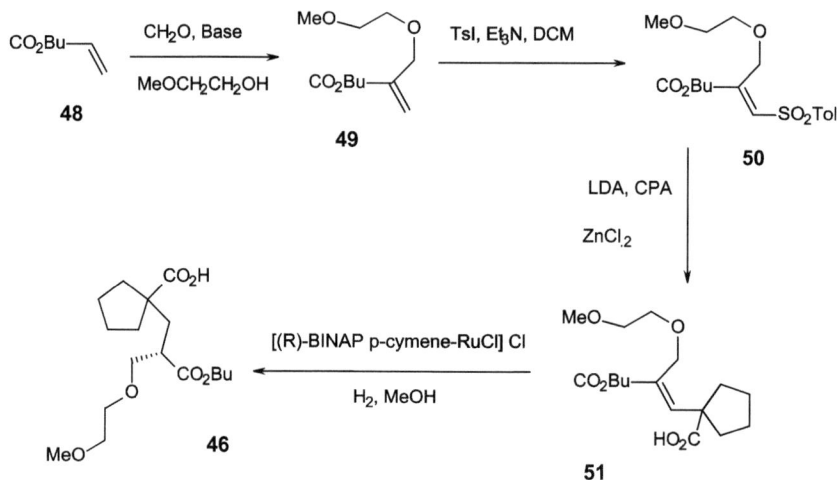

Scheme 5.12 Synthesis of candoxatril using asymmetric hydrogenation (reproduced with permission from American Chemical Society).

in the API specification and through working to cGMP (current good manufacturing practice). A key challenge for the process chemist is to scale-up the process reproducibly and without adversely affecting the quality of the API. Control of API quality is achieved through control of chemical and physical parameters in the process. When assessing a synthetic route for control issues, the process chemist should identify the following: (1) non-selective reactions (chemo-, regio- and stereo-) and other side reactions that are likely to generate process related impurities; (2) the chemical stability and physical properties of each intermediate and reagent (in particular, labile functional groups and chiral centres; stability towards heat, moisture and oxygen; hygroscopicity; viscosity; and crystallinity); (3) the number and efficiency of potential purification points in the route.

Homochiral pyridinediol **56** (Scheme 5.13) is a key intermediate in the bulk manufacture of an AstraZeneca drug candidate. The medicinal chemistry route[23] to this intermediate involved acetonide protection and periodate oxidation of gluconolactone **52** to generate 2,3-*O*-isopropylidene-L-glyceraldehyde (**53**). While it might be possible to find more economic routes to **53**, the stability of this aldehyde is variable and presents a significant concern for commercial manufacturing. Several reports in the literature cite difficulties with epimerisation, although work on the D-isomer[24] was not adversely impacted by instability.

Extensive polymerisation of **53** occurs within a few days, but exclusion of moisture and air retards this polymerisation to a rate of 10% per week. The option of learning how to manage the stability issues with **53** was considered. However, given the high cost of gluconolactone, it would still be necessary to evaluate new routes at some point in development. It was therefore attractive to design an alternative route to **56** using a stable intermediate in which the

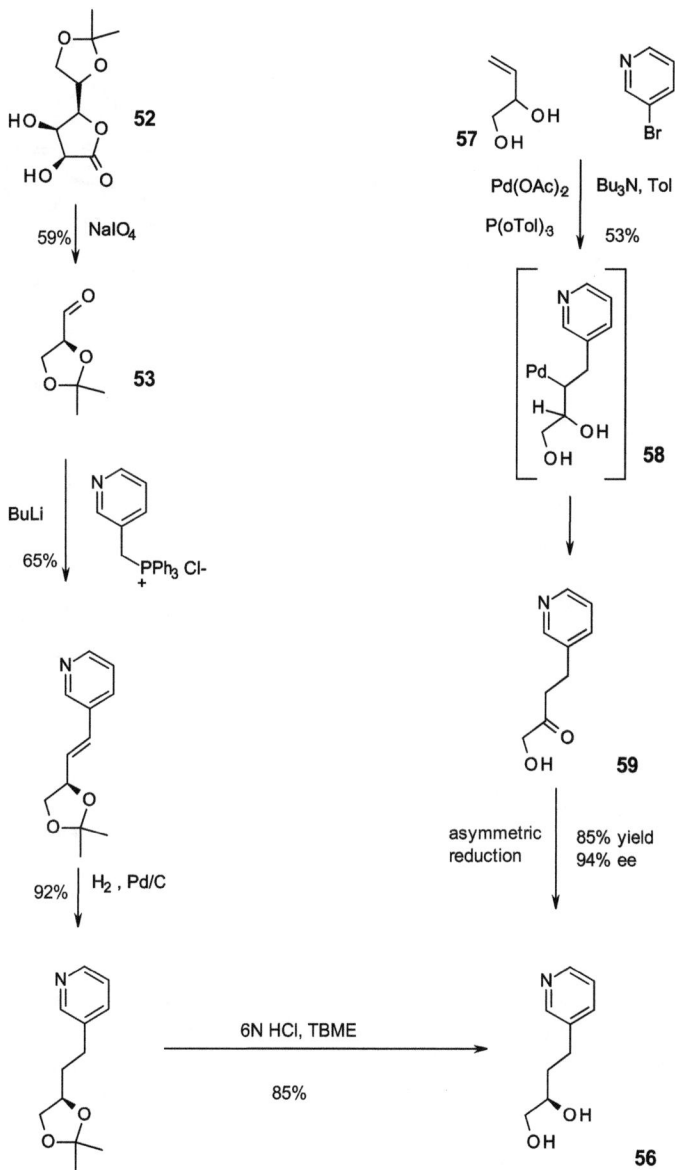

Scheme 5.13 Routes to chiral diol intermediate (reproduced with permission from American Chemical Society).

chiral centre could be controlled without loss of optical purity. Asymmetric reduction of ketone **59** gave diol **56** in high yield and enantiomeric excess. A Heck reaction of 3-bromopyridine with but-3-ene-1,2-diol (**57**) provided a simple route to **59** whereby the dehydropalladation of **58** favoured the enol product.

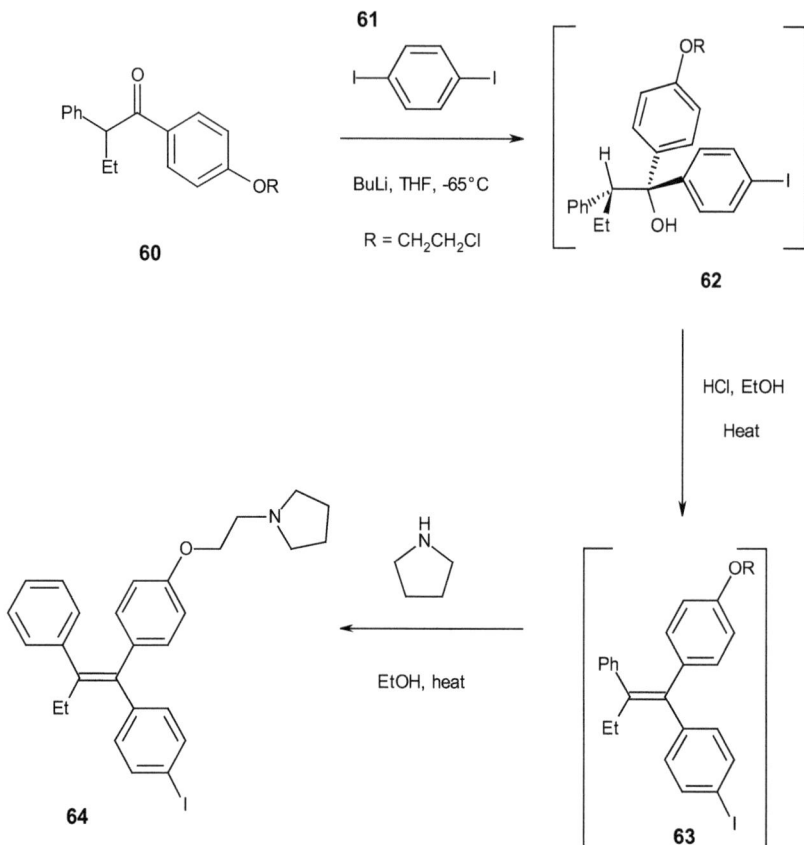

Scheme 5.14 Synthesis of idoxifene *via* dehydration (reproduced with permission from American Chemical Society).

Idoxifene (**64**) (Scheme 5.14), a selective estrogen receptor modulator developed by GSK, contains a tetra-substituted double bond. The initial route[25] to supply this compound involved preparation of the alcohol **62**, which was dehydrated under acidic conditions to give the desired *E*-alkene **63** as a mixture of *E/Z* isomers (70:30). The alkene mixture was then taken forward to **64** by substitution with pyrrolidine and purification by crystallisation. This gave a 35% yield from **60** and a 24% overall yield. The lack of selectivity with respect to the alkene geometry meant that a new long-term route was required. It was noted that much of the stereochemical information was lost in the acid-catalysed dehydration reaction, and it was believed that an improved *E/Z* ratio could be obtained using a concerted *syn*-elimination of a diastereomerically enriched substrate. The preparation of alcohol **67** was therefore developed *via* a Felkin–Anh controlled Grignard addition to **65** (Scheme 5.15). This alcohol was converted to the pivalate ester **68**, which was in turn converted into **64** *via* a selective *syn*-elimination by refluxing with HMDS in 1,2,4-trimethylbenzene. This

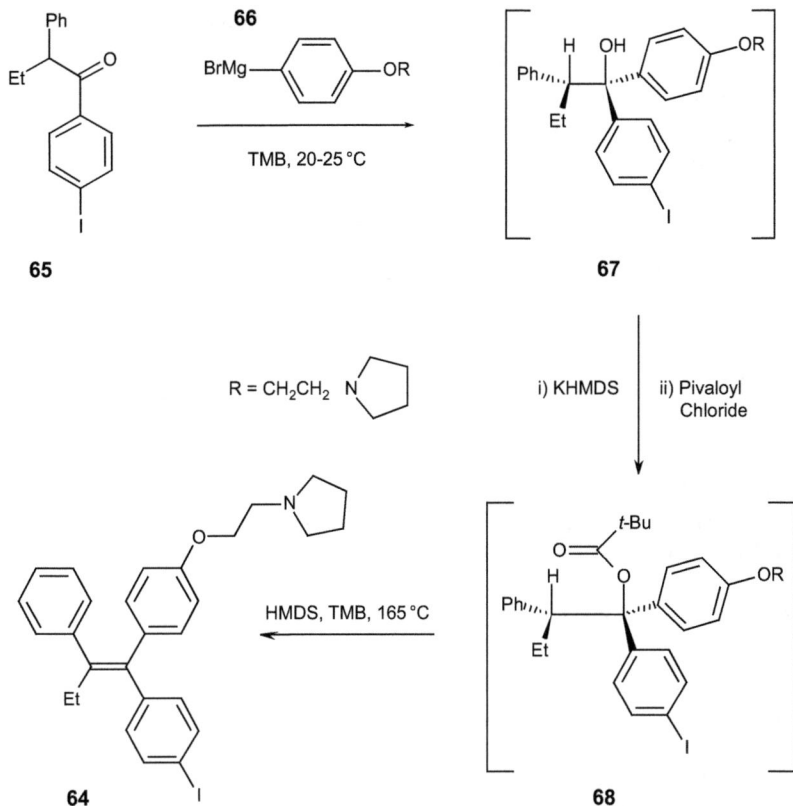

Scheme 5.15 Synthesis of idoxifene *via* thermal elimination (reproduced with permission from American Chemical Society).

methodology gave excellent control and a considerably improved *E/Z* alkene ratio of 93:7. This allowed formation of **64** in a 70% yield from **65** and an overall 66% yield and was subsequently selected as the route of manufacture.

5.3.6 Throughput

The throughput of a process defines the weight of material (in grams, kilograms or tonnes) that can be manufactured in unit time. Throughput issues may not be identified until late in development, potentially only upon transfer to manufacturing. However, consideration of throughput issues earlier than transfer to manufacturing can benefit process chemists in several ways, including delivery time and plant availability. Much of the literature[26] on batch process design and operation focuses on plant-related aspects of throughput; rarely are the process constraints considered. A recent survey (considered in more detail elsewhere[1]) of processes (using first cGMP batches) in AstraZeneca established a clear relationship between the number of synthetic steps and the

Effect of Sequence Length of Throughput

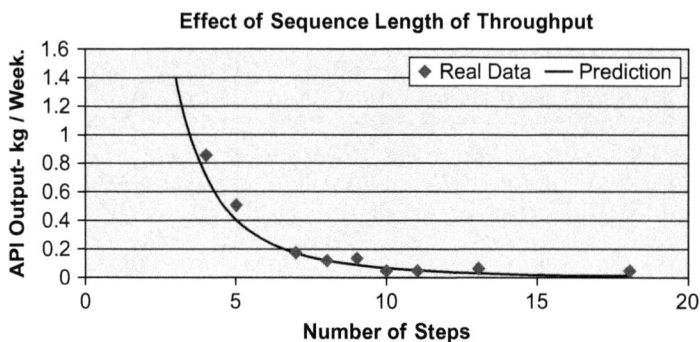

Figure 5.5 Effect of sequence length on throughput (reproduced with permission from American Chemical Society).

amount of API that can be manufactured in a unit time. A simple model (Figure 5.5) has been devised which supports these data.

Options to improve the various elements of throughput include:

(1) Chemical yield can often be improved through a deeper understanding of the kinetics and mechanism. Screening of alternative solvents, reagents and catalysts is also an important approach to yield improvement.

(2) The capacity and types of processing vessels as well as their availability is a limiting factor, and it might be attractive to transfer the campaign to an alternative plant.

(3) In general, reducing the number or the length of the most time-consuming unit operations will improve throughput (such as reaction time, number of solvent replacements, extraction time, crystallisation, filtration and drying time, and cleaning activities). A popular approach is to "telescope" two or more reactions to avoid extended isolation and drying operations.

(4) Poor solubility (limiting concentration) is a common issue that impacts throughput. This can be difficult to overcome, and it may be necessary to change solvent systems or make soluble derivatives of either the starting material or product.

(5) The number of unit operations is linked to the number of chemical steps and convergency.

(6) Some specialist techniques, such as chromatography, can be very time-consuming or limiting on throughput. On the other hand, throughput can be improved by using continuous processing techniques.

(7) High molecular weight protecting groups or salt forms can unnecessarily increase the size of a campaign, so lower molecular weight alternatives should be considered.

(8) Avoidance of raw materials whose poor availability will lead to supply bottlenecks.

The proton pump inhibitor omeprazole (**70**) (Scheme 5.16) is a racemic mixture used in the treatment of gastric reflux disease[27] and was the industry's largest selling

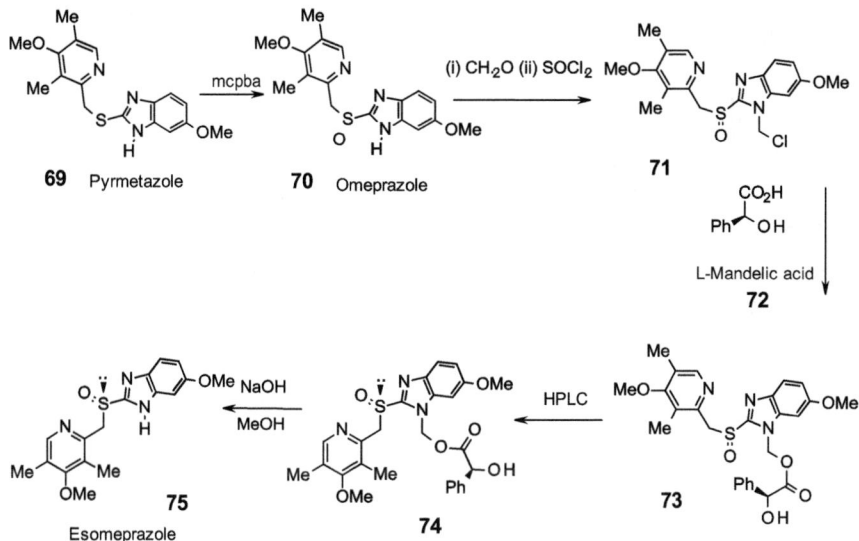

Scheme 5.16 Synthesis of esomeprazole *via* diastereomer separation (reproduced with permission from Nature Publishing Group).

drug in 1999 with sales of approximately $6 billion. Esomeprazole (**75**),[28] the (*S*)-enantiomer of **70**, entered development in 1993, and the chemical synthesis involved separation of the diastereoisomers of **73** by HPLC. The first development campaign for **75** converted 40 kg of **70** to supply only 500 g of pure enantiomer **75**. It took six weeks to perform the first three steps on a 250–500 L pilot scale, providing 5.5 kg of unresolved mandeloyl derivative **73**. A low-throughput separation of diastereomers was achieved using 430 injections on a 15 cm × 100 cm HPLC column, taking at least one week with continuous operation. After processing the column fractions and cleaving the mandelate auxiliary of **74**, crystallisation gave 500 g of **75** as the sodium salt. The next supply requirement was 5 kg of bulk esomeprazole. Extrapolation of the HPLC method would require 60 000 L of eluent to support such a campaign, with an unacceptable campaign time. This serious throughput issue triggered a research programme to identify an enantioselective synthesis to support development of the product. In principle, the most attractive approach to the homochiral sulfoxide **75** would involve catalytic asymmetric oxidation of the sulfide precursor **69**. Applying Kagan conditions unfortunately gave a near racemic mixture of isomers. An enantioselective process (Scheme 5.17) was discovered[29] when the catalyst system was modified using ethyldiisopropylamine (Hunig's base), giving **75** in 94% ee and 92% conversion. This new route to **75** offered a significant improvement in throughput.

The availability of starting materials clearly has an impact on the potential throughput of a process and can be a major factor requiring the identification of a new route. This was the case in the development of the water-soluble prodrug fosfluconazole (**80**) (Scheme 5.18). The first route[30] to this compound involved reaction of fluconazole (**76**) with phosphoramidite **77** to yield the corresponding phosphite **78**. Oxidation to the phosphate **79** and

Esomeprazole
>94%ee / 92% conversion

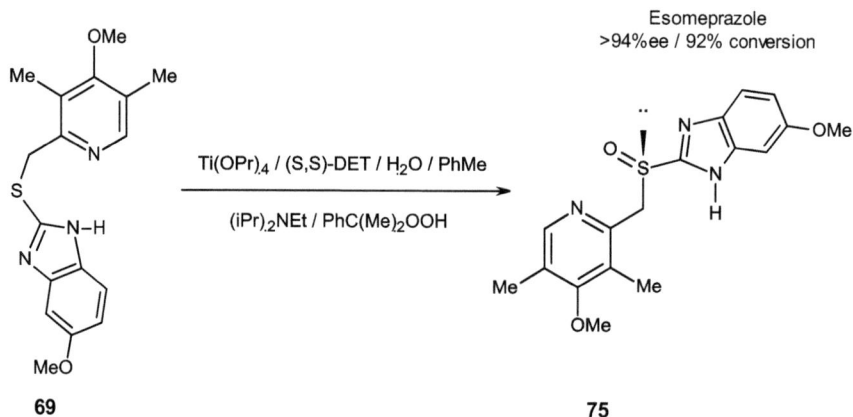

69

Ti(OPr)$_4$ / (S,S)-DET / H$_2$O / PhMe

(iPr)$_2$NEt / PhC(Me)$_2$OOH

75

Scheme 5.17 Synthesis of esomeprazole *via* asymmetric oxidation (reproduced with permission from Wiley-VCH Verlag GmbH & Co).

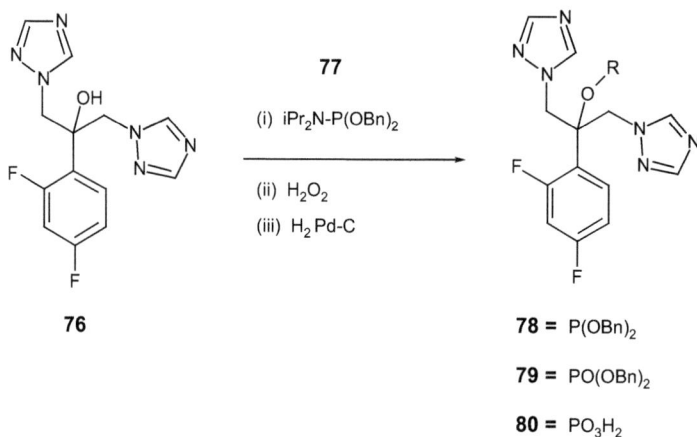

77

(i) iPr$_2$N-P(OBn)$_2$

(ii) H$_2$O$_2$

(iii) H$_2$ Pd-C

76

78 = P(OBn)$_2$

79 = PO(OBn)$_2$

80 = PO$_3$H$_2$

Scheme 5.18 Synthesis of fosfluconazole using a diisopropylphosphoramidite reagent (reproduced with permission from American Chemical Society).

removal of the benzyl groups by hydrogenolysis then completed the synthesis of **80**. While this route was acceptable for providing early bulk for the programme, the dibenzyl diisopropylphosphoramidite **77** had limited commercial availability from catalogue suppliers. Attempts to source this material from custom manufacturers were hampered by the instability of the material toward moisture and difficulties in purification of this high-boiling liquid on a kilogram scale. As a result, a new route was identified (Scheme 5.19) in which the dibenzyl phosphate moiety of **79** was installed by sequential reaction of **76** with phosphorus trichloride, benzyl alcohol and

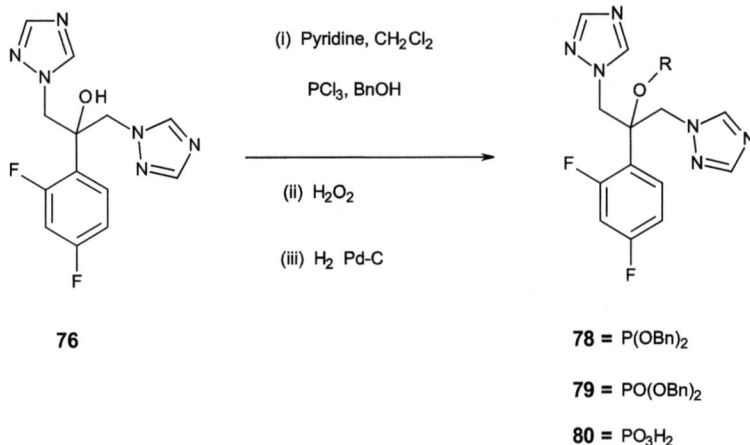

Scheme 5.19 Synthesis of fosfluconazole using phosphorus trichloride (reproduced with permission from American Chemical Society).

then hydrogen peroxide. All of these reagents are commercially available on a large scale, and the dibenzyl phosphate ester **79** has now been produced on an approximately 1 tonne scale using this route.

5.4 Generation and Prioritisation of Ideas for Route Design

As a discipline, organic synthesis requires an ability to recognise bond-forming opportunities for the construction of a target molecule and to relate these opportunities to available synthetic methodology. Using the logic of the disconnection approach, it is now possible to design routes on paper with a good level of predictability. However, the chemical literature contains a vast number of possible reaction types and it is easy to overlook possibilities that might offer a superior approach. Route design is a creative activity and success depends upon the capability of the process chemist to generate ideas of high potential. It is unusual for the final sequence to be conceived at the outset; many types of ideas are combined together over the course of a project. These ideas might include: new starting materials, adaptation of known routes to similar targets, new routes to/from a key intermediate, speculative reactions with limited precedence. It is therefore important to generate as many ideas as possible and a variety of chemical databases can be used to help:

- *SciFinder* (CAS)[31] or *Beilstein* (Reaxys)[32] offer searching based upon known compounds, citations, key words, substructure similarity or reactions based on conversion of substructures.

- Retrosynthesis software is helpful in terms of generating options for route design; examples include ICSYNTH[33] and LHASA.[34]
- *Science of Synthesis*[35] is a database of reliable synthetic methods for 2700 different types of compounds.
- Patent literature (*Derwent World Patent Index*)[36] for keyword, compound name and structure searching based on the Markush system.
- *Spresi* (Infochem)[37] is a structure-based database for commercially available compounds.
- *Directory of World Chemical Producers* (Chemical Information Services)[38] is a database of worldwide producers of chemical raw materials.
- *Electronic Laboratory Notebooks* (Symyx)[39] is software to enable planning and recording of experimentation; enables structure-based searching of "in-house" knowledge.

Brainstorming meetings can generate more sophisticated ideas and will often involve academic consultants. There are four basic rules in brainstorming: (i) focus on quantity; (ii) withhold criticism; (iii) welcome unusual ideas; and (iv) combine and improve ideas. Another useful technique is to further develop each idea using a list of "chemistry keywords" (such as oxidation, reduction, rearrangement, and so on) as a prompt to stimulate diversity of thinking.

After the ideas-generation exercise, it is important to sort and group ideas into an "ideas map". The benefit of the "ideas map" is to identify common intermediates and gaps.

In situations where there is limited resource and an excessive numbers of ideas, it is helpful to develop a shortlist and decide the order of priority for investigation. Kepner Tregoe (KT) techniques[40] are sometimes a useful way to achieve this and can lead to objective decision making in a team environment through the "weighting" of each criterion. Given there will be limited data on "paper routes", the KT criteria at this stage should include things like (i) steps to reach the key step; (ii) literature precedent; (iii) total number of steps in longest linear sequence; (iv) estimated time for manufacture of next campaign; (v) patent issues; and (vi) projected cost of raw materials. Deciding upon the "end-game" (often the final three steps) is usually the highest priority when designing a commercial route. After commitment to a particular "end-game", deciding on the optimum route to late intermediates can be postponed to a later point in the drug development programme, thereby delaying resource investment.

5.5   Maximising the Value of Experimental Work

Experimentation is time consuming and it is important to maximise the value gained and ensure that the right data are obtained to support evaluation and decision making. To do this effectively, a continuous and regular plan–do–review cycle should be operated and the team should work flexibly and collaboratively so that highest priorities are addressed. In the planning phase, work plans are geared to answer key questions and care should be exercised to avoid wasting time on developing chemistry that ultimately may not be needed. Sometimes it is possible to convert samples of an API into putative late intermediates and thus rapidly test various "end-game" options. Most paper routes will have a "key step" representing a point of particular uncertainty. Designing rapid experiments to verify a key step is especially attractive.

The use of computational techniques can be helpful to predict the selectivity of a paper transformation. The outcome of certain reaction types (such as aromatic electrophilic substitution or metallation) can now be predicted with a high degree of reliability. Chapter 9 provides helpful guidance on the use of methods to understand and optimise a process. These techniques can also be useful when the process chemist is unable to prove a key step in a sequence. It is also productive to screen a variety of reagents, catalysts and solvent systems. If the initial objective is to prove a synthetic sequence, the process chemist can afford to ignore some of the large-scale constraints (such as cryogenic conditions), as these can often be developed-out later on. Reaction and product

Table 5.2 Investigation of reaction parameters.

Factor	Impact on desired product	Impact on product B	Impact on product C
Stoichiometry Temperature & time Concentration Catalysts Order/rate of addition Reagents & solvents pH Leaving group	Profile of product formation, rate of reaction and selectivity?	Profile of product formation, rate of reaction and selectivity?	Profile of product formation, rate of reaction and selectivity?
Work-up	Partition (solubility) of product in each phase	Partition (solubility) of product in each phase	Partition (solubility) of product in each phase
Phases in reaction	Profile of product formation *versus* changing number and types of phases (*e.g.* heterogeneous)	Profile of product formation *versus* changing number and types of phases (*e.g.* heterogeneous)	Profile of product formation *versus* changing number and types of phases (*e.g.* heterogeneous)
Rate of reaction	Selectivity *versus* unwanted products at fast or slow reaction rates		
Analytical method	Specificity for product, stability of product during analysis	Specificity for product, stability of product during analysis	Specificity for product, stability of product during analysis
Starting material purity	Influence of organic and inorganic impurities upon reaction profile	Influence of organic and inorganic impurities upon reaction profile	Influence of organic and inorganic impurities upon reaction profile

characterisation can be surprisingly difficult but is vital to success. Access to high-field NMR, LCMS and GCMS is therefore important, as well as use of preparative chromatography tools like Biotage.

In the "review" part of the cycle, the aim is to (i) draw conclusions from the experimental data, (ii) eliminate routes from further investigation, (iii) decide how to further exploit the findings and (iv) consider avenues to improve problematic chemistry. Despite high confidence in a particular idea, the initial experimental results may prove frustrating. As a first step, it can be useful to construct an inventory (Table 5.2) of what is/is not known about the problematic reaction. Prompting these questions may improve the focus on experimental planning, and should improve speed to understanding the issues.

5.6   Conclusions: Route Selection

The final aim of a route design project is to identify and select a suitable (hopefully the best) synthetic route for API manufacture. The process chemist is faced with making many decisions along the way, and often relies upon technical judgement. To make effective decisions in selection of a commercial route, it is important to consult with a variety of colleagues in the pharmaceutical company and potentially a contract manufacturer. In many instances, selection of the best route will be an obvious decision to make and this is especially the case where there is a strong economic driver. The KT technique discussed above can sometimes help where decision making is less straightforward.

Acknowledgements

I would like to thank the following people for their help and advice in preparing material for this chapter: A. Boss, D. Catterick, P. Cornwall, A. Craig, A. Curzons, D. Dale, L. Diorazio, D. Ennis, A. Gillmore, S. P. Green, D. Lathbury, J. Leonard, I. Marziano, B. Moss, T. Rein, J. P. Sherlock, M. Sulur, M. Thelin, C. Thomson, A. S. Wells and M. T. Williams.

References

1. M. Butters, D. Catterick, A. Craig, A. Curzons, D. Dale, A. Gillmore, S. P. Green, I. Marziano and J. P. Sherlock, *Chem. Rev.*, 2006, **106**, 3002.
2. S. A. Lee, *J. Fluorine Chem.*, 2001, **109**, 55.
3. A. C. Barker, K. A. Boardmanz, S. D. Broady, W. O. Moss, B. Patel, M. W. Senior and K. E. H. Warren, *Org. Process Res. Dev*, 1999, **3**, 253.
4. Vinyl bromide properties: *Dangerous Prop. Ind. Mater. Rep.*, 1989, **9**, 80; *Chem. Abstr.*, 1989, **110**, 198295.
5. L. J. Goldwater, *Sci. Am.*, 1971, **224**(5), 15.

6. I. P. Andrews, R. J. Atkins, G. F. Breen, J. S. Carey, M. A. Forth, D. O. Morgan, A. Shamji, A. C. Share, S. A. C. Smith, T. C. Walsgrove and A. S. Wells, *Org. Process Res. Dev.*, 2003, **7**, 655.

7. R. J. Atkins, A. Banks, R. K. Bellingham, G. F. Breen, J. S. Carey, S. K. Etridge, J. F. Hayes, N. Hussain, D. O. Morgan, P. Oxley, S. C. Passey, T. C. Walsgrove and A. S. Wells, *Org. Process Res. Dev.*, 2003, **7**, 663.

8. (a) http://www.unodc.org/unodc/index.html; (b) http://www.incb.org/incb/index.html; (c) http://www.opsi.gov.uk/; (d) http://www.usdoj.gov/dea/index.htm (last accessed August 2010); (e) http://www.whitehouse-drugpolicy.gov/index.html (last accessed August 2010); (f) http://www.justice.gov/dea/index.htm (last accessed August 2010).

9. http://www.hse.gov.uk/comah (last accessed August 2010).

10. http://www.hse.gov.uk/index.htm (last accessed August 2010); details of EUSES can be found at the European Chemicals Bureau at http://ecb.jrc.it/ (last accessed August 2010).

11. H. Jackson, *Basic Intellectual Property Concepts, in Patent Strategy: For Researchers and Research Managers*, Wiley, Chichester, 2nd edn, 2001, ch. 1.

12. W. Cabri and R. Di Fabio, *From Bench to Market: The Evolution of Chemical Synthesis*, Oxford University Press, Oxford, 2000, p. 168.

13. H.-J. Federsel, *Chirality in Industry II*, Wiley, Chichester, 1997, p. 225.

14. G. S. Bulteau, Afr. Pat. 6802593, 1968; *Chem. Abstr.*, 1969, 71, 30354b.

15. *Pharmaceutical R&D, Costs, Risks and Rewards*, Office of Technology Assessment, Congress of the U.S., Washington, 1993, appendix G, p. 302.

16. J. D. Moseley, W. O. Moss and M. J. Welham, *Org. Process Res. Dev.*, 2001, **5**, 491.

17. S. Challenger, A. Derrick, C. P. Mason and T. V. Silk, *Tetrahedron Lett.*, 1999, **40**, 2187.

18. M. J. Burk, F. Bienewald, S. Challenger, A. Derrick and J. A. Ramsden, *J. Org. Chem.*, 1999, **64**, 3290.

19. http://www.fda.gov/AboutFDA/CentersOffices/CDER/default.htm (last accessed August 2010).

20. *Specifications: Test Procedures And Acceptance Criteria For New Drug Substances And New Drug Products: Chemical Substances*, http://www.ich.org/LOB/media/MEDIA422.pdf-Q3A(R) (last accessed August 2010); *ICHQ3C: Impurities: Guideline For Residual Solvents*, http://www.ich.org/LOB/media/MEDIA430.pdf (last accessed August 2006); http://www.ich.org/LOB/media/MEDIA433.pdf ICH Q7A GMP for active pharmaceutical ingredients; http://www.ich.org/LOB/media/MEDIA431.pdf (last accessed August 2006).

21. http://www.emea.europa.eu/ema (last accessed August 2006).

22. http://www.pharmtech.com/pharmtech/data/articlestandard/pharmtech/062003/45872/article.pdf (last accessed August 2010).

23. D. Ainge and L.-M. Vax, *Org. Process Res. Dev.*, 2002, **6**, 811.

24. C. R. Schmid, J. D. Bryant, M. Dowlatzedah, J. L. Phillips, D. E. Prather, R. D. Schantax and C. S. Vianco, *J. Org. Chem.*, 1991, **56**, 4056.
25. K. W. Ace, M. A. Armitage, R. K. Bellingham, P. D. Blackler, D. S. Ennis, N. Hussain, D. C. Lathbury, D. O. Morgan, N. O'Connor, G. H. Oakes, S. C. Passey and L. C. Powling, *Org. Process Res. Dev.*, 2001, **5**, 479.
26. (a) *Handbook of Batch Process Design*, ed. P. N. Sharratt, Kluwer, New York, 1999; (b) H.-M. Ku, D. Rajagopalan and I. A. Karimi, *Chem. Eng. Prog.*, 1987, **83**(8), 35; (c) J. L. Manganaro, *Chem. Eng. Prog.*, 2002, **98**(8). 70.
27. P. Lindberg, A. Brandstrom, B. Wallmark, B. Mattsson, L. Rikner and K. J. Hoffman, *Med. Res. Rev.*, 1990, **10**, 1.
28. L. Olbe, E. Carlsson and P. Linberg, *Nat. Rev. Drug. Discov.*, 2003, **2**, 132.
29. H.-J. Federsel and M. Larsson, *Asymmetric Catalysis on Industrial Scale*, Wiley, Chichester, 2004, p. 413.
30. A. Bentley, M. Butters, S. P. Green, W. J. Learmonth, J. A. MacRae, M. C. Morland, G. O'Connor and J. Skuse, *Org. Process Res. Dev.*, 2002, **6**, 109.
31. www.cas.org (last accessed August 2010).
32. www.reaxys.com (last accessed August 2010).
33. http://infochem.de (last accessed May 2011).
34. http://www.lhasa-llc.com (last accessed May 2011).
35. www.science-of-synthesis.com (last accessed August 2010).
36. www.thomsonreuters.com (last accessed August 2010).
37. www.spresi.com (last accessed August 2010).
38. www.chemicalinfo.com (last accessed August 2010).
39. www.symyx.com (last accessed August 2010).
40. www.kepner-tregoe.com (last accessed August 2010).

CHAPTER 6

The Importance of Green Chemistry in Process Research & Development

PETER J. DUNN

Pfizer Worldwide Research and Development, Sandwich Laboratories, Kent, CT13 9NJ, UK

6.1 Introduction

Green chemistry or sustainable chemistry is defined by the Environmental Protection Agency[1] as "the design of chemical products and processes that reduce or eliminate the use or generation of hazardous substances".[†] In recent years there has been a greater societal expectation that chemists and chemical engineers should produce greener and more sustainable chemical processes, and it is likely that this trend will continue to grow over the next few decades.

Green chemistry came into prominence through the work and publications of Anastas,[2] Clarke,[3] Sheldon,[4] Trost,[5] Warner[2] and others during the 1990s. Of course, efficient, low-waste chemistry existed before these publications and indeed one of the best examples of green chemistry in the pharmaceutical industry is the Upjohn synthesis of cortisone in 1952,[6] a full 40 years before the term "green chemistry" was actually coined. In his book *Laughing Gas,*

[†]The full definition is: Green chemistry, also known as sustainable chemistry, is the design of chemical products and processes that reduce or eliminate the use or generation of hazardous substances. Green chemistry applies across the life cycle of a chemical product, including its design, manufacture and use.

RSC Drug Discovery Series No. 9
Pharmaceutical Process Development: Current Chemical and Engineering Challenges
Edited by A. John Blacker and Mike T. Williams
© Royal Society of Chemistry 2011
Published by the Royal Society of Chemistry, www.rsc.org

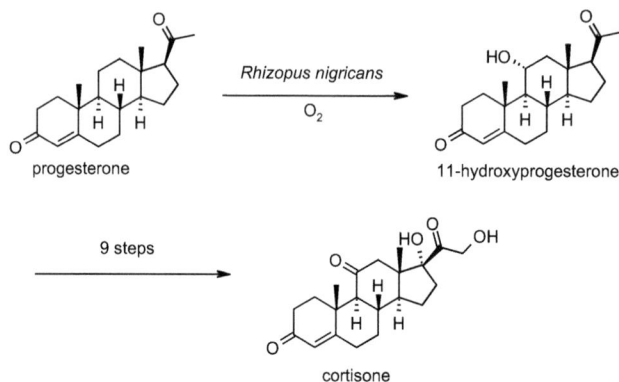

Scheme 6.1   The Upjohn synthesis of cortisone.

Viagra and Lipitor,[7] Jie Li says the Upjohn company "stunned the world" when they reported a 10-step synthesis of cortisone involving a key hydroxylation reaction at the 11-position of progesterone (Scheme 6.1). This is probably an over-exaggeration, but it certainly must have stunned the Merck company who had discovered cortisone and whose own 31-step synthesis of cortisone (itself an outstanding scientific achievement)[8] was made redundant overnight.

Clearly, low-waste, efficient chemistry existed well before the 1990s, but in the main the strong focus on green chemistry in process R&D (particularly in the pharmaceutical industry) has occurred in the last decade and this has resulted in the discovery and publication of green syntheses of pharmaceuticals such as sertraline hydrochloride,[9] sildenafil citrate,[10] paclitaxel[11] and sitagliptin phosphate monohydrate.[12] Examples from the fine chemical industry include the BASF route to citral[13] and the Lonza synthesis of biotin.[14,15]

6.2   Solvents and Solvent Selection

Solvents typically make up more than 80% of the material usage for active pharmaceutical ingredient (API) manufacture.[16] Solvent use also consumes about 60% of the overall energy and accounts for 50% of the post-treatment greenhouse gas emissions;[17] hence solvent selection is a major consideration in the design of a chemical synthesis. A number of solvent selection guides have been reported by GSK,[17] Pfizer,[18‡] the Pharmaceutical Roundtable[19] and by academic groups.[20] The GSK approach is to assess the solvents in a variety of categories such as health, safety and lifecycle assessment (LCA) and then to report a score for the various categories to the end user using a colour-coded table. An example of that assessment for the alcohol class of solvents is shown in Figure 6.1. The Pfizer approach is to do a similar, detailed, multi-category assessment but then

---

‡Note that the detailed information to classify these 39 solvents and a further 50 solvents are held in a more detailed solvent selection tool (the solvent bundle book), which is made available to all Pfizer scientists through various internal websites.

	Solvent	Environ-mental Waste	Environ-mental Impact	Health	Safety	LCA ranking
Alcohols	Ethylene glycol	4	9	8	9	9
	1-Butanol	5	8	8	8	5
	Diethylene glycol butyl ether	5	7	10	9	7
	Isoamyl alcohol	7	7	7	8	6
	2-Ethylhexanol	9	6	8	7	6
	2-Butanol	4	7	7	7	6
	1-Propanol	3	7	5	8	7
	Ethanol	3	8	10	7	9
	2-Propanol	3	9	9	7	5
	1-Butanol	3	10	7	7	8
	Methanol	3	10	5	8	9

Figure 6.1 An illustrative portion of the GSK solvent selection guide.

Preferred	Usable	Undesirable
Water	Cyclohexane	Pentane
Acetone	Heptane	Hexane(s)
Ethanol	Toluene	Di-isopropyl ether
2-Propanol	Methylcyclohexane	Diethyl ether
1-Propanol	TBME	Dichloromethane
Ethyl Acetate	Isooctane	Dichloroethane
Isopropyl acetate	Acetonitrile	Chloroform
Methanol	2-MeTHF	NMP
MEK	THF	DMF
1-Butanol	Xylenes	Pyridine
t-Butanol	DMSO	DMAc
	Acetic Acid	Dioxane
	Ethylene Glycol	Dimethoxyethane
		Benzene
		Carbon tetrachloride

Figure 6.2 The Pfizer solvent selection guide.

report a much simpler summary of those assessments to the end user; this is shown in Figure 6.2. Of course, there are pros and cons to both approaches, but Pfizer reports that its simple selection guide is particularly effective when used in the medicinal chemistry environment and has resulted in reduced chloroform usage of 98.5% and diisopropyl ether usage by 100% over the past few years.[21]

Pfizer have also reported a replacement table for all solvents in the undesirable red category (Table 6.1). However, one group of solvents for which

Table 6.1 The Pfizer solvent replacement table.

Undesirable solvents	*Alternative*
Pentane	Heptane
Hexane(s)	Heptane
Diisopropyl ether or diethyl ether	2-Methyltetrahydrofuran or *tert*-butyl methyl ether
Dioxane or dimethoxyethane	2-Methyltetrahydrofuran or *tert*-butyl methyl ether
Chloroform, dichloroethane, carbon tetrachloride	Dichloromethane
Dimethylformamide, dimethylacetamide, *N*-methylpyrrolidinone	Acetonitrile
Pyridine	Triethylamine (if pyridine used as base)
Dichloromethane (extractions)	Ethyl acetate, *tert*-butyl methyl ether, toluene, 2-methyltetrahydrofuran
Dichloromethane (chromatography)	Ethyl acetate/heptane
Benzene	Toluene

there are few good replacements are the dipolar aprotic solvents such as dimethylacetamide (DMAc), dimethylformamide (DMF) and *N*-methylpyrrolidinone (NMP). These solvents have reprotoxicity issues and are strongly regulated, particularly in the European Union. However, they have many features which are desirable in chemistry; for example, they are very good solvents for metal-catalysed cross-coupling reactions and nucleophilic substitution reactions. Sometimes they are the only solvents that will dissolve polar materials (known by chemists as "brick dust") or salts. Given a choice between using one of these three solvents, many companies recommend NMP. This solvent carries the same reprotoxicity hazards as DMF and DMAc but is less volatile and hence the risk of exposure is reduced.

6.3   Green Chemistry Metrics

The green chemistry movement has developed a series of metrics to support and reinforce behaviour change on both industry and academia in the move towards greener and more sustainable chemistry. Four of the most common metrics that are used are atom economy, the *E*-factor, reaction mass efficiency (RME) and process mass intensity (PMI). These are defined and discussed in turn.

6.3.1   Atom Economy

The concept of atom economy was first proposed by Trost in 1991.[5] At about the same time, Sheldon proposed the very similar concept of atom utilisation,[22] but it is the atom economy name that has become widespread in use. Atom economy is simply defined (see eqn 6.1) as:

$$\text{Atom economy} = \frac{\text{Molecular weight of desired product}}{\text{Molecular weight of all products}} \times 100\% \qquad (6.1)$$

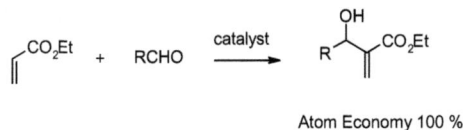

Atom Economy 100 %

Scheme 6.2 An example of the Baylis–Hillman reaction.

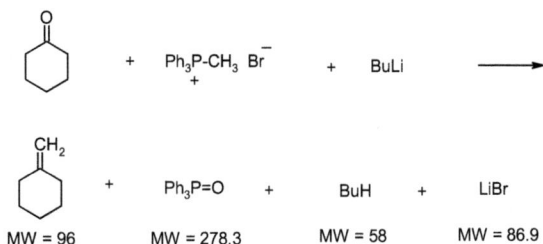

| MW = 96 | MW = 278.3 | MW = 58 | MW = 86.9 |

Scheme 6.3 An example of the Wittig reaction.

An example of a reaction with perfect atom economy is the Baylis–Hillman reaction, shown in Scheme 6.2. In this reaction, all of the atoms from the two starting materials are incorporated into the product; there are no co-products and so the atom economy is 100%.

In contrast, a reaction with poor atom economy is the Wittig reaction. The example shown in Scheme 6.3 has an atom economy of only 18.5%.

The atom economy calculation for the Wittig reaction shown in Scheme 6.3 is given in eqn (6.2):

$$\text{Atom economy} = \frac{96}{(96 + 278.3 + 58 + 86.9)} \times 100\% = 18.5\% \qquad (6.2)$$

While the concept of atom economy is simple and can be calculated directly from the reaction scheme, it has a number of downsides. It does not take into account reaction yield or stoichiometry or make any allowance for the amount of solvents or other reagents used in either the reaction or the work-up.

6.3.2 Environmental Factor

The simplest concepts are often the most effective, and this can certainly be said for the environmental factor (*E*-factor), which was proposed by Sheldon in 1992.[4] The *E*-factor is defined as the ratio of waste over product. Sheldon originally proposed *E*-factor ranges for different branches of the chemical industry (shown in Table 6.2), but of course they can be applied to any product, so it is equally possible to calculate an *E*-factor for producing a laptop or a mobile phone.

By coincidence, before 1992 when the *E*-factor was first proposed, most synthetically produced chiral drugs were marketed as the racemate. However,

Table 6.2   *E*-Factors in different chemical industry segments.

Industry segment	Volume $(t\ y^{-1})$	E-Factor
Bulk chemicals	10^4–10^6	<1–5
Fine chemicals industry	10^2–10^4	5–50 +
Pharmaceutical industry	10–10^3	25–100 +

after 1992 many chiral drugs were marketed as the single enantiomer. Hence the range of 25–100 proposed by Sheldon at the end of the "racemic era" is probably now more realistically 25–200 in the "chiral era". That said, some pharmaceutical companies are now routinely measuring their *E*-factors and are setting reduced targets. By 2008, three companies, GSK, Lilly and Pfizer, had published their *E*-factor (or equivalent) targets,[23] so a drive to lower numbers should now be expected. The same can be said for some fine chemical companies, who are also setting lower targets and driving improvement.[24]

E-Factors can be calculated with or without consideration of process water; as this is a very large contributing factor, it is important to state which method is being used. In general, it is usual to calculate *E*-factors without water. As Sheldon says "when considering an aqueous waste stream, only the inorganic salts and the organic compounds contained in the water are counted; the water itself is excluded. Inclusion of water used in the process can lead to exceptionally high *E*-factors ... and can make meaningful comparisons of processes difficult".[4]

6.3.3   Reaction Mass Efficiency

The concept of reaction mass efficiency (RME) was first published by GSK in 2002.[25] The idea was to keep the simplicity of the atom economy concept but avoid the high impact of solvents which are found in the *E*-factor. RME takes into account reaction yield, stoichiometry and the use of catalysts or other reagents. It is defined (see eqn 6.3) as:

$$\text{Reaction mass efficiency} = \frac{\text{Mass of desired product}}{\text{Mass of all reactants}} \times 100\% \qquad (6.3)$$

6.3.4   Process Mass Intensity

Process mass intensity (PMI) is very similar to the *E*-factor but measures the mass of materials used to make 1 kg of the API. Like the *E*-factor it can be measured with or without the inclusion of water. It is defined (see eqn 6.4) as:

$$\text{Process mass intensity} = \frac{\text{Mass of all materials used to make the product}}{\text{Mass of product}} \times 100\%$$

$$(6.4)$$

One argument for using PMI is that it is easier to calculate using the inputs into a reaction (taken from a batchcard or laboratory notebook) than waste measurements, which can be less readily available. Proponents of the *E*-factor argue that the perfect process (which does not produce any waste) has an *E*-factor of zero, which better reflects the goal of zero waste, whereas the perfect process has a PMI of 1. The two measures are related by eqn (6.5):

$$E\text{-factor} = \text{Process mass intensity} - 1 \qquad (6.5)$$

The Pharmaceutical Roundtable (a group of 13 pharmaceutical companies and the American Chemical Society Green Chemistry Institute, who promote green chemistry) recently published some PMI figures derived from data submitted by several companies. For 21 products, which at that time were either in phase 3 or commercial products, the average solvent usage was 55 kg per kg of product (with a range of 10–170) and the average PMI (excluding water) was 77 kg per kg of product (range 25–240).[26]

6.4 The Importance of Biocatalysis in Green Chemistry

Biocatalysis has many attractive features in the context of green chemistry: reactions are often performed in water under mild conditions of temperature, pressure and pH. In addition, the catalyst (an enzyme) is biodegradable and derived from renewable raw materials. The enzyme often affords high chemo-, regio- and stereoselectivities and avoids the need for protection and deprotection sequences that are required in traditional syntheses.

An outstanding example of the application of biocatalysis to yield environmental benefits is the synthesis of 6-aminopenicillinic acid (6-APA). More than 10 000 tonnes of penicillin G is enzymatically hydrolysed to 6-APA every year for subsequent conversion to semi-synthetic penicillins such as amoxicillin and ampicillin (Scheme 6.4).

Converting penicillin G to 6-APA sets a significant chemical challenge; it requires the hydrolysis of a stable amide bond while leaving the reactive β-lactam bond untouched. An ingenious chemical solution is first to protect the acid as the silyl ester with chlorotrimethylsilane, then react the amide group with phosphorus pentachloride to give the imino chloride **1** (Scheme 6.5). Hydrolysis of both the imino chloride and the silyl ester groups gives 6-APA.

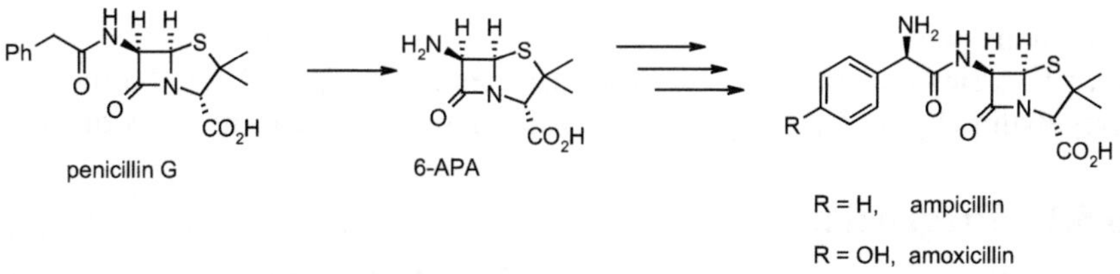

R = H,   ampicillin

R = OH, amoxicillin

Scheme 6.4   The synthesis of semi-synthetic penicillins.

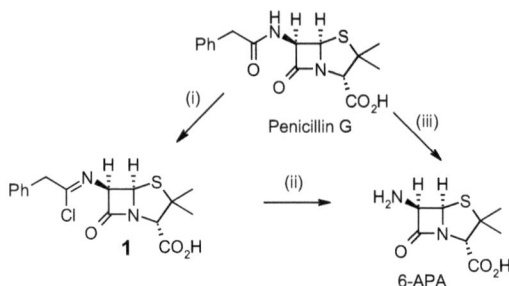

(i) TMSCl then PCl$_5$, PhNMe$_2$, CH$_2$Cl$_2$, -40°C (ii) n-BuOH, -40°C, then H$_2$O, 0°C (iii) Pen-acylase, H$_2$O, 37°C

Scheme 6.5   Enzymatic and chemical deacylation of penicillin G.

This chemical route of synthesis provided society with access to semi-synthetic penicillins until the mid-1980s, but it had a number of drawbacks from a green chemistry perspective. These included the use of highly reactive and hazardous phosphorus pentachloride, the use of an undesirable chlorinated solvent (CH$_2$Cl$_2$) and the need to employ a reaction temperature of –40 °C, leading to high energy use. In contrast, the biocatalysis process proceeds in a single step in water at a temperature of 37 °C, leading to major environmental savings.[27]

Enzymatic processes have been widely used for many years in the fine chemical industry and a fairly recent review provides information on more than 130 commercial processes that are being carried out with biocatalysis;[28] however, until very recently the use of biocatalysis in pharmaceutical manufacturing was somewhat limited. Fortunately, this is now changing and there are three factors driving that change:

1. There is a growth in the numbers of enzymes that are available, particularly through bioinformatics and cloning.
2. Reactions can now be screened using this increased enzyme pool by means of automated technology and with minimal amounts of substrate.[29]
3. Once a lead enzyme has been identified, it can be optimised (if required) using forced evolution techniques pioneered by Stremmer,[30] Arnold,[31] Reetz[32] and others.

6.5   Case Histories

For many scientists, green chemistry comes alive through a consideration of case histories and three recent case histories are presented for the medicines pregabalin (Lyrica®), sitagliptin (Januvia®) and rosuvastatin (Crestor®).

6.5.1   Pregabalin

Pregabalin is a lipophilic γ-aminobutyric acid (GABA) analogue that was developed for the treatment of several nervous system disorders, including

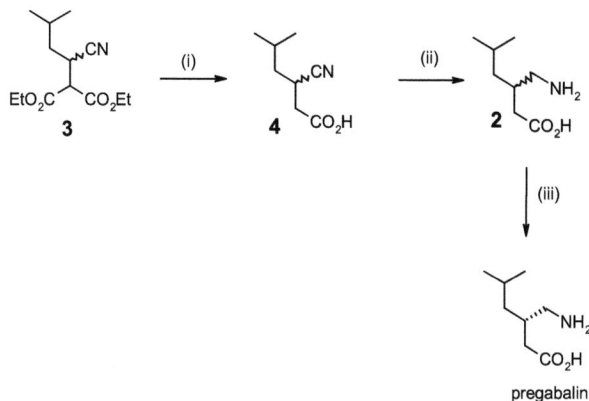

(i) KOH, MeOH, H₂O, reflux (ii) H₂ RaNi, EtOH, H₂O; then HOAc: then IPA wash (iii) (*S*)-mandelic acid, IPA, H₂O, recrystallise from IPA, H₂O, salt break, recrystallise from IPA, H₂O

Scheme 6.6   The classical resolution route to pregabalin.

epilepsy, neuropathic pain, anxiety and social phobia.[33,34] The medicine was launched as Lyrica® in the U.S. in September 2005. The first commercial synthesis, developed by the Parke-Davis company, is shown in Scheme 6.6. The β-cyano diester **3** was prepared by the condensation of isovaleraldehyde (3-methylbutanal) with diethyl malonate, followed by the addition of potassium cyanide. The cyano diester **3** was hydrolysed and decarboxylated to give the β-cyano acid **4**. Reduction with Raney nickel gave racemic pregabalin **2**, which was resolved with (*S*)-mandelic acid.[35]

In summary, this was an efficient synthesis of the racemic pregabalin, but by using a final stage resolution, half of the synthetic materials used to prepare the racemate were by definition thrown away. In this particular example the classical resolution was not particularly efficient (the yield from racemic pregabalin to pregabalin was only around 30%) and so some 70% of the materials were lost in this classical resolution step. When an *E*-factor analysis was performed on the classical resolution chemistry it was found that 86 kg of waste was being produced for every kilogram of the desired product and this inspired a search for more efficient chemistry.[36]

An enzymatic screen was performed by Pfizer on the cyano diester **3** to see if a biocatalytic resolution could be found. This would have the additional advantage that the resolution would be performed early in the synthesis. The screen identified two enzymes which hydrolysed the desired (*S*)-cyano diester 200 times faster than the (*R*)-cyano diester **5**. These enzymes were *Thermomyces lanuginous* lipase (sold commercially as Lipolase) and *Rhizopus delemar* lipase. Although both enzymes were equally selective, the Lipolase had higher specific activity and hence was selected for scale-up as the specific activity would translate directly into lower biocatalyst loadings and hence less waste. The process was skilfully designed so that every process step was performed in water; the full synthesis can be seen in Scheme 6.7. In September 2006, Pfizer

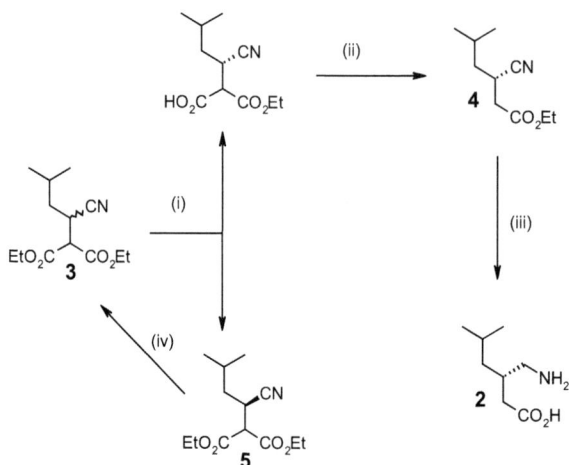

(i) Lipolase catalysed enzymatic resolution performed in water (ii) thermal decarboxylation performed in water (iii) hydrolysis (KOH) and hydrogenation (H$_2$ RaNi) both reactions performed in water (iv) base catalysed epimerisation

Scheme 6.7   The biocatalytic route to pregabalin.

started commercial manufacture with the new process, which had a much improved *E*-factor of 17.[37]

The next major step in the development of the process was to recycle the wrong enantiomer [the (*R*)-cyano diester **5**]. This is achieved using a base-catalysed epimerisation which is performed in a continuous reactor. The introduction of this process involved designing, building and validating a new continuous processing plant at a Pfizer manufacturing facility and filing the new chemistry with the FDA and other regulatory agencies. The new process produces significantly less waste and has a much higher overall yield from the cyano diester **3** to pregabalin. The FDA approved the new process in December 2009 and Pfizer has started manufacturing with the latest most environmentally friendly process.

A summary of key materials used by the three processes is shown in Table 6.3.

A summary of the process changes and the *E*-factors calculated for each process is shown in Table 6.4. In addition, the energy use for the classical resolution and biocatalysis routes have been calculated and reported.[38]

6.5.2   Sitagliptin

Sitagliptin is a medicine that was discovered by Merck for the treatment of type II diabetes.[39,40] The initial route of synthesis that was used to prepare the supplies for early safety and clinical studies is shown in Scheme 6.8. The acid **6** was converted through to the β-keto ester **7** using Masamune conditions.[41] The asymmetric hydrogenation of **7** was carried out using a modified (*S*)-Binap

Table 6.3 Comparison of key material inputs for the manufacture of 1000 kg of pregabalin using the classical resolution and biocatalytic routes.

Inputs	Kilograms per 1000 kg pregabalin		
	Classical resolution route	Biocatalytic route	
		No recycling of 5	With recycling of 5[a]
Cyano diester 3	6212	4798	2810
Enzyme	0	574	574
(S)-Mandelic acid	1135	0	0
Raney nickel	531	80	55
Solvents	50 042[c]	6230[b]	2620[b]
Total	57 920	11 682	6059

[a]Plus some other process improvements.
[b]After solvent recycle.
[c]Some solvents recycled.

Table 6.4 Summary of pregabalin processes and *E*-factors.

Year	Process used to supply major markets	E-Factor[a]
2005–2006	Classical resolution route	86
2006–2009	Biocatalysis route	17
2010–present	Biocalysis route with recycle of the (R)-cyano diester	8

[a]Process water not included in the calculations.

catalyst. Hydrolysis of the resulting chiral ester gave the acid **8** in excellent yield and 94% ee. Coupling of acid **8** with $BnONH_2 \cdot HCl$ using *N*-(3-dimethylaminopropyl)-*N'*-ethylcarbodiimide hydrochloride (EDC) gave the hydroxamate **9**. Cyclisation to form the β-lactam **10** was carried out using diisopropyl azodicarboxylate (DIAD) and triphenylphosphine.[42] The β-lactam was isolated in 81% yield (from compound **8**) by crystallising from aqueous methanol and this crystallisation also upgraded the optical purity to greater than 99% ee.

The synthesis of sitagliptin was finished by hydrolysing the β-lactam **10** to the amino acid **11**, which was coupled to the triazole **12** with EDC. Hydrogenation gave the free amine **13**, which was converted to the phosphate salt **14**. The triazole **12** is common to all three routes presented in this section and after optimisation was prepared by Merck with an *E*-factor of 68.

This synthesis proceeded in 45% overall yield from the acid **6**. The route was rather long, but each step proceeded in high yield. Merck made more than 100 kg of the API using this route and believed that the route could have been commercialised. Nevertheless, from a green chemistry perspective there are a number of downsides, which included:

- Four steps with some high molecular weight reagents were used to convert the hydroxy group to the amino group. This made the synthesis rather lengthy.

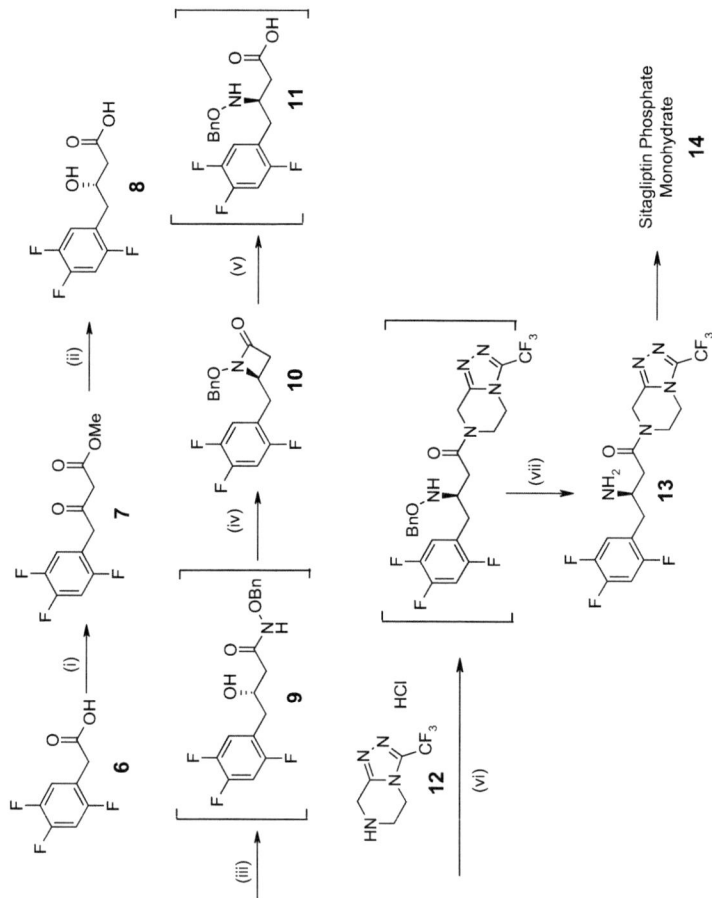

Scheme 6.8 Initial large-scale synthesis of sitagliptin phosphate (β-lactam route).

(i) CDI, potassium methyl malonate, MgCl₂, Et₃N (86 %) (ii) (S)-BinapRuCl₂, HBr, 90 psi H₂, MeOH, 80°C, then NaOH MeOH/H₂O (83 %) (iii) BnONH₂·HCl, EDC, LiOH, THF/H₂O (iv) DIAD, PPh₃, THF (81 %) (v) LiOH, THF/H₂O (25°C) (vi) EDC, N-methylmorpholine, CH₃CN, 0°C (vii) H₂Pd/C then H₃PO₄, 78 % yield from **10**.

- The Masamune reaction had a high dilution of 30 L kg^{-1}, the most dilute point in the process.
- The use of EDC in two coupling steps. EDC has poor atom economy and generates a urea by-product with a molecular weight of 209.7, whereas the theoretical ideal by-product of the coupling reaction should just be water with a molecular weight of 18.
- The use of energetic reagents such as DIAD.

The commercial process that Merck invented addresses all of these weaknesses and is shown in Scheme 6.9.[43]

The acid **6** and the triazole **12** are intermediates which are common with the earlier route. The acid **6** was activated with pivaloyl chloride (2,2-dimethylpropanoyl chloride) and then reacted with Meldrum's acid to form the adduct **15**. Meldrum's acid was used as a two-carbon synthon. The adduct **15** was not isolated but converted to the ketoamide **16** by adding the triazole **12** and a catalytic quantity of trifluoroacetic acid. The ketoamide **16** was also not isolated but converted through to the enamine **17** by the addition of a methanolic solution of ammonium acetate. The overall yield from the acid **6** to enamine **17** was an impressive 84%.

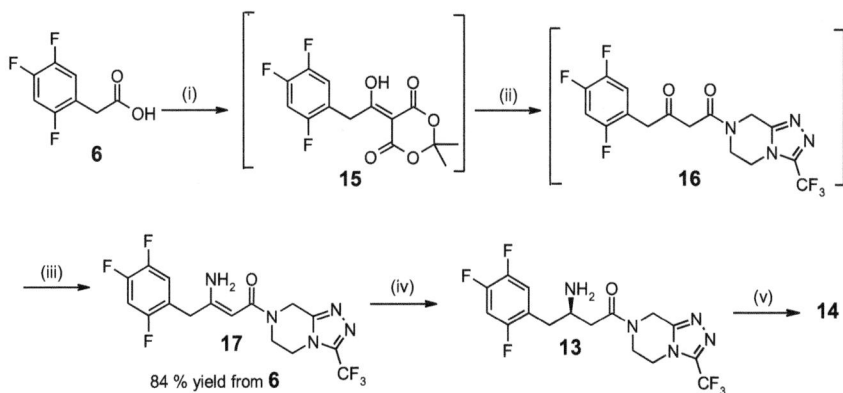

(i) Meldrum's acid, iPr$_2$NEt, DMAP (8 mol %), CH$_3$CN, (ii) triazole **12**, TFA (cat.) (iii) NH$_4$OAc, MeOH (iv) 0.15 mol % [Rh(COD)Cl]$_2$, 0.31 mol % *t*-Bu Josiphos, 90–100 psig H$_2$, MeOH 50°C, Ecosorb C-941, IPA/heptane crystallisation (79%) (v) H$_3$PO$_4$, H$_2$O, IPA (95%).

Scheme 6.9 The first commercial synthesis of sitagliptin (asymmetric hydrogenation route).

The enamine **17** was then subjected to an asymmetric hydrogenation reaction using a catalyst generated from Rh(COD)Cl$_2$ and *t*-Bu-Josiphos. Particularly of note was that no protecting groups were required for this chemistry. Following the reduction, 90–95% of the rhodium was recovered for recycling using an Ecosorb C-941 adsorbant.[44] The free amine **13** was then converted to the final phosphate salt **14**.

Table 6.5   Environmental assessments of the β-lactam route and asymmetric
hydrogenation route to sitagliptin.

	β-Lactam route	*Asymmetric hydrogenation route*	*Reduction*
Aqueous waste	60 kg	2 kg	97%
Organic waste	205 kg	65 kg	68%

Merck report[45] some significant environmental savings for the asymmetric
hydrogenation route (Scheme 6.9) when compared with the β-lactam route
(Scheme 6.8) and these are summarised in Table 6.5.

This synthesis was successfully used to launch the product, but in spite of the
major environmental savings, one significant environmental issue remained.
Rhodium is an extremely rare metal that is only present in the Earth's crust at
4 ppb. This scarcity is also reflected in its price volatility: between early 2006
(when Januvia® was being brought to the market) and July 2008 (before the
global recession) the price of rhodium rose from $3000 per ounce to $10 000 per
ounce.[46] Hence for both environmental and financial reasons, Merck started to
look at other options.

An alternative to the two-step process to convert the ketoamide **16** to the
sitagliptin free base **13** is to perform this transformation as a single step using a
transaminase enzyme (Scheme 6.10). Merck have recently reported the direct
conversion of **16** to **13** using a transaminase enzyme and with isopropylamine
as the nitrogen source. Merck have also indicated that they plan to file the new
synthesis to replace the asymmetric hydrogenation route with the transaminase
route for commercial production.[47]

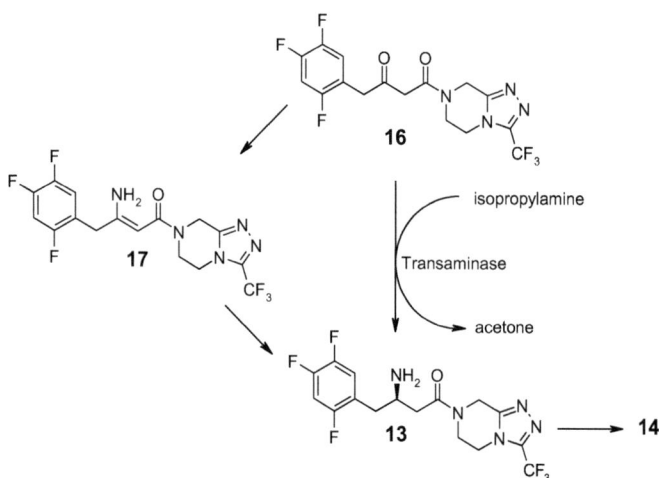

Scheme 6.10   The transaminase route to sitagliptin.

6.5.3 A Rosuvastatin Intermediate

6.5.3.1 Introduction to Statins

Pharmaceutical medicines that inhibit the enzyme HMG-CoA reductase, which is the rate-limiting enzyme in cholesterol biosynthesis, have become the standard of care for the treatment of hypercholesterolemia.[48] Lovastatin (**19**), simvastatin (**20**) and pravastatin (**21**) are naturally occurring fungal metabolites or their semi-synthetic derivatives (see Figure 6.3), whereas fluvastatin (**22**), atorvastatin (**23**) and rosuvastatin (**24**) are totally synthetic inhibitors.

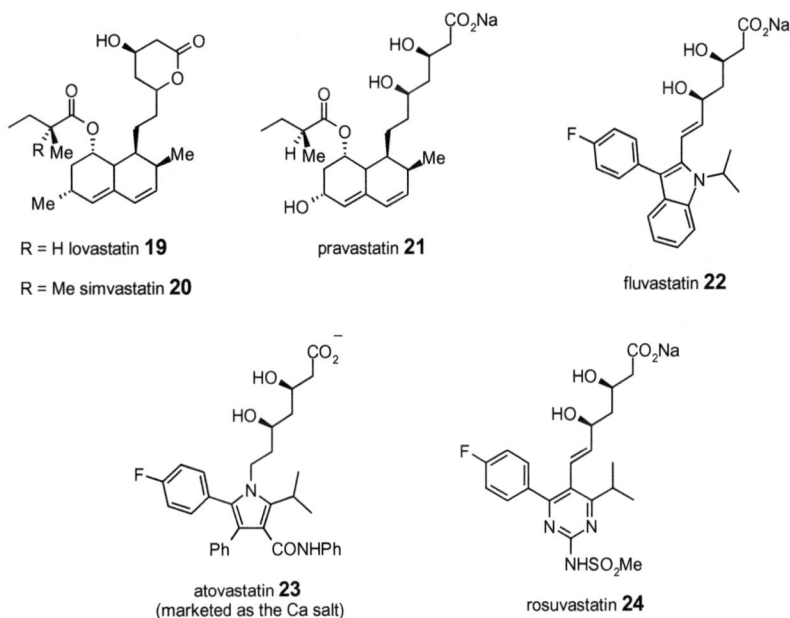

R = H lovastatin **19**

R = Me simvastatin **20**

pravastatin **21**

fluvastatin **22**

atovastatin **23**
(marketed as the Ca salt)

rosuvastatin **24**

Figure 6.3 Commercially available statins.

Fluvastatin was developed and marketed as a racemic mixture.[49] Atorvastatin and rosuvastatin are marketed as single enantiomers. Rosuvastatin was the last of these commercial statins to be introduced to the market and its sales are still growing: they were $3.6 billion, $4.5 billion and $5.7 billion in 2008, 2009 and 2010, respectively. The chemistry that was initially used to prepare the rosuvastatin side-chain is shown in Scheme 6.11.[50] Typically, a β-keto ester was asymmetrically reduced to set the first chiral centre; this was followed by a Claisen condensation to give the hydroxy ketone **25**. Following introduction of the desired R-group functionality, the second chiral centre was obtained by reduction with sodium borohydride and a boron complex at –78 °C, based on conditions that were initially developed by Pfizer/Parke-Davis for atorvastatin.[51]

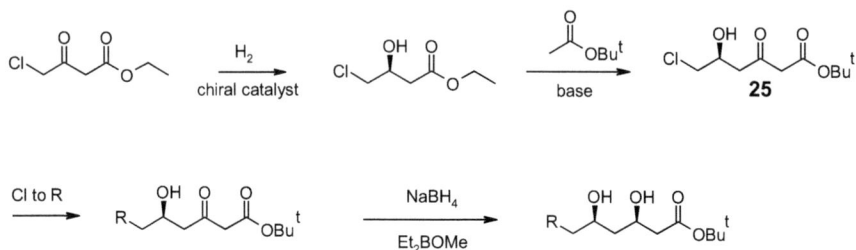

Scheme 6.11   Typical early chemistry to the rosuvastatin side-chain.

6.5.3.2   Biocatalytic Routes to Statins

Although initially prepared by chemical means (shown in Scheme 6.11), the statin side-chain is "made" for biocatalysis. An early example was the enantioselective reduction of the 3,5-diketo carboxylate **26** using an NADP-dependent alcohol dehydrogenase of *Lactobacillus brevis* (Scheme 6.12).[52,53] The NADP co-factor is recycled using isopropanol. The second chiral centre can also be introduced using a biocatalytic reduction.[54,55] Sheldon and co-workers have recently reported detailed E-factors (with and without water) and an environmental analysis based on the 12 principles of green chemistry[2] to make the statin side-chain for atorvastatin using a three-enzyme approach that was developed by the Codexis company.[56]

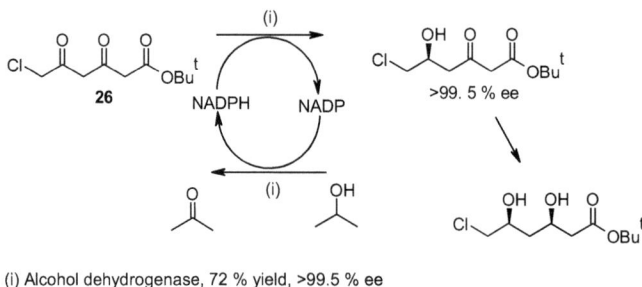

(i) Alcohol dehydrogenase, 72 % yield, >99.5 % ee

Scheme 6.12   An early biocatalytic route to the statin side-chain.

However, the most exciting development in this area came when Wong and co-workers reported that statin side-chains could be prepared in enantiopure form and in a single enzymatic step from basic chemicals using deoxyribose aldolase (DERA) enzymes.[57–60] It is very unusual when working in process R&D that an academic discovers and publishes exactly the reaction you want to help you with your project, but this certainly happened in the case of the rosuvastatin side-chain. The reaction sequence discovered by Wong and co-workers at the Scripps Institute is shown in Scheme 6.13.

The academic work formed the basis of the commercial process to make the rosuvastatin side-chain,[61] but there was still much to be done. Early DERA

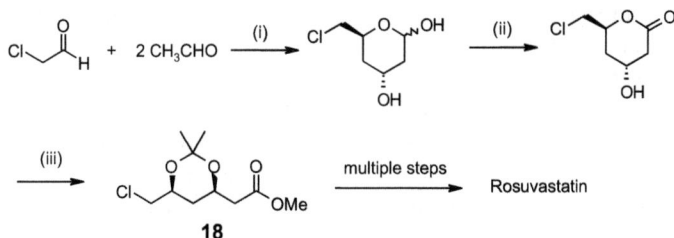

(i) DERA (ii) Br₂, H₂O, BaCO₃ (iii) 2,2-dimethoxypropane, cat. H⁺

Scheme 6.13 The synthesis of the rosuvastatin side-chain using the DERA enzyme.

enzymes had very low affinity for chloroacetaldehyde and hence very high enzyme levels were required. The problems were exacerbated as chloroacetaldehyde also caused rapid and irreversible deactivation of the enzyme. These problems were reduced to acceptable levels by outstanding scientific work in which the DERA enzyme was improved by a combination of extensive directed evolution and site-specific mutagenesis.[62]

The real power of the DERA chemistry is that two carbon–carbon bonds and two chiral centres are formed in a single process step; if you compare this with the Codexis chemistry[56,63] or the chemistry shown in Scheme 6.12, those transformations are occurring in multiple chemical steps.[§]

6.6   Future Trends

The importance of sustainability in process R&D seems set to continue and grow for the foreseeable future. In fine chemical and pharmaceutical manufacturing, solvents contribute heavily to the waste burden. Currently, much academic effort is directed at finding new solvents and indeed whole conferences are directed at this goal. However, many common oxygenated solvents, such as acetone, methyl ethyl ketone, methanol, ethanol, isopropanol, ethyl acetate, tetrahydrofuran and 2-methyltetrahydrofuran, can all be produced from renewable feedstocks rather than from petroleum; hence, in the author's view, more research into finding more efficient, cost-effective syntheses of existing oxygenated solvents from renewable feedstocks would be a very useful objective on the road to sustainability and may prove more useful than finding new solvents.

However, it appears that rare metals such as rhodium, iridium, palladium and platinum will become very scarce long before petroleum becomes scarce[64,65] and the goal of finding new chemical methodologies either using more abundant metals or, even better, using enzymatic methodologies also seems to have a high value for society.

[§]The next steps of the Codexis synthesis are a Claisen condensation followed by another keto-reductase reduction to set the second chiral centre.

Finally, there seems to be much opportunity to improve the teaching of more sustainable chemistry in our academic universities. Students are taught what a wonderful, selective and versatile reagent the Dess–Martin periodinane is for the oxidation of alcohols to aldehydes,[66] but how many are taught about the downsides of this methodology in terms of its poor atom economy and that it is a highly energetic reagent which is problematic for scale-up?

That said, the stunning ingenuity of chemists and chemical engineers within process R&D departments continues to produce a stream of outstanding chemical processes and it is hoped that the case histories presented in this chapter will inspire process chemists to target even more sustainable chemical processes. In the words of Nobel Laureate Professor Ryoji Noyori "Green chemistry is not just a mere catch phrase; it is the key to the survival of mankind".[67]

References

1. www.epa.gov/greenchemistry (last accessed 25 January 2011).
2. *Green Chemistry Theory and Practice*, ed. P. T. Anastas and J. C. Warner, Oxford University Press, Oxford, 1998.
3. J. H. Clarke, *Green Chem.*, 1999, **1**, 1–8.
4. R. A. Sheldon, *Chem. Ind.*, 1992, 903–906; *Green Chem.*, 2007, **9**, 1273–1283.
5. B. Trost, *Science*, 1991, **254**, 1471–1477.
6. D. H. Peterson and H. C. Murray, *J. Am. Chem. Soc.*, 1952, **74**, 1871–1872.
7. J. J. Li, *Laughing Gas, Viagra and Lipitor*, Oxford University Press, Oxford, 2006.
8. S. H. Pines, *Org. Process Res. Dev.*, 2004, **8**, 708–724.
9. G. P. Taber, D. M. Pfisterer and J. C. Colberg, *Org. Process Res. Dev.*, 2004, **8**, 385–388.
10. P. J. Dunn, S. Galvin and K. Hettenbach, *Green Chem.*, 2004, **6**, 43–48.
11. P. G. Mountford, in *Green Chemistry in the Pharmaceutical Industry*, ed. P. J. Dunn, A. S. Wells and M. T. Williams, Wiley-VCH, Weinheim, 2010, pp. 145–160.
12. K. B. Hansen, Y. Hsiao, F. Xu, N. Rivera, A. Clausen, M. Kubryk, S. Krska, T. Rosner, B. Simmons, J. Balsalls, N. Ikemoto, Y. Sun, F. Spindler, C. Malan, E. J. J. Grabowski and J. D. Armstrong III, *J. Am. Chem. Soc.*, 2009, **131**, 8798–8804.
13. A. Chauvel, A. Delmon and W. F. Hoelderich, *Appl. Catal. A: Gen.*, 1994, **115**, 173–215.
14. J. McGarrity, F. Spindler, R. Fux and M. Eyer (Lonza), EP 624587, 1994.
15. R. Imwinkelried, *Chimica*, 1997, **51**, 300.
16. D. J. C. Constable, C. Jimenez-Gonzalez and R. K. Henderson, *Org. Process Res. Dev.*, 2007, **11**, 133–137.
17. C. Jimenez-Gonzalez, A. D. Curzons, D. J. C. Constable and V. L. Cunningham, *Clean Techn. Environ. Policy*, 2005, **7**, 42–50.
18. K. Alfonsi, J. Colberg, P. J. Dunn, T. Fevig, S. Jennings, T. A. Johnson, H. P. Kleine, C. Knight, M. A. Nagy and M. Stefaniak, *Green Chem.*, 2008, **10**, 31–36.

19. C. R. Hargreaves, *A Collaboration to Deliver a Solvent Selection Guide for the Pharmaceutical Industry*, presented at the American Institute of Chemical Engineers Annual Meeting, Philadelphia, 17 November 2008.
20. C. Capello, U. Fischer and K. Hungerbuhler, *Green Chem.*, 2007, **9**, 917–934.
21. P. J. Dunn, A. S. Wells and M. T. Williams, in *Green Chemistry in the Pharmaceutical Industry*, Wiley-VCH, Weinheim, 2010, pp. 333–355.
22. R. A. Sheldon, presented at the International Symposium on Catalytic Chemistry for the Global Environment, Sapporo, Japan, July 1991 and published in ref. 4.
23. (*a*) M. E. Kopach, T. Zhang, S. Coffey, A. Borghese, M. Korbeirski and W. Tranke, *A Practical and Green Chemistry Approach for the Manufacture of NK1 Antagonist LY686017*, presented at the 12th Annual Green Chemistry and Engineering Conference, Washington, 25 June 2008; (*b*) http://www.gsk.com/responsibility/downloads/GSK-CR-2008-full.pdf (pp. 221 and 222) (last accessed 25 January 2011); (*c*) C. Boswell, *ICIS Chem. Bus. Mag.*, 2008, 16–17.
24. M. Philippe, presented at the Third International Conference on Green and Sustainable Chemistry, Delft, Holland, July 2007.
25. D. J. C. Constable, A. D. Curzons and V. L. Cunningham, *Green Chem.*, 2002, **4**, 521–527.
26. R. K. Henderson, J. Kindervater and J. M. Manley, presented at the Third International Conference on Green and Sustainable Chemistry, Delft, Holland, July 2007.
27. R. A. Sheldon, I. Arends and U. Hanefeld, *Green Chemistry and Catalysis*, Wiley-VCH, Weinheim, 2007; see also R. A. Sheldon, *Green Chem.*, 2007, **9**, 1273–1283.
28. A. J. J. Straathof, S. Panke and A. Schmid, *Curr. Opin. Biotechnol.*, 2002, **13**, 548–556.
29. D. R. Yazbeck, J. Tao, C. A. Martinez, B. J. Kleine and S. Hu, *Adv. Synth. Catal.*, 2003, **4**, 524–532.
30. W. P. Stremmer, *Nature*, 1994, **370**, 389–391.
31. O. Kucher and F. H. Arnold, *Trends Biotechnol.*, 1997, **15**, 523–530.
32. M. T. Reetz, L.-W. Wang and M. Bocola, *Angew. Chem. Int. Ed.*, 2006, **45**, 1236–1241.
33. B. A. Lauria-Horner and R. B. Pohl, *Expert Opin. Invest. Drugs*, 2003, **12**, 663–672.
34. I. Selak, *Curr. Opin. Invest. Drugs*, 2003, **2**, 828–834.
35. M. S. Hoekstra, D. M. Sobieray, M. A. Schwindt, T. A. Mulhern, T. M. Gote, B. K. Huckabee, V. S. Hendrickson, L. C. Franklyn, E. J. Granger and G. L. Karrick, *Org. Process Res. Dev.*, 1997, **1**, 26–38.
36. For an asymmetric hydrogenation route to pregabalin, see M. J. Burk, P. D. de Koning, T. M. Grote, M. S. Hoekstra, G. Hoge, R. A. Jennings, W. S. Kissel, T. V. Le, I. C. Lennon, T. A. Mulhern, J. A. Ramsden and R. A. Wade, *J. Org. Chem.*, 2003, **68**, 5731–5734.
37. C. A. Martinez, S. Hu, Y. Dummond, J. Tao, P. Kelleher and L. Tully, *Org. Process Res. Dev.*, 2008, **12**, 392–398.

38. P. J. Dunn, K. Hettenbach, P. Kelleher and C. A. Martinez, in *Green Chemistry in the Pharmaceutical Industry*, ed. P. J. Dunn, A. S. Wells and M. T. Williams, Wiley-VCH, Weinheim, 2010, pp. 161–177.
39. A. E. Weber, *J. Med. Chem.*, 2004, **47**, 4135–4141.
40. P. E. Wiedeman and J. M. Trevillyan, *Curr. Opin. Invest. Drugs*, 2003, **4**, 412–420.
41. D. W. Brooks, L. D. Lu and S. Masamune, *Angew. Chem. Int. Ed.*, 1979, **18**, 72–74.
42. M. J. Miller, P. G. Mattingly, M. A. Morrison and J. F. Kerwin, *J. Am. Chem. Soc.*, 1980, **102**, 7026–7032.
43. K. B. Hansen, Y. Hsiao, F. Xu, N. Rivera, A. Clausen, M. Kubryk, S. Krska, T. Rosner, B. Simmons, J. Balsells, N. Ikemoto, Y. Sun, F. Spindler, C. Malan, E. J. J. Grabowski and J. D. Armstrong III, *J. Am. Chem. Soc.*, 2009, **131**, 8798–8804.
44. The commercial source of Ecosorb C-941 is Graver Technologies, Glasgow, DE 19702, USA.
45. J. Balsells, Y. Hsiao, K. B. Hansen, F. Xu, N. Ikemoto, A. Clasuen and J. D. Armstrong III, in *Green Chemistry in the Pharmaceutical Industry*, ed. P. J. Dunn, A. S. Wells and M. T. Williams, Wiley-VCH, Weinheim, 2010, pp. 101–126.
46. www.kitco.com (last accessed 25 January 2011).
47. (*a*) G. Hughes, *A Greener Biocatalytic Manufacturing Route to Sitagliptin*, presented at the 13th Annual Green Chemistry and Engineering Meeting, College Park, MD, June 2009; (*b*) for an excellent publication by Merck and Codexis, see C. K. Saville, J. M. Janey, E. C. Mundorff, J. C. Moore, S. Tam, W. R. Jarvis, J. C. Colbeck, A. Krebber, F. J. Fletz, J. Brands, P. N. Devine, G. W. Huisman and G. J. Hughes, *Science*, 2010, **328**, 305–309.
48. J.-J. Li, D. S. Johnson, D. R. Sliskovic and B. D. Roth, in *Contemporary Drug Synthesis*, Wiley-Interscience, Hoboken, NJ, 2004, pp. 113–124.
49. O. Repic, K. Prasad and G. T. Lee, *Org. Process Res. Dev.*, 2001, **5**, 519–527.
50. H. Kierkels, S. Panke, M. Schuermann and M. Wolberg, in *Green Chemistry in the Pharmaceutical Industry*, ed. P. J. Dunn, A. S. Wells and M. T. Williams, Wiley-VCH, Weinheim, 2010, pp. 127–144.
51. P. L. Browner, D. E. Butler, C. F. Deering, T. V. Le, A. Millar, T. N. Nanninga and B. D. Roth, *Tetrahedron Lett.*, 1992, **33**, 2279–2282.
52. M. Woberg, W. Hummel, C. Wandrey and M. Muller, *Angew. Chem. Int. Ed.*, 2000, **39**, 4306–4308.
53. M. Woberg, M. V. Filho, S. Bode, P. Geilenkirchen, R. Feldmann, A. Liese, W. Hummel and M. Muller, *Bioprocess Biosyst. Eng.*, 2008, **31**, 183–191.
54. R. N. Patel, A. Banerjee, C. G. McNamee, D. Brzozowski, R. L. Hanson and L. J. Szarka, *Enzyme Microb. Technol.*, 1993, **15**, 1014–1021.
55. R. A. Holt, A. J. Blacker and C.D. Reeve, WO 01/85975, 2001.
56. S. K. Ma, J. Gruber, C. Davis, L. Newman, D. Gray, A. Wang, J. Grate, G. W. Huisman and R. A. Sheldon, *Green Chem.*, 2010, **12**, 81–86.

57. L. Chen, D. P. Dumas and C.-H. Wong, *J. Am. Chem. Soc.*, 1992, **114**, 741–748.
58. H. J. M. Gijsen and C.-H. Wong, *J. Am. Chem. Soc.*, 1994, **116**, 8422–8423.
59. C.-H. Wong, E. Garcia-Junceda, L. Chen, O. Blanco, H. J. M. Gijsen and D. H. Steensma, *J. Am. Chem. Soc.*, 1995, **117**, 3333–3339.
60. J. Liu, C.-C. Hsu and C.-H. Wong, *Tetrahedron Lett.*, 2004, **45**, 2439–2441.
61. N. Ran, L. Zhao, Z. Chen and J. Tao, *Green Chem.*, 2008, **10**, 361–372.
62. S. Jennewein, M. Schuermann, M. Wolberg, I. Hilker, R. Luiten, M. Wubbolts and D. Mink, *Biotechnol. J.*, 2006, **1**, 537–548.
63. (a) R. J. Fox, S. C. Davis, E. C. Mundorff, L. M. Newman, V. Gavrilovic, S. K. Ma, L. M. Chung, C. Ching, S. Tam, S. Muley, J. Grate, J. Gruber, J. C. Whitman, R. A. Sheldon and G. W. Huisman, *Nat. Biotechnol.*, 2007, **25**, 338–344; (b) J. Grate, *presented at the Third International Conference on Green and Sustainable Chemistry*, Delft, Holland, July 2007.
64. J. Johnson, E. M. Harper, R. Lifset and T. E. Graedel, *Environ. Sci. Technol.*, 2007, **41**, 1759–1765.
65. R. B. Gordon, M. Bertram and T. E. Graedel, *Proc. Natl. Acad. Sci.*, 2006, **103**, 1209–1214.
66. D. B. Dess and J. C. Martin, *J. Org. Chem.*, 1983, **48**, 4155–4156.
67. R. Noyori, *Tetrahedron*, 2010, **66**, 1028.

CHAPTER 7

Kinetic Approaches for Faster and Efficient Process Development

SUJU P. MATHEW

Chemical Research and Development, Pfizer Global Research and
Development, Sandwich, CT13 9NJ, UK

7.1 Introduction

Developing efficient chemical processes, and reducing the cost and time factors
involved in the development work for transferring the process from bench to
manufacturing scale, is becoming key to the implementation of any successful
process in the current environment of stringent patent laws, environmental
concerns and stiff competition. The pharmaceutical process development
roadmap and approaches are being continuously improved in an effort to
achieve an organisation and a path which can deliver quality active pharmaceutical ingredients (APIs) in the most cost effective, greenest and fastest
manner. An in-depth understanding of scale-up and robustness issues, together
with a detailed understanding of the process parameters that determine the
process efficiency, are key factors in this effort. A multidimensional analysis of
the process from synthetic, physical organic chemistry and reaction engineering
viewpoints, coupled with the adoption of state-of-the-art process optimisation
and manufacturing technologies and methodologies, is essential to face these
challenges. Synthetic organic chemistry will remain the main force in small-
molecule pharmaceutical API synthetic route selection and development.

RSC Drug Discovery Series No. 9
Pharmaceutical Process Development: Current Chemical and Engineering Challenges
Edited by A. John Blacker and Mike T. Williams
© Royal Society of Chemistry 2011
Published by the Royal Society of Chemistry, www.rsc.org

However, in changing process development scenarios, it is important to develop process development skills enabling the problem to be analysed from multiple angles. Understanding how the interplay of equilibria, kinetics and mass/heat transfer properties influence process robustness, efficiency and scalability forms the key part of process development. We frequently try to interpret experimental results based on reaction mechanisms and try to come up with different reagents or bond-forming steps, whereas solving the actual problem might involve dealing with a pre-equilibrium step or a kinetic or mass transfer related issue. Considering the timelines involved, this is quite challenging, as we deal with a wide range of complex organic transformations and processing scenarios.

Understanding the kinetic parameters affecting the formation of the desired product and any undesired impurities, and identifying their dependency on mass and heat transfer effects (and if needed decoupling these two effects), is necessary for the successful scale-up of the process. In the petrochemical and bulk chemical industries, where longer process development time lines are involved, kinetics and reactor design play a prominent role. These industries also employ more reaction engineers and give them a bigger responsibility for developing the process, unlike the pharmaceutical industry where synthetic organic chemists constitute the main work force. The traditional way of understanding reaction kinetics, using approaches like initial rate analysis, and predicting reactor performance involves detailed mathematical treatment. Many synthetic chemists find such traditional kinetic approaches and model development very "mathematical" and time consuming. To deal with most day-to-day process development problems, use of such mathematical approaches is not usually necessary. Simple approaches, or treatments using basic kinetic and reaction engineering principles (coupled with appropriate kinetic methodology and data gathering methods), are found to be very effective in most cases. Such approaches can be applied across various process development phases, including reagent/condition selection, process understanding, robustness and scale-up analysis, and also in defining the edges of failure for various operating conditions. Detailed mathematical treatments like reactor modelling can be of more significance when predicting the scale-up performance of reactions in which mixing and mass and/or heat transfer properties of the reactor can have a significant effect on the reaction behaviour on larger scales. This chapter highlights some of the basic kinetic approaches we have found highly effective in our day-to-day process development efforts, such as reagent selection, process optimisation, robustness and scale-up analysis and understanding mass transfer effects.

7.2 Reagent and Processing Condition Design Based on Kinetic Experimentation

Owing to the high early candidate attrition in the pharmaceutical industry, detailed process development efforts are often initiated at later stages when more clarity on the candidate success is available. Once the commercial route is identified, development effort focuses on reagent selection and determining the processing conditions. An in-depth understanding of the process from mechanistic and parametric dependencies is the key in achieving the right reagent and operating condition selection. Identifying the rate-determining step or steps involved in a chemical reaction is critical for this understanding as the processing conditions are developed. Well designed and interpreted kinetic experiments can give valuable information about the rate-determining step and the reaction mechanism involved. This information can then be used to design reagents and/or conditions to obtain the desired result. This approach allows us both to gain very early process understanding and also to "design" the right reagents, rather than just "picking up" available ones.

Any particular chemical transformation can involve multiple steps. Even if multiple steps are involved, the reaction rate and selectivity are mainly determined by the rate-determining step (which is usually the slowest step) involved in that chemical transformation. Let us consider a general reaction in which A and D react in the presence of reagents B and F (eqn 7.1):

$$A + D \xrightarrow[\text{solvent,temperature}]{\text{B,F}} \text{Product} \tag{7.1}$$

For such a chemical reaction involving many reagents, identifying the right reagents and processing conditions can be challenging. Usually this involves a detailed screen of reagents and conditions, which can be a quite laborious and time-consuming process. If the reaction involves equilibria and multiple steps as shown below (eqns 7.2–7.4), it is important to understand the rate-limiting step and the role that reagents play in this step. Identifying the rate-determining step involved helps us to identify the right reagent and operating conditions for the process. For this case, let us assume that the following individual steps are involved:

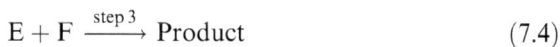

$$A + B \underset{}{\overset{\text{step 1}}{\rightleftharpoons}} C \tag{7.2}$$

$$C + D \xrightarrow{\text{step 2}} E \tag{7.3}$$

$$E + F \xrightarrow{\text{step 3}} \text{Product} \tag{7.4}$$

This breakdown of steps can be carried out using mechanistic knowledge or speculation. Once you have split the chemical transformation into possible individual steps, kinetic experiments should be carefully designed to understand the rate-limiting step. This can be carried out by singling out one step at a time and varying the concentration of the components to understand its effect on the reaction rate. For example, if we assume step 2 is the rate-limiting step in the above case, the concentrations of E and F will have little influence on the net reaction rate, and experiments with different initial concentrations of E and F will confirm this. Experiments with varying concentration of C (will depend on A and B concentrations and the equilibrium position) and D should show a

$$\text{R-OH} \quad + \quad \text{R}_1\text{-Cl} \quad \xrightarrow[\text{Solvent}]{\text{Base}} \quad \text{R-O-R}_1$$

A1 B1 C1

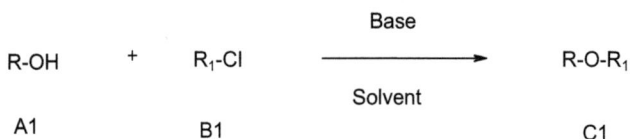

Scheme 7.1 Reaction between a hydroxy and a chloro compound.

considerable effect of the net reaction rate observed. Once you identify the rate-limiting step, the properties of the reagent required and/or the processing conditions can be chosen. Such an approach helps us to quickly identify the critical process parameters of the process, to systematically design reagents required for the process and to gain an early in-depth understanding of the process.

A reaction examined in our laboratories involved the reaction of a hydroxy compound ROH (A1) with a chloro compound R_1Cl (B1) in the presence of a base to generate the desired product C1 (Scheme 7.1).[1] Initial conditions identified for this transformation, guided by the literature, involved using a weaker organic base and an organic solvent. At 80 °C the reaction conversion was around 95% in 7 hours, and achieving a faster reaction rate was considered desirable.

To identify the critical parameters that could be influencing the reaction rate, the reaction was first split into the possible individual steps involved: (1) deprotonation of the hydroxy compound to generate the anion (eqn 7.5) and (2) displacement of the chlorine by the anion (eqn 7.6):

$$\text{ROH} + \text{Base} \underset{}{\overset{\text{step 1}}{\rightleftharpoons}} \text{RO}^- \tag{7.5}$$

$$\text{RO}^- + \text{R}_1\text{Cl} \xrightarrow{\text{step 2}} \text{R--O--R}_1 \tag{7.6}$$

The preferred situation for this reaction would be conditions under which step 2 (eqn 7.6) was rate limiting, thus minimising the process parameters that would need to be focused on for development. Under such conditions, the rate of product formation would be determined by the concentration of the anion and the chloro compound in the solution. Two experiments were designed to check this: (1) a standard experiment with 1.05 equivalents of chloro compound (0.73 M) and (2) a higher concentration of chloro compound (1.0 M), while keeping the concentration of the limiting reagent, hydroxy compound and other parameters constant. The product formation as a function of time for these two experiments is shown in Figure 7.1; with weaker base the initial rate of product formation is not significantly different at different chloro compound concentrations, suggesting that step 2 might not be the rate-determining step (another likely possibility of zero-order dependency for chloro compound concentration also exists).

The possibility of step 2 not being rate-determining points to the potential that the step 1 (eqn 7.5) equilibrium might be playing a crucial role in determining the reaction rate. Since the concentration of the anionic species is

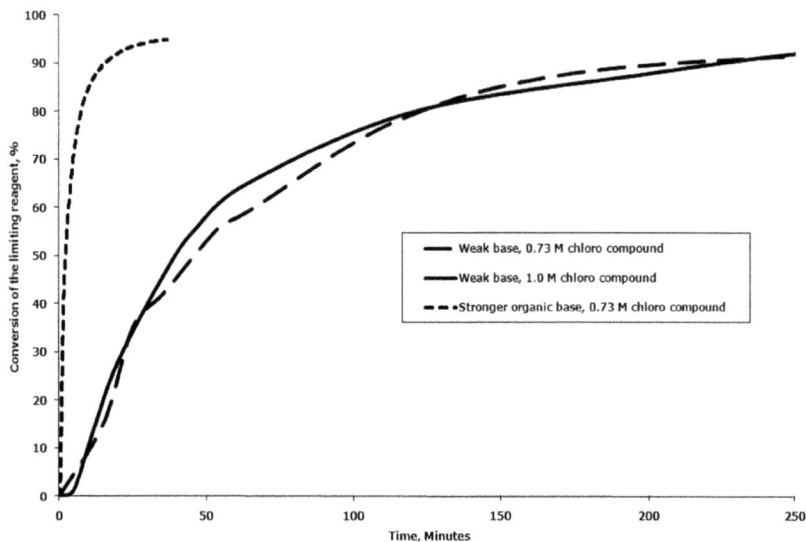

Figure 7.1   Product formation profile for C1.

determined by the position of the hydroxy group deprotonation equilibrium, the net observed rate of formation of the product could be determined by either step 1 or step 2. If step 1 is the rate-limiting step, the reaction rate will be determined by the position of the equilibrium in step 1, and the influence of chloro compound concentration on the reaction rate will be minimal. In this scenario, the important parameters affecting the reaction could be those that affect the equilibrium position, like strength of the base, concentration of the base (not only equivalents) and common parameters like temperature and solvent. A very small screen with stronger bases (based on theoretically predicted pK_a of the hydroxylic proton) showed that with a stronger (but expensive) organic base the reaction was completed in less than 30 minutes even at lower temperatures (Figure 7.1, curve with stronger organic base). This proves that when a weaker base is used, the rate was mainly determined by the pre-equilibrium involved, whereas with a stronger base the rate-determining step is shifted to the desired displacement reaction. Now having understood that shifting the pre-equilibrium involved is the key, we could go back and speed up the reaction with the cheaper weaker base by shifting to a more polar solvent and using a higher temperature (other factors that determine the equilibrium).

When assessing potential new routes to an API, the chances of success for high-value multi-step routes will often depend on key synthetic steps ("kill steps") which offer considerable cost advantage, but have a limited chance of success. Literature precedents may exist for such "kill step" reactions in simpler molecules. The likelihood of the key transformation working with the API of interest, which is often more complex and has very different electronic properties, needs to be rapidly assessed. The same kinetic approach can be used to identify the important parameters influencing the rate-determining steps, based on the possible mechanisms or reported kinetic investigations, and then designing a targeted screen. Carrying out reactions with reactants having the same functional group but different electronic properties can change the rate-determining step, and care should be taken in interpreting results in such cases; further experimentation might be required to confirm trends.

When designing such kinetic experiments, care should be taken to avoid using reagents in high excess (or as a solvent) or using multiple-fold concentrations, as this might lead to pseudo-zero-order behaviour. High reagent excesses can also considerably alter physical properties like the polarity of the system, which will be reflected in the reaction rate and can be misinterpreted as dependency of the reagent concentration on the reaction rate. Another important point that should always be kept in mind is the zero-order kinetics shown by certain reagents in certain steps.

7.3 Robustness and Process Understanding from Kinetic Trends and Rate Analysis

Process robustness is one of the most important areas of concern in pharmaceutical process development. Most common sources of process robustness issues are related to:

- Substrate instability or decomposition.
- Side reactions involving one or more substrates.
- Product instability, impurity formation or increased reaction or processing periods and/or downstream processing (such as distillation or crystallisation).
- Mixing issues associated with fast reactions.
- Controlled reaction termination.
- Mass and/or heat transfer issues.

Such factors can lead to increased levels of certain impurities or to product purity not meeting specification, resulting in batch failures. In all these cases it is extremely important to understand the robustness and scale-up behaviour by designing the right small-scale laboratory experiments. Most of these robustness issues (other than mixing, mass and heat transfer related) are directly related to the concentration of the components, processing periods and temperature. These three factors constitute the main components of reaction rates and kinetic trends, and an understanding of rate and kinetic trends is key for addressing any robustness issues.

7.3.1 Reaction Profiling for Robustness and Process Understanding

Simple techniques like kinetic profiling of reactions and interpreting and understanding the rate behaviour of both substrate disappearance and product/impurity formations (as well as any decompositions) are very powerful

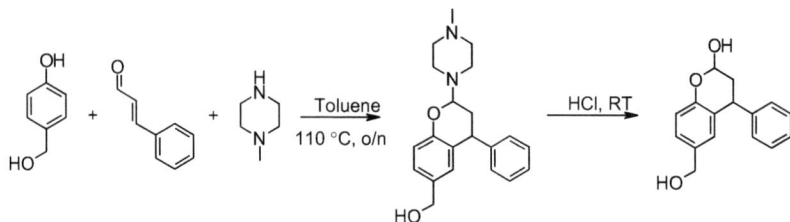

Scheme 7.2 Amine-mediated Friedel–Crafts alkylation of 4-hydroxybenzyl alcohol.

tools for fast and efficient process development. Since reaction rate is the rate of change of concentration, it is reflected as the slopes of the curve, and the steepness of the slope will tell whether the rate of depletion is slow or fast. So the key is looking at the shape of the concentration profiles and the steepness of the slopes of the curves at different periods, which is the rate of formation/ depletion over that reaction period.

One of the reactions we had been investigating in our laboratory was the amine-mediated Friedel–Crafts alkylation reaction involving 4-hydroxybenzyl alcohol, cinnamaldehyde and N-methylpiperazine (Scheme 7.2).[2]

The reaction involved two parts: (1) the main annulation reaction in refluxing toluene conditions for ~8 hours and (2) the product–base adduct hydrolysis using aqueous hydrochloric acid solution to liberate the free product. The downstream processing involved a phase separation followed by solvent swap for crystallisation. The main issues we faced with this reaction were that the product formation was stalling (~60% isolated product yield) and that a brown oily third phase was formed which was sticking to the glass wall, making both the phase separation and reactor cleaning difficult. HPLC analysis showed only the formation of one impurity, benzaldehyde, at the relatively low level of <5%, along with the product. In order to understand the kinetic trends involved, a concentration–time profile of a typical reaction is shown in Figure 7.2.

As can be seen from the graph, at the beginning of the reaction the cinnamaldehyde concentration is in 1.3 fold excess compared to the limiting reagent hydroxybenzyl alcohol. If the desired product formation was the only significant reaction happening and contributing to the mass balance, this excess should have been maintained throughout the reaction course, based on known reaction stoichiometry. However, as can be seen from Figure 7.2, even though cinnamaldehyde is the excess reagent at the beginning of the reaction, during the course of the reaction it becomes the limiting reagent. During the entire reaction course, the rate of product formation matched reasonably well with the rate of disappearance of the hydroxybenzyl alcohol, whereas the rate of cinnamaldehyde disappearance was considerable higher. The shape of the concentration profiles and the steepness of the slopes of the curves at different periods highlight these trends. The instability of the cinnamaldehyde with temperature might be the contributing factor towards the formation of the oily polymeric impurity, and also explains the loss of mass balance from cinnamaldehyde.

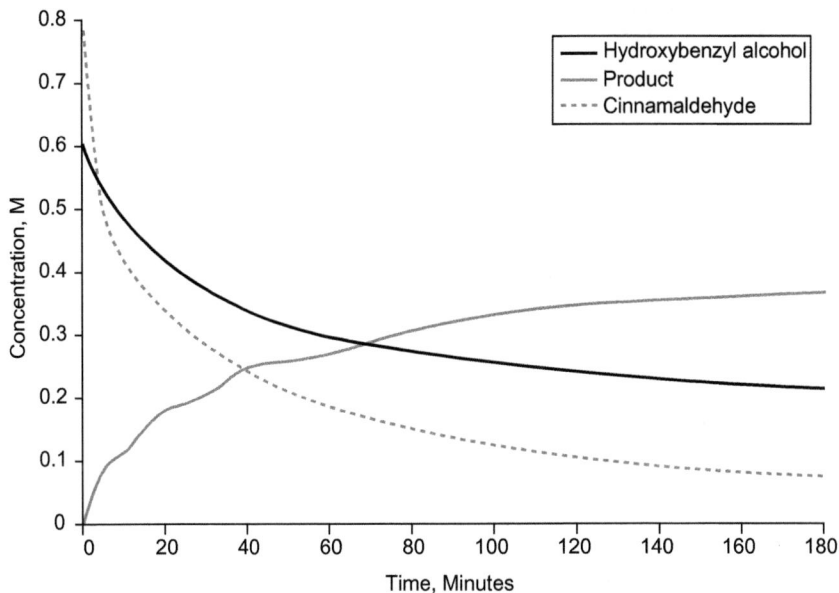

Figure 7.2   Concentration–time profile of reaction components of Scheme 7.2. Reaction conditions: [hydroxybenzyl alcohol] = 0.6 M; [cinnamaldehyde] = 0.78 M; [NMP] = 1.5 M; solvent, toluene; temperature, 110 °C.

We found that the rate of formation of the benzaldehyde impurity formation was too small to account for this high rate of depletion in cinnamaldehyde concentration, suggesting an undetectable impurity formation from the cinnamaldehyde, as shown in the reaction pathway (Scheme 7.3). Apart from the cinnamaldehyde concentration, the brown oily impurity formation was found to be strongly dependent on the reaction period and temperature, and shorter processing time was preferable. The reaction profile also showed that the reaction rate drops considerably with time, and that the rate is extremely slow after ~120 minutes. Most of the product formation happens within this reaction time. Thus, terminating the reaction after 2 hours, and distillation under vacuum, considerably reduced the oily impurity formation as shown in Figure 7.3.

The manner in which such concentration–time kinetic profiling is carried out is also critical in identifying robustness-related issues. The frequency and number of samples required depends on the nature of the system and the components of interest. If the concern is the stability of the initially charged reagents, or an intermediate species formed initially from the reagents, then more sample points in the beginning will be critical in understanding the rate trends. If the concern is certain impurities forming towards the end of the reaction and/or product decomposition, then sufficient samples are also required at the end of the reaction. The frequency of sampling should be decided based on how fast the undesired reaction takes place. A good starting

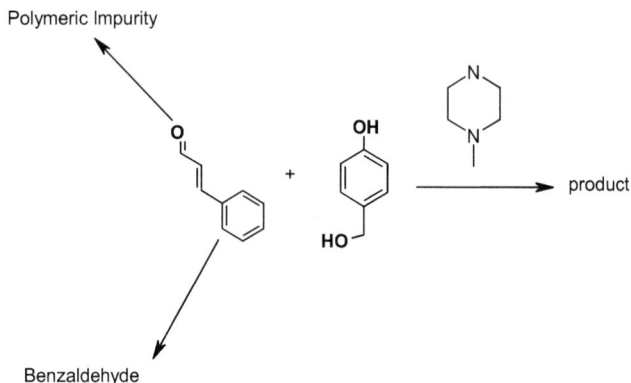

Scheme 7.3   Proposed reaction pathway for the reaction Scheme 7.2.

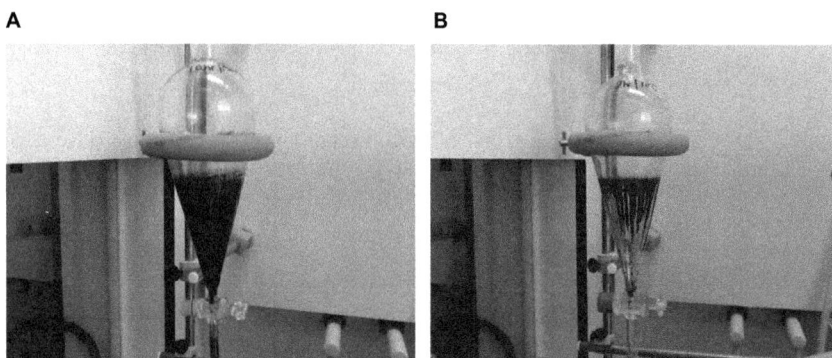

Figure 7.3   Effect of processing conditions on impurity formation: (A) 10 h reaction time and distillation at reflux temperature; (B) 2 h reaction time followed by distillation at lower temperature.

point to determine sampling frequency and timing is to run a reaction with extended reaction time with similar sampling frequency, and then identify the reaction periods where the rates of reaction involving the desired components are important. Monitoring the rate behaviour of the species formation during the entire course of the reaction period gives a great deal of information regarding reaction termination, robustness and relative reaction rates. What needs to be compared is how fast substrates and intermediate species are becoming depleted, and whether this is reflected in the formation of product and/or certain impurities. This will also highlight mass balance and robustness issues, the reaction pathways involved, the available reaction termination window and interdependencies of the rate on various reaction components.

Even though concentration profiles can be applied commonly, in many cases converting these concentration data into rate data can give clearer insights into the reaction trends. This can be done either manually, by determining the slope at various time intervals during the reaction course, or by fitting the concentration–time profile using a mathematical function (like a polynomial fit) and taking the derivative of that. However, in some cases such a single mathematical fit might not be good enough to represent the entire reaction period, and might need splitting into various time periods within the reaction. Such mathematical manipulation can be reduced by directly using a differential technique like heat flow, which is a direct representation of the rate and will be discussed in Section 7.3.2. A Raney nickel catalysed aqueous phase hydrogenation (eqn 7.7) of a nitrile compound was investigated in Argonaut Endeavor hydrogenation equipment which had a multiple reaction facility:[3]

$$RCN \xrightarrow[\text{aqueous basic conditions}]{H_2, \text{Raney Ni, } 30\,°C} RCH_2NH_2 \tag{7.7}$$

Hydrogen uptake curves for low pressure (50 psig = 2885 Torr) and higher pressure (100 psig = 5770 Torr) runs are shown in Figure 7.4.

From the hydrogen uptake curves for the above two cases it can seen that the high-pressure reaction is considerably faster and finished before the low-pressure reaction. This is as expected for normal hydrogenation reactions, as the rate is usually proportional to the hydrogen partial pressure. Such a hydrogen profile curve indicates the dependency of the reaction on the hydrogen partial pressure, and also helps to identify the amount of processing time required. Now let us take the derivative of the two curves and see whether we can gather any further information. The rates of hydrogen consumption calculated by taking the derivative of the hydrogen consumption curves of the two reactions over the reaction period are shown in Figure 7.5.

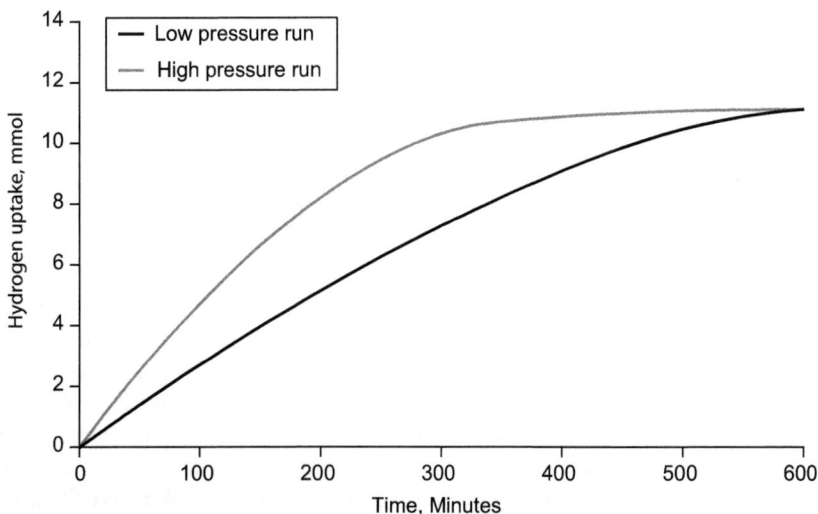

Figure 7.4 Hydrogen uptake curves for nitrile reduction.

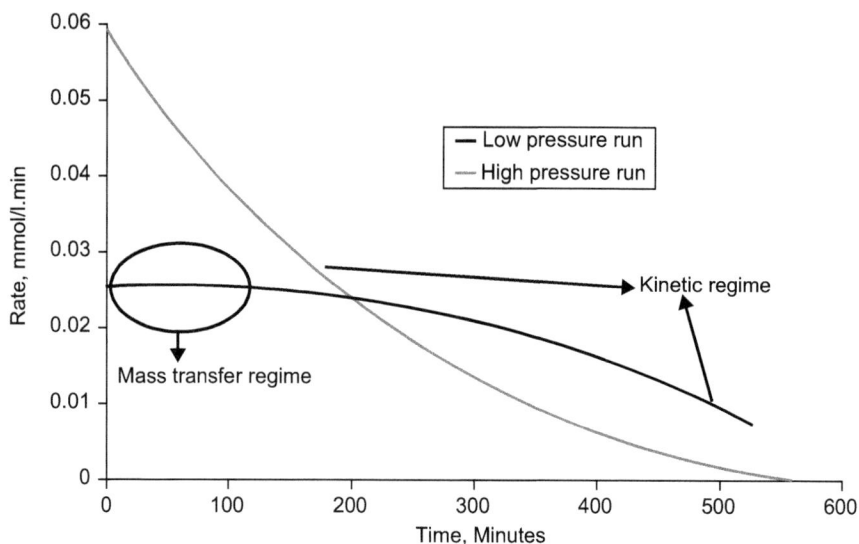

Figure 7.5   Rate of hydrogen consumption for the nitrile reduction.

Now more information about the reaction can be extracted. Comparing the rates at similar hydrogen consumption data points can give information about the order of the reaction on hydrogen concentration. Comparing rates at two different points within the same curve, and correlating that to the nitrile concentration from hydrogen consumption, tells you the dependency of rate on the nitrile concentration. The rate curve for the higher pressure run in this case shows approximately first-order behaviour (not quite first-order as it follows Langmuir–Hinshelwood kinetics) with respect to the nitrile substrate; as the substrate is consumed, the rate slows down. No effect for hydrogen will be observed during the reaction period in this case, as the reaction was carried out in a semi-batch mode with constant hydrogen pressure and hence almost constant hydrogen liquid-phase concentration. By comparison, the rate curve for the lower pressure run clearly shows a strong effect for hydrogen concentration. Comparing initial reaction rates for the two runs is difficult, because if we look at the lower pressure reaction run the initial phase of the curve is almost a straight line, but as the reaction progresses it shows a falling trend. This suggests that the initial part of the reaction might be in a mass transfer controlled regime (high reaction rate due to higher nitrile concentration), and that as the nitrile is consumed (as the reaction progresses) its concentration falls, the rate slows down and the reaction moves into a kinetic regime. Thus interpreting the reaction profile in terms of rates gives considerably more information about reaction behaviour and kinetic dependencies, and also gives a very good indication of the mass transfer issues involved. We have found this approach very useful, not only for hydrogenation reactions but also for a wide range of stoichiometric and catalytic reaction systems.

Since pharmaceutical process development efforts involve wide-ranging processing conditions, we often come across reactions where mass transfer efficiency between the component phases can influence the process behaviour. Early application of reaction engineering principles during process development, in cases where mass transfer can be contributing, is extremely important to make sure that the scaled-up process runs as predicted, or as found in laboratory scale studies. In cases where the desired reaction is the only major product, and no concerns exist with impurity formation, the same product purity and yield could be achieved even under mass transfer limited operating conditions, but the processing time required might vary. For reactions with robustness issues and which require accurate reaction termination, it is very important to understand the mass transfer behaviour and its influence on the reaction performance. In such cases, understanding kinetic parameters affecting the reaction rates and adjusting the parameters in such a way that the reaction is "pushed" into the kinetic regime might be preferred. This can be achieved either (1) by adjusting the kinetic parameters (such as temperature and concentration of reagents) to slow the rate down so that the reaction is no longer controlled by mass transfer efficiency, or (2) by improving the mass transfer efficiency by varying the physical parameters (for example, mixing or adding a component which facilitates better mass transfer) so that the mass transfer rate is higher than the kinetics. Defining the limits of those parameters where the reaction can shift between kinetic and mass transfer regimes is also important. If you still have concerns about the dependency of such limits on mixing efficiency, computational fluid dynamics based or theoretical calculations can be used to predict the mixing characteristics of the reactor in which the process was developed in the laboratory scale; then theoretical prediction can be used to show how these mixing characteristics can be translated in terms of choosing the right equipment on different scales. This also highlights the importance of using the right reactor equipment for well defined and reproducible laboratory scale process development efforts.

7.3.2 Heat flow as a Tool for Rate Analysis and Process Development

Reaction calorimetry is another powerful tool that can be used for fast process development efforts, as the thermal data obtained are a direct representation of the reaction rate at any particular moment. Hence by using heat data, the mathematical treatment of taking the derivative of the concentration can be avoided. The heat observed at any particular point within the reaction period is related to the rate at the point as shown in eqn (7.8):

$$q = \text{Rate} \times V_{\text{Rn}} \times \Delta H_{\text{Rn}} \qquad (7.8)$$

Where q is the heat flow at any particular reaction time, V_{Rn} the reaction volume and ΔH_{Rn} is the heat of reaction. Since the heat of reaction is constant, if you carry out the reaction at constant volume (or by taking into account the

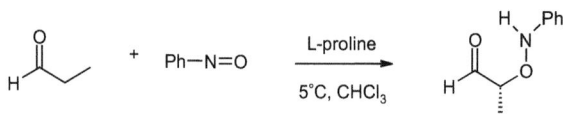

Scheme 7.4 Catalysed oxyamination of propionaldehyde with nitrosobenzene (Coypright Wiley-VCH Verlag GmbH & Co. KGaA. Reproduced with permission).

change in volume), heat flow provides a visual understanding of the rate behaviour of the process during the course of the reaction time (differential measurement). Hence the shape of the heat flow curve can be used directly to understand reaction behaviour and also to identify the kinetic trends involved, as discussed before for the nitrile hydrogenation case.

We found that calorimetric data can help to quickly identify anomalous reaction behaviour, such as induction periods involved in reactions, substrate or product inhibited kinetic trends, and auto inductive and auto catalytic reaction behaviour. Monitoring the rate or heat flow is like "fingerprinting" a reaction and helps to quickly identify anomalous reaction behaviour and give insights into reaction mechanisms, without any further time-consuming mathematical manipulation. One example where we found such "reaction fingerprinting" was very useful was for a proline-catalysed oxyamination reaction. Proline-catalysed aldol reactions were reported in the literature to be slow reactions and to require high catalyst loadings. However, two nearly simultaneous reports by Zhong and MacMillan on proline-catalysed oxyamination reactions involved short reaction periods and lower catalyst loadings.[4,5] In order to understand the reaction, the proline-catalysed oxyamination reaction between propionaldehyde and nitrosobenzene (Scheme 7.4) was investigated in a small reaction calorimeter.[6]

For most of the reactions we usually come across, such as first- or second-order reactions, the reaction rate (heat flow) will be highest at the beginning of the reaction (as shown later, in Figure 7.8) when the concentrations of the reactants are at their maximum, and will then slow down as the reactants are consumed. However, the oxyamination reaction showed a quite different trend, with initial low rates gradually increasing with reaction progress and maximising at the end (Figure 7.6, reaction 1).

Such behaviour is usually possible under three scenarios: (1) reactions with substrate inhibition, where a higher concentration of substrate lowers the reaction rate, and as it becomes consumed the rate increases; (2) autocatalytic reactions, in which the product formed acts as a catalyst, driving up the reaction rate; (3) auto-inductive reactions, where the product formed leads to "catalyst activation" or "conversion of a pre-catalyst" and the catalyst interacts to form an improved catalyst, or a structure making the reaction faster. To identify the cause in this case we first carried out a reaction without proline, but with added product along with the substrates, to check whether the product was acting as the catalyst. This run showed no reaction, ruling out autocatalytic

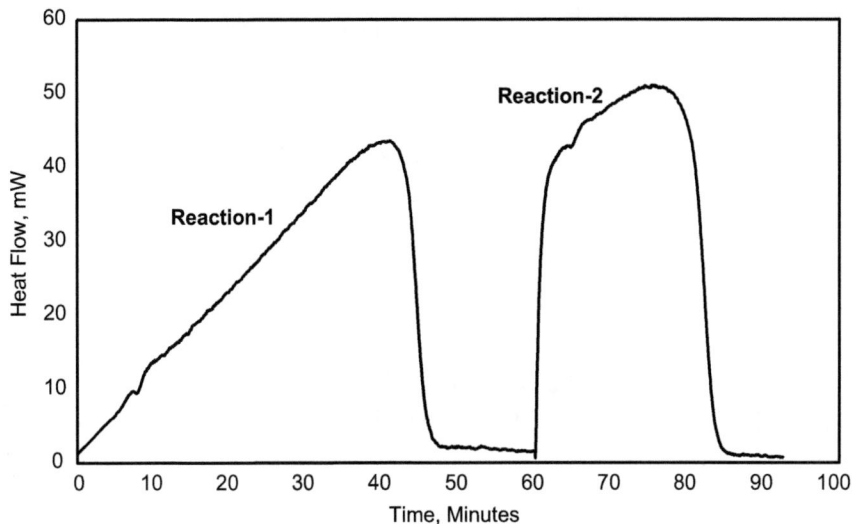

Figure 7.6 Proline-catalysed oxyamination reaction heat flow.

behaviour. In another experiment, we started the reaction by taking excess propionaldehyde with catalyst and solvent in the reaction calorimeter, and one reaction was carried out by injecting nitrosobenzene solution into the calorimeter. Once the heat flow came to a base line, a second reaction was initiated by adding a second aliquot of nitrosobenzene solution, and the heat curve observed for this second reaction is shown in the curve for reaction 2 of Figure 7.6. As can be seen for the second reaction, the rate picked up from where it stopped after the first reaction, ruling out the involvement of substrate inhibition, and suggesting auto inductive rate behaviour. This rapid insight into the reaction mechanism and process behaviour from just a few experiments highlights the importance of reaction rate "fingerprinting" for fast and efficient process development, without the need for any detailed mathematical treatment.

Reaction "fingerprinting" can also be used as a powerful tool for rapidly identifying kinetic trends, enabling the chemist to decide upon and design appropriate experiments to identify optimal processing conditions. With modern small-scale multiple calorimeter equipment available on the market, many kinetic experiments can be carried out simultaneously using very little material, and the data can be interpreted to provide rate and kinetic trends. Calorimetric heat flow also provides information regarding thermodynamic properties of the system, which can be used to understand the thermal hazards involved and modify the processing conditions accordingly.

Another example of interest is the catalytic amination reaction of aryl halides, a very important category of bond-forming reaction in the pharmaceutical industry (see Chapter 3).[7] The rapid insights, gained directly from reaction calorimeter heat curves, into the kinetic trends of this amination reaction are well

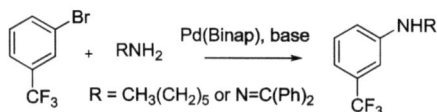

Scheme 7.5   Pd(binap)-catalysed amination of 3-bromobenzotrifluoride.

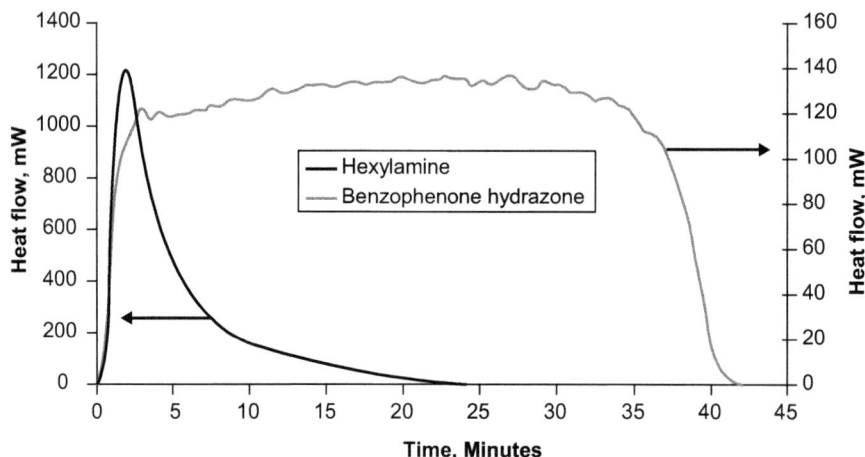

Figure 7.7   Calorimetric heat flow curve for Pd(binap)-catalysed amination reactions
with hexylamine and benzophenone hydrazone.

demonstrated in a paper from the Blackmond group.[8] Pd(binap)-catalysed
amination of 3-bromobenzotrifluoride [1-bromo-3-(trifluoromethyl)benzene]
with two different amines, hexylamine and benzophenone hydrazone (Scheme
7.5), was carried out simultaneously in a multiport small-scale reaction calorimeter (using 500 mg of limiting reagent, 5 mL reaction volume).

The heat flows from the two small-scale reactions observed are shown in
Figure 7.7.

These data-rich experiments tell us a lot about the reaction and its kinetic
behaviour. Even though the catalyst and other processing parameters are
similar, as can be seen from the graphs, the kinetics of the reaction for these two
different amines are completely different and reveal much from both a process
understanding and a process development viewpoint. When benzophenone
hydrazone was used as the base, the reaction showed net zero-order kinetics,
whereas when hexylamine was used as the base the reaction showed overall first-
order reaction kinetics, and the rate of reaction was considerably higher. This
clearly shows that even though the reaction mechanism and the catalytic cycle
might be similar in the two cases, the rate-determining step within the catalytic
cycle might be different. From a process development viewpoint, for a process
with benzophenone hydrazone as the amine, changing the concentration of

either of the reactants will have no influence on the reaction rate, and hence these concentrations cannot be varied to increase the productivity. However, in the case of the reaction with hexylamine, two further experiments varying the concentration of each substrate will determine whether the observed net first-order kinetics is due to the first-order dependence on one of the reactants and zero-order dependence on the other, or partial order dependence on both reactants. Once this is identified, we can obtain a clear picture of how to optimise the reactant concentrations to obtain the best operating conditions. These experiments will also give valuable information about the heat generation and how the feed rate and heat transfer parameters will need to be controlled to run the process in a safe and efficient manner in the plant.

7.4 Reaction Progress Kinetic Analysis in Process Development

Further mathematical manipulation of the heat and rate data can give more accurate and detailed kinetic information, as well as mechanistic insights. Reaction progress kinetic analysis (RPKA) developed by Blackmond and her group is achieving considerable attention and interest due to its effectiveness in providing insights into reaction kinetics and mechanism, from far fewer experiments than from traditional kinetic analysis techniques like initial rate analysis. Reaction calorimetry is one of the main experimental tools used for this approach, even though this could be carried out from the concentration–time data obtained by various techniques. In RPKA, mathematical correlations and reaction stoichiometry are utilised to interpret the rate data as a function of the component concentrations throughout the entire course of the reaction. This makes RPKA visually powerful, as we are able to correlate rates and concentrations which are the major components of reaction kinetics. When we collect data from a reaction calorimeter, the rate obtained is seen as a function of time, as shown in Figure 7.8.

In the rate expression for a general chemical reaction, the rate of any particular reaction is normally expressed as a function of the concentration of the reagent involved, as shown in eqn (7.9):

$$\text{Rate} = k\,[\text{substrate}]^x \tag{7.9}$$

where k is the rate constant, [substrate] is the concentration of the substrate and x is the order dependency of rate on the substrate. Time does not appear explicitly in the rate equation, making it difficult to understand the concentration dependency on the reaction rate from the plot of rate as a function of time. If we want to understand the influence of concentration of the reagents on the reaction rate, it is necessary to plot rate against concentration of the reagent. If we know the initial concentration of the substrates, from the

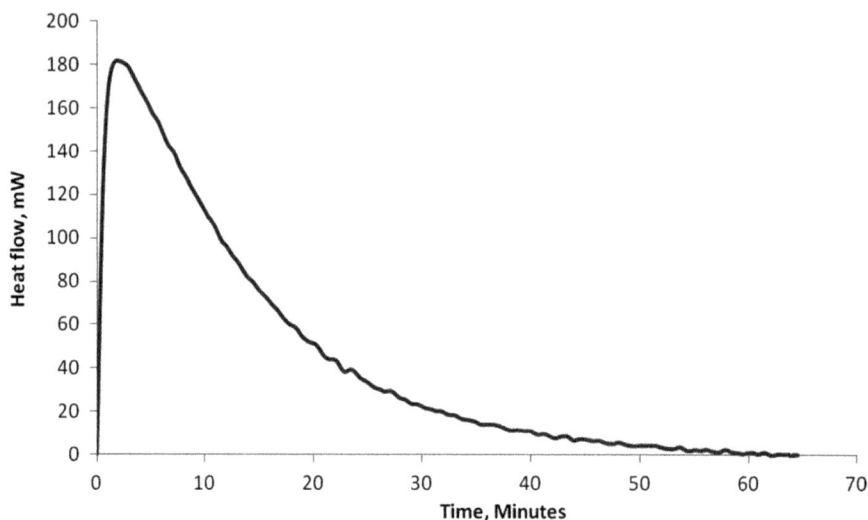

Figure 7.8   Reaction heat plotted against reaction time for a typical reaction.

stoichiometry of the reaction, the concentration of the reagents at any point can be determined from the conversion values calculated from the reaction heat flow using eqns (7.10) and (7.11):

$$\text{fraction conversion} = f = f_{\text{final}} \frac{\int_0^t q(t)\,dt}{\int_0^{t(\text{final})} q(t)\,dt} \tag{7.10}$$

$$[\text{substrate}] = [\text{substrate}]_0(1-f) \tag{7.11}$$

where f is the fractional conversion, f_{final} is the observed final conversion and $q(t)$ is the heat flow at any reaction time t.

Now the rate can be plotted as a function of the concentration of the reactant involved, giving the reaction rate behaviour plot as a function of the substrate concentration (shown in Figure 7.9), which is a representation of the rate equation shown above (eqn 7.10). Such a graphical representation of the rate gives a direct visual picture of the reaction rate variation as the reagent concentration changes during the reaction course.

For typical bimolecular reactions, if you want to understand the reaction rate dependence of individual substrates, the rate can be normalised with one of the substrate concentrations and then plotted as a function of the second substrate concentration. This analysis of reactions, using graphical rate equations, provides a quick understanding of the driving forces in terms of the reagent concentrations, as well as good insights into the reaction mechanism.

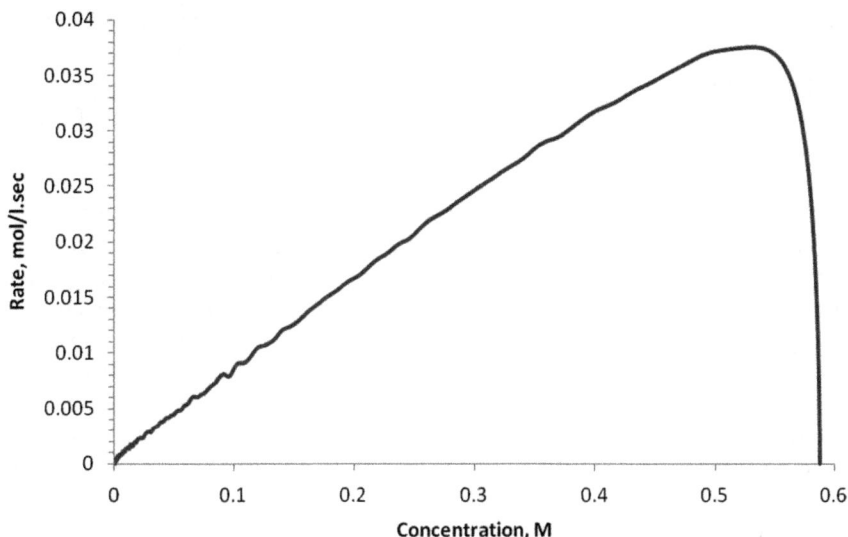

Figure 7.9 Rate behaviour as a function of the substrate concentration.

Detailed descriptions of such an approach and its applications are available in the literature.[8,9]

Another area where RPKA has been found to be highly effective in process development has been to check the robustness of catalytic reactions, with graphical rate equations used to check for catalyst deactivation/activation, substrate/product influence on the reactions, and in-depth mechanistic understanding. A detailed description of the approach, covering the theoretical background, experimental tools and data collection, mathematical manipulation of the data acquired and its application in various cases for process understanding, has been very well documented in reviews by Blackmond.[9–12] The use of calorimetric data and the corresponding graphical rate equation can become complicated when the target reaction is not the only major contributing factor towards the net heat observed, and side reactions and/or impurity formation also contribute significantly. In such cases, alternative analytical tools (like *in situ* FTIR and HPLC analysis of the samples during the reaction course) need to be coupled with heat data to understand the mass balance distribution and its effect on the reaction stoichiometry.

7.5 Role of Theoretical Kinetic and Reactor Modelling in Process Development

The kinetic methodologies discussed above have mainly focused on treatments without using any theoretical predictive model development. Kinetic or reactor modelling, which involves the development of mathematical models to predict kinetic trends and/or reactor performance, requires much more complex mathematical treatment. Theoretical kinetic model development based on various possible reaction mechanisms, and validation against experimental data by fitting, can be used to validate reaction mechanisms, which in turn could be utilised for process improvements. Assuming a kinetic regime, such kinetic models can predict the reaction behaviour as a function of various kinetic parameters. If your process is found to be influenced by other physical parameters like mass/heat transfer effects, and/or needs to take into account reactor characteristics and other processing parameters, you might need to develop a detailed reactor model taking into account all these parameters, together with the reaction kinetics. Using kinetic modelling and reactor designing methodology to theoretically predict the process and reactor performance needs considerable mathematical calculations and a good understanding of reaction engineering principles. Reactor design software now available on the market is advanced and can be used for this purpose. Some of this software even comes with a good built-in library of commercially available large-scale reactors (with their characteristics) required for scale-up prediction. We have found that application of such theoretical scale-up prediction for the process can be very useful for identifying key reactor characteristics (such as heating/cooling capabilities and feed rates) that need to be kept in mind when defining the robustness range and edge of failure. Reactions that particularly benefit from the use of predictive reactor models include very fast and highly exothermic ones, which require well-controlled operating conditions to maintain the necessary purity. A well-validated model can also be used to predict the operating conditions under which the process can be run with maximum efficiency.

7.6 Choosing the Right Experimental Tool for Kinetic Studies

Choosing the right kit, for experimentation and accurate data collection to monitor the desired reactions during the entire course of a reaction, is critical for good kinetic understanding. This selection should be based on process characteristics like product/impurity distribution, the number of phases and physical properties of the phases involved, mass and heat transfer characteristics, and reaction period. Modern *in situ* monitoring techniques can be very effective, but the selection of the technique should be done carefully, taking account of the needs of the process under investigation. Some important points to keep in mind while making this choice include: (1) how easy it is to determine the concentration from the measurable property, (2) the sensitivity of the technique to changes in physical properties of the reaction mixture during the reaction course, (3) the chances of overlap with other species (and how easy it is to de-convolute if needed), (4) the sensitivity of the technique to monitor low-level impurities if the reaction of interest involves these and (5) the suitability to the scale of the investigation and material availability. For example, if you are interested in understanding the kinetic trends involved in the formation of certain impurities which are observed in low levels, *in situ* FTIR might not be the best tool for monitoring impurity formation, as it would be out of its accurate detection limits. In such a case, manual sampling or auto-sampling techniques, coupled with analysis by accurate chromatographic techniques, would be appropriate.

Most traditional reaction calorimeters that are capable of collecting calorimetric data require reaction volumes of 100 mL or more. However, more small-scale reaction calorimeters are now entering the market, which are very useful for early process development scenarios and also when faster process development with lower reagent quantities is required. One apparatus we use is a 10-port reaction calorimeter, in which eight reactions can be carried out simultaneously using lower reaction volumes (1–12 mL scale) with high sensitivity (~ 10 μW).

To undertake process development studies involving mass transfer sensitive reactions, it is really useful to dedicate one particular reactor set-up to carrying out the initial development studies, such as defining the operating ranges for kinetic and mass transfer regimes in terms of kinetic and operating parameters. By dedicating one reactor set-up, variations of reactor parameters can be avoided, and the reactor characteristics (such as the type of stirrer used, stirrer blades, baffles and clearance from the bottom) should be set to achieve uniformity throughout the experimental investigation.

7.7 Analytics: the Backbone of Kinetic Approaches

Efficient and accurate analytical techniques constitute the backbone of good process development efforts (as discussed in Chapter 12). The first and most important issue that needs to be addressed, before full-scale process development is initiated, is the development of a suitable analytical method that can identify and track the various components within the reaction mixture. The transformation map, or the reaction network built from the quantification of the various components, forms the foundation for understanding the mass balance distribution, which is essential for good process development. Impurity identification, separation and synthesis are time consuming tasks, and on many occasions understanding the mass balance by taking into account the impurities forms the key to understanding the process and solving robustness related issues. In cases where impurities or components are not detected by chromatographic techniques, gaining an understanding of the key process parameters can be very challenging.

Use of internal standards can be particularly useful in understanding reactions with mass balance issues and uncharacterised or non-quantifiable impurities. Internal standards can be used for the quantitative concentration determination of components, or to understand the relative rates of the main reaction, impurity formation and decomposition. The concentration determination can be done by well-established internal standard quantification methods in which a known amount of an inert standard is added to the reaction, and from the chromatographic analysis of the reaction samples the concentration of any known compound can be determined using eqn (7.12):

Concentration of component $=$

$$\frac{\text{Area of component} \times \text{RF of component} \times \text{Concentration of standard}}{\text{Area of standard} \times \text{RF of standard}}$$

$$(7.12)$$

Where RF of the component is the response factor of the component that needs to be quantified, and RF of standard is the response factor of the internal inert standard used. Even though this information can be obtained by other quantitative analytical methods, the main advantage with the use of internal standards is that samples from the reaction mixture can be directly used for analysis after dilution, whereas other techniques usually involve collecting accurate volumes of samples and then carefully diluting to required known volumes, which can be a time consuming process.

In cases where the response factors of components are not known, or reaction components involve unknown impurities, inert compounds can be added to the reaction mixture. Monitoring the depletion and formation of various components with respect to this inert molecule can then be used as a very simple and efficient way of understanding the rates of formation and depletion of various components. This method does not involve any mathematical manipulation, and monitoring the ratio of the peak areas of the components to the inert peak area tells how the component composition varies with time in the reaction, and can also be directly used for rate assessment. We have found this a very efficient way to interpret trends when compounds with differing response factors are present and the chromatographic peak area cannot be directly used for analysing the trends. This ratio analysis of the chromatographic areas of various components with the inert compound will give a relative rate comparison of the desired reaction and the side reaction/decomposition, even if details of the all the components are not known.

7.8 Summary

The various kinetic approaches that have been discussed are suitable for different stages of process development, from choosing the reagents to making the process ready for manufacturing. Well-designed kinetic experiments to determine the rate-determining steps can be used to design reagents and/or conditions and allows us to gain good process understanding. Analysis of the rate of formation and depletion of various reaction components throughout the entire course of a reaction, and comparing this with rates of other competing reactions, can give very valuable information on kinetic trends, mass transfer and robustness issues, along with process understanding. Reaction progress kinetic analysis is another powerful methodology which can be applied across the process development spectrum for driving force analysis using graphical rate equations, and also in studying catalytic reactions from robustness and mechanistic viewpoints. The reactor modelling approach can be useful for predicting the scale-up behaviour of a process, and also for identifying key reactor characteristics that could influence the performance of the process on scale.

Acknowledgements

The author would like to thank Pfizer CRD colleagues and management. Timely support from Mike Williams with manuscript preparation and valuable discussions with Professor Donna Blackmond are gratefully acknowledged.

References

1. S. P. Mathew, S. Howard-Field and F. Susanne, unpublished Pfizer work.
2. S. P. Mathew, C. Burns, S. Field, C. Mason and O. Dirat, unpublished Pfizer work.
3. F. Rawlinson, S. P. Mathew and D. Henderson, unpublished Pfizer work.
4. S. Brown, M. Brochu, C. J. Sinz and D. W. C. MacMillan, *J. Am. Chem. Soc.*, 2003, **125**, 10808.
5. G. Zhong, *Angew. Chem. Int. Ed.*, 2003, **42**, 4247.
6. S. P. Mathew, H. Iwamura and D. G. Blackmond, *Angew. Chem. Int. Ed.*, 2004, **43**, 3317.
7. J. S. Carey, D. Laffan, C. Thomson and M. T. Williams, *Org. Biomol. Chem.*, 2006, **32**, 464.
8. A. C. Ferretti, J. S. Mathew, I. Ashworth, M. Purdy, C. Brennan and D. G. Blackmond, *Adv. Synth. Catal.*, 2008, **350**, 1007.
9. D. G. Blackmond, *Angew. Chem. Int. Ed.*, 2005, **44**, 4302.
10. N. Zotova, S. P. Mathew, H. Iwamura and D. G. Blackmond, in *Process Chemistry in the Pharmaceutical Industry*, vol. 2, ed. K. Gadamasetti and T. Braish, CRC Press, Boca Raton, FL, 2008, p. 455.
11. J. S. Mathew, M. Klussmann, H. Iwamur, F. Valera, A. Futran, E. Emanuelsson and D. G. Blackmond, *J. Org. Chem.*, 2006, **71**, 4711.
12. T. Rosner, A. Pfaltz and D. G. Blackmond, *J. Am. Chem. Soc.*, 2001, **123**, 4621.

CHAPTER 8
The Design of Safe Chemical Reactions: It's No Accident

DAVID J. DALE

The Briars, Clavertye, Elham, Kent, CT4 6YE, UK

8.1   Introduction

Pharmaceutical manufacture by its very nature involves the use of reactive chemicals to enable chemical reactions to take place and hence deliver desired intermediates or products. Consequently, there are inherent hazards associated with pharmaceutical manufacture, making it essential to fully evaluate the hazards that may exist. However, as this book explains, safety is not the only demand placed on chemical process development. Reactions must be investigated, developed, optimised for both yield and purity and, ultimately (hopefully), performed on a manufacturing scale. The overall cost of development and the requirement for a fast time to market need to be balanced against the development of a safe chemical process. Many pharmaceutical and chemical companies have invested heavily in specialist process safety laboratories capable of assessing and making recommendations on chemical reaction hazards. These laboratories were traditionally set up as "add-ons" to the development process where safety experts performed an assessment on the process in question, often just before introduction into a scale-up facility. This linear approach can be successful, but if safety issues are discovered at this stage the process may have to be supplemented by additional risk-reducing measures, or even redesigned. This can often lead to costly delays in both time and money. It is therefore prudent to ensure that safety is an integral part of chemical

RSC Drug Discovery Series No. 9
Pharmaceutical Process Development: Current Chemical and Engineering Challenges
Edited by A. John Blacker and Mike T. Williams
© Royal Society of Chemistry 2011
Published by the Royal Society of Chemistry, www.rsc.org

development, with consideration being given to potential safety issues at all stages of development. However, it is vitally important that the safety assessment is fully commensurate with the stage and scale of the development process. Clearly it makes no sense to carry out a very detailed hazard study on a process only used to produce small amounts of materials in a laboratory environment. However, at the same time it is clear that the early identification of hazards can reduce incidents even at the laboratory scale and result in a more efficient development process. The purpose of this chapter is to outline the basic principles of process safety in a way easily understood by non-safety specialists. The aim is to give an appreciation of process safety requirements in pharmaceutical development activities to non-specialists in this field.

8.2 Background

Understanding chemical reaction hazards is a basic requirement of the chemical development process and the consequences of runaway reactions can be devastating. The world's worst chemical incident occurred at Bhopal in India in 1984,[1] and even today serves as a reminder of what can happen if hazards are not fully understood, control measures not put in place, and training and control systems are inadequate. There are many examples of incidents in the literature and the list unfortunately keeps growing. Some of the most frequent contributing factors to thermal runaways are:

- Inadequate understanding of process chemistry and thermochemistry
- Inadequate design for heat removal
- Inadequate control and safety systems
- Inadequate operational procedures, including training.

8.3 Reaction Hazards

There are essentially four major hazards in chemical processing:

- Chemical reaction hazards
- Fire and explosion hazards
- Health hazards
- Environmental hazards.

All of the above need to be assessed for any chemical process. Explosion and health hazards are touched upon briefly in Chapter 5 (Section 5.3.1), while environmental issues are covered in Chapter 6. This chapter is principally concerned with chemical reaction hazards, which generally fall into the three main areas of runaway reactions (Section 8.3.1), thermal instability issues (Section 8.3.2) and gas evolution (Section 8.3.3).

8.3.1 Runaway Reactions

Chemical reactions either release heat (exothermic) or absorb heat (endothermic). The majority of useful reactions performed in industry are exothermic and this heat release should be controlled in order to ensure safety and, often, product quality. If the rate of heat production by the chemical reaction exceeds the rate of heat removal available, then a thermal runaway may result. In this runaway scenario, as the surplus heat cannot be removed this heat will raise the temperature of the reaction mass. This in turn will accelerate the reaction even further (an approximate rule of thumb suggests that reaction rate doubles with every 10 °C rise in temperature) and thus increase the rate of heat production. In simple diagrammatic terms the result is a runaway cycle as depicted in Figure 8.1.

The scenario described above, where reaction control is lost, is not helped by some fundamental thermodynamic heat transfer principles. In particular, as the temperature of the reaction increases the rate of heat production will increase in an exponential fashion whilst the rate of heat removal only increases linearly. This is illustrated in Figure 8.2.

The reaction can be controlled whilst the heat removal line remains above the heat production line, but the upper intersection point represents a "point of no return", as this is where the reaction enters thermal runaway.

It is also important to remember that as a reaction is increased in scale, the problems are exacerbated. This is because the reaction heat generation rate is a function of the reaction volume (that is a cubic term), whereas the cooling capacity of the reaction vessel is a function of the available cooling area (that is a squared term). Therefore as the reaction scale and the ratio of reactor volume to surface area increases, the cooling capacity may well become inadequate, and hence the result may well be a runaway reaction.

Once control of the reaction as described above has been lost, the reaction vessel may be in danger of over-pressurisation due to violent vaporisation or

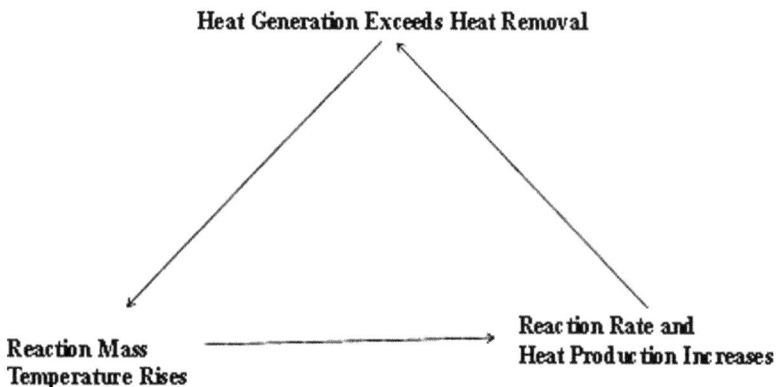

Figure 8.1 Runaway reaction cycle.

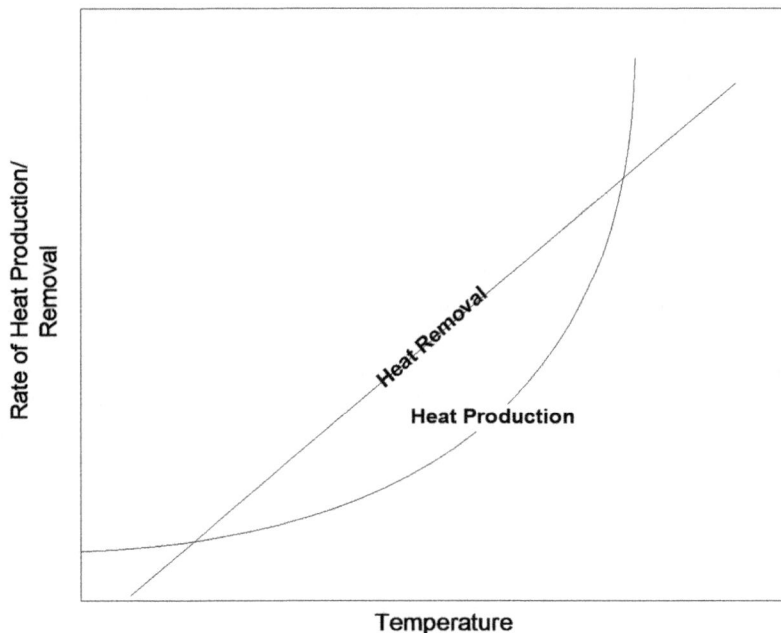

Figure 8.2   Heat production *versus* heat removal.

rapid gas generation, and high energy secondary decomposition reactions may also be triggered.

A useful illustration of the effect of heat transfer in relation to vessel volume is comparing the time taken for a 1 °C drop in water temperature from an initial temperature of 80 °C by natural cooling only (surroundings at 20 °C):

- 10 mL test tube = 10 s
- 100 mL glass beaker = 20 s
- 2.5 m^3 reactor = 20 min
- 25 m^3 reactor = 230 min.

It can clearly be seen that as the scale increases, the reaction heat losses are drastically reduced and tend towards adiabatic (no heat loss) behaviour. This fact is demonstrated in reactions performed in laboratory glassware where no temperature excursions may be observed, but when scaled up into larger vessels temperature control problems due to heat release could result. This has important implications for the scale-up of processes (and the potential for runaway) and the way in which we study reaction hazards.

8.3.2   Thermal Instability Issues

Hazards can also arise from the thermal instability of process materials such as starting materials, intermediates, products, reaction mixtures and waste streams.

These decompositions can often be of very high energy and liberate gas capable of over-pressurising reactors. To illustrate this fact, consider a decomposition value of 2000 J g^{-1} for a high-energy material (for example, an undiluted dinitro aromatic compound), which could promote an adiabatic temperature rise in the order of 1000 K (based on a heat capacity of 2 J K^{-1} g^{-1}).

8.3.3   Gas Evolution

Many chemical reactions generate a non-condensable gas as they progress. As with heat output, it is important to quantify both the amount and the rate of the gas being formed so as to design a safe chemical process. This information helps to ensure that the gas can be vented from the reactor without over-pressurisation, to aid calculations regarding the concentrations of flammable gases, and to assist in the sizing of scrubber units used to treat hazardous or environmentally unfriendly evolved gases. These three areas will be discussed further during the course of this chapter.

8.4   Lifecycle Approach to Process Safety

In order to deal with any chemical reaction hazards, you need to identify them, decide how likely they are to occur and how serious the consequences may be; that is, you need to carry out a risk assessment for the process.[2] As pharmaceutical process development goes through various stages of evolution, a convenient way of describing an effective process safety hazard assessment programme is to link the chemical development lifecycle activities with a similar process lifecycle approach to reaction hazards testing. To explore this approach it is convenient to break down the development lifecycle into distinct phases and then describe the tools, techniques and procedures applicable that allow us to define a "basis of safety" for any chemical process and scale.

Each of these phases is now discussed in turn, along with experimental assessment techniques frequently employed. Figure 8.3 shows how testing requirements change across the development lifecycle.

However, it must be remembered that whilst it is convenient to break down testing requirements based on scale, there are no hard and fast rules to be applied. A hazardous process may require sophisticated and detailed testing even at a relatively small scale, whilst some simple and intrinsically safe processes can be operated at large scales with a reduced level of safety data. A mention should also be made here of continuous processes, as clearly they do not readily fit into a scale-driven testing paradigm, and for these types of reaction a clear understanding of kinetics, start-up and shut-down issues, and possible upset scenarios, is required in order to drive an effective testing strategy.

It is also wise to bring in the concept of "inherent" safety at this stage.[3] An inherently safe process is largely a theoretical concept as it implies no hazard

Discovery	Development	Commercial Manufacture

Increasing Scale

Desk Screening.

Functional group analysis

Reaction heat prediction.

Adiabatic temperature rise prediction.

Thermal stability screening.

Reaction mixture screening

Reaction heat measurement.

Gas generation quantification.

Adiabatic temperature rise calculation.

Adiabatic testing.

Maloperations,

Relief Calculations

HAZOP

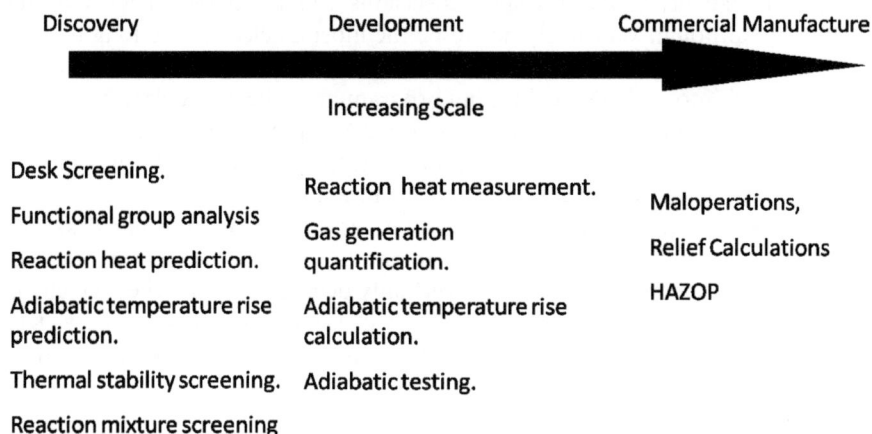

Figure 8.3 Development lifecycle and safety testing.

exists, which is difficult to achieve for the majority of chemical processes. However, the concept of making the process as inherently safe as possible is a useful one and involves removing hazards completely where possible, or reducing their magnitude sufficiently to avoid the need for elaborate safety systems and procedures. Approaches to the design of inherently safer processes can be grouped into four hierarchical strategies:

- Substitute: replace a material with a less hazardous substance or a hazardous reaction with a less hazardous reaction.
- Minimise: use smaller quantities of hazardous substances in the process at any one time.
- Moderate: use less hazardous conditions, a less hazardous form of a material or facilities which minimise the impact of a release of a hazardous material or energy.
- Simplify: design facilities and procedures that eliminate unnecessary complexity and make operating errors less likely with a greater degree of error tolerance.

It is considerably easier to factor in inherent safety to a process during the early stages of the development lifecycle, and hence the strategies above should be an integral part of chemical development activities.

8.4.1 Initial Route Assessment

Pharmaceutical chemical development activities really begin when a potential new compound has been identified *via* chemical discovery programmes. The new target compound will until then have been produced only on a small scale, and the synthesis may have involved many steps using many exotic or hazardous chemicals. At this stage, decisions will be made on whether the chemistry, although very likely not to be the most efficient in the long term, is sufficiently safe and robust to allow some rapid scale-up work to take place in order to feed into early clinical testing programmes. There is a clear balance to be had here between investing heavily in the development of the process before clear positive clinical results are available, and the requirement to operate in a safe environment at all stages of development activity.

At this stage, techniques for process safety assessment tend to be "desktop" screening and rapid small-scale tests; there may well be limited amounts of material for testing at this stage and certainly decisions on whether the chemistry is viable, at least in the short term, will need to be made quickly. Key components of the process safety risk assessment at this stage are desk screening and basic screening tests.

8.4.1.1 Desk Screening

Are there any potentially highly energetic materials or reactions involved? The existing literature makes a good starting point here and *Bretherick's Handbook of Reactive Chemical Hazards*,[4] along with the National Fire Protection Association's *Manual of Hazardous Chemical Reactions*,[5] are two good sources of information. However, it should always be remembered that lack of information on a particular material or reaction does not always imply that no hazards exist. Of particular use at this early screening stage is a list of functional groups known to exhibit high-energy decomposition, such as the one shown in Figure 8.4.

Early identification of molecules containing these groups is important as it means consideration can be given to making changes to the chemistry to avoid such materials, or hazard testing can be undertaken to examine the magnitude of the potential hazard and as a result appropriate control measures put in place.

Thermochemical calculations are often also useful at this stage. Indications of heats of reactions can be determined by subtracting the heats of formation of the reactants from those of the products.[6] Estimation of the heat of reaction gives an idea of the total energy being released and the maximum possible adiabatic temperature, ΔT_{ad} (calculated from the expression shown in eqn 8.1):

$$\Delta T_{ad} = \frac{\Delta H_r \cdot n}{m C_p} \tag{8.1}$$

where n = number of moles of limiting reagent (mol), m = total mass of the reaction mixture (kg) and C_p = heat capacity of the reaction mixture (J kg^{-1} K^{-1})

However, the estimated heat of reaction does not give an indication of how fast this energy will be released. Obviously a large energy released over a

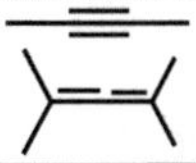 (acetylenic structure)	acetylenic, metal acetylides, haloacetylene derivatives, allenes
 (epoxide structure)	epoxides
—N=N— —N≡N	all substances with N-N double or triple bonds, i.e. azo, diazonium salts, azides, diazirines and other high nitrogen containing compounds like triazoles, triazenes, tetrazoles etc.
—O–O—	all substances with O-O bonds, i.e. peroxides, peroxyacids and their salts, hydroperoxides, peroxyesters
—N=O —N–O	all substances with N-O bonds, like nitro, nitroso, hydroxylamines, nitrite, nitrate, fulminates, oximes, oximates
—N–X	halogen azides, N-halogen compounds, N-haloimides
—O–X	alkyl perchlorates, aminium perchlorates, chlorite saltes, halogen oxides, hypohalites, perchloryl compounds (bromates, iodates, as well)
—N–M	metal nitrides, amides, hydrazides, imides, cyanamide. main concern is the pyrophoric nature of the pure solid material. dilute solutions of metal amides and substituted amides (i.e. LDA, LiHMDS) are generally acceptable depending on use and fate of excess quantities
Ar—M—X X—Ar—M	non-catalytic use of haloarylmetals, haloarenemetal π–complexes Note: Grignards of concern are only halo-phenyl Grignards containing trifluoromethyl moeities.

Figure 8.4 High-energy functional groups.

short period of time is potentially more dangerous than a small energy released over a longer period, and hence an indication of the rate of the reaction from laboratory experiments is of value at this stage. However, if we add the calculated maximum possible adiabatic temperature rise (calculated from eqn 8.1) to the operating temperature, we obtain the maximum temperature of the synthesis reaction (MTSR). This is an important value as if this is below the temperature at which additional chemistry or physical transitions occur (such as decompositions, side reactions, boiling, gas generation), then again informed safety decisions can be made even at this early stage. This concept has been explored in great detail by Stoessel,[7] who categorised reactions according to the potential hazard based on operating temperatures, heats of reaction, boiling points and the calculated or measured potential temperature rise due to the desired reaction. A simplified version of this categorisation is shown in Figure 8.5.

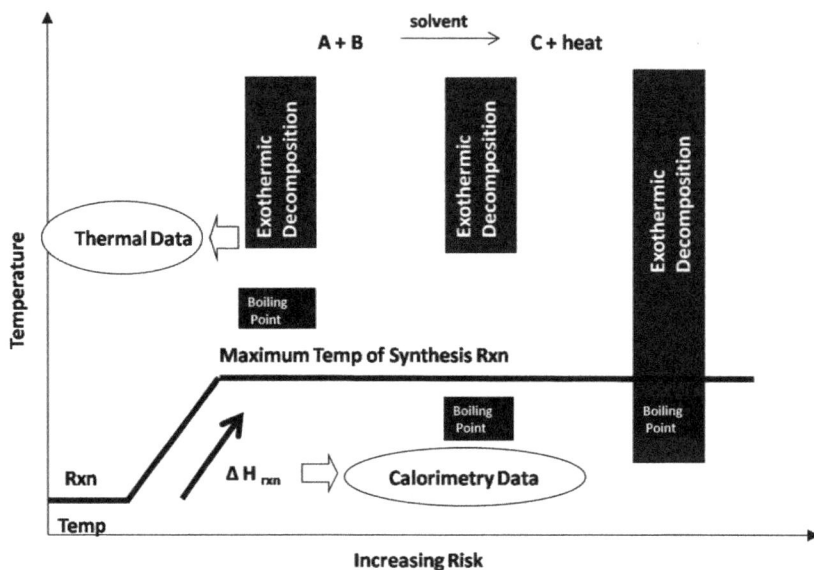

Figure 8.5   Simplified Stoessel diagram.

8.4.1.2   Basic Screening Tests

The typical early testing strategy is to perform desktop calculations as described above, followed by basic screening tests, and then more sophisticated and detailed testing if required. Screening tests refer to those types of test that are less expensive, have quicker turnaround times, are small scale and are less complicated to perform and interpret. Differential scanning calorimetry (DSC) is perhaps the most common[8] of these types of test. There are commercially available instruments that basically involve the linear programmed heating of a small amount (typically 10 mg) of a material and reference. The material is sealed in a capsule (capable of withstanding high pressures) and generally heated from ambient to 400 °C at around 5 °C min^{-1}. The resulting energy trace yields peaks and troughs for exothermic (for example, decomposition or reaction) or endothermic (for example, melting) activity. The areas under the resulting peaks give a value for the heat released, and the point where the event occurs gives an indication of the onset of the event. Whilst DSC is a useful technique for low-cost material, sparing examination of the thermal properties of a material (or reaction), it can sometimes be difficult to obtain representative samples and it gives no indication of gas evolution as no pressure data are recorded. There is also an important point to be made about onset temperatures, which are sometimes quoted as absolute values, *e.g.* material X exothermically decomposes at Y °C. Extreme caution should be taken with quoted values as the onset point is not an intrinsic property of the material in question, unlike a melting point, and therefore should not be treated as an absolute number. The onset point is in fact the temperature at which the thermal activity is first observed in a dynamic thermal measurement. Its value is a result of a balance between the sensitivity and scan rate of the instrument, the mass of the sample, the mass and conductivity of the sample cell and the kinetics and energy of the decomposition reaction. For the above reasons a safety margin of around 100 °C is usually applied to DSC test results.[9]

Other basic screening tests which may be employed at this stage to assess the thermal stability of materials and reactions include techniques such as the TS[U] (thermal screening unit)[10] and Carius[11] tube tests. These techniques are both similar in their requirements and data obtained. They consist of heating the material(s) in question in a glass tube or cell ($\sim$ 10–30 mL) which is attached to a thermocouple and a pressure transducer. The sample cell is then heated in a temperature-programmed oven, usually at a ramp rate of around 2 °C min^{-1} up to 250–400 °C, depending on the pressures being generated in the glass cell. The output traces from such a test are usually three plots consisting of a temperature–pressure–time trace, a temperature *versus* a differential temperature between oven and sample and an Antoine plot of reciprocal of absolute temperature *versus* the natural logarithm of the pressure (this helps identify permanent gas formation, as a plot just resulting from vapour pressure will be a straight line). From these graphs, an onset for both temperature and pressure can be obtained, but as with DSC onsets described earlier, it must be remembered that this is not an absolute value and safety margins of around 60–70 °C are usually applied to these tests.

8.4.2 Chemical Route Identification

In the majority of cases the initial discovery route will only ever be used for very early material requirements (often not even for this), and hence will be modified or completely changed as we progress through the pharmaceutical development process. Factors that are of high importance in new route development include safety, cost, intellectual property issues, throughput and robustness, as discussed in Chapter 5. As potential new routes are investigated by development project teams, so the safety assessment of these potential new routes will continue, much in the same way as described above for the initial scale-up routes. During this stage the safety of various route options can readily be screened, and this should be used as one of the criteria in final route selection. The aim must be to make the targeted new route as inherently safe as possible, as previously discussed in this chapter, and thus many of the early screening techniques already described will be employed during this phase of the lifecycle. An example of this approach is shown in Schemes 8.1 and 8.2.

In this example, early DSC screening highlighted an unexpected problem with the thermal stability of the mesyl compound UK-81,752 (Scheme 8.1). The compound was found to have a large exothermic decomposition

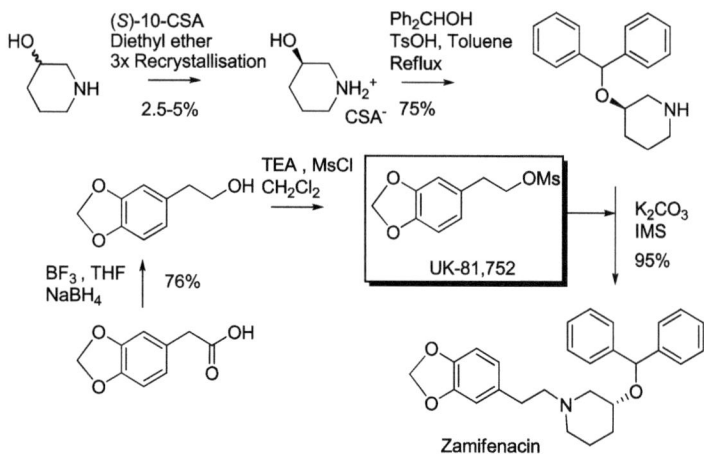

Scheme 8.1   Original zamifenacin route.

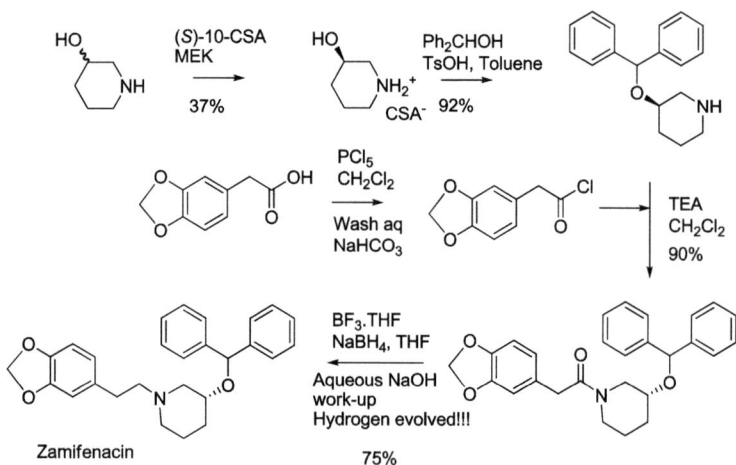

Scheme 8.2   Modified zamifenacin route.

occurring at $<200\,^{\circ}C$ in a DSC scan and it was also observed to be auto-catalytic in nature (time and temperature dependent). However, as this was found early in the development process, the chemistry steps were easily reordered to avoid this compound. The original process involved acid reduction followed by mesylation and coupling, whereas the modified route (Scheme 8.2) involved acid chloride preparation followed by coupling and finally reduction.

8.4.3 Process Development and Optimisation

With the desired manufacturing route selected, development activities then enter a phase of development and optimisation, where the aim is to maximise yield, quality, throughput and minimise the associated hazards. It is important at this that stage process safety remains an integral part of the development process and the chemical development activities are closely aligned with the safety studies. The mutually beneficial outcome of this synergy is that safe processes are developed in an efficient manner, along with the collection of good quality safety data which can often help identify and contribute to areas where quality, yield or throughput can be improved.

The three key areas requiring safety evaluation at this stage of development are:

- Thermal stability of all process materials (raw materials, intermediates and products). If high-energy materials have not been designed out at the route selection stage, more detailed studies may now be required.
- A more detailed characterisation of the normal process, including heat release magnitude and rate, and gas evolution issues.
- Full assessment of the explosivity potential of process materials that contain high-energy functional groups.

8.4.3.1   Thermal Stability

The assessment should now include thermal stability data on all components of the reaction. DSC data and other screening data may already have been obtained for some of the materials in question, and if after the use of appropriate safety margins the onsets are close to normal process temperatures, further more detailed testing may be required. Testing on other materials screened as having no issues of concern can be ceased at this stage.

For those materials or mixtures requiring more detailed testing, the use of an adiabatic calorimeter is recommended. Examples of these calorimeters include the accelerating rate calorimeter (ARC),[12] Phi-Tec[13] and vent sizing package (VSP).[14] Adiabatic calorimeters can examine the exothermic potential of individual materials and reactions. They generally consist of a ~ 10 mL sample "bomb" which is surrounded by a set of electrical heaters. The sample will normally be heated in a "heat–wait–search" mode, which involves heating up in steps of between 5 and 10 °C and then waiting for a period of time to see if any exothermic activity is detected. If none is detected, the sample will be a heated up by another step increment and the process repeated until activity is detected. Once thermal activity is detected (and these instruments can generally detect temperature rise rates of around $0.02 \,°C \, min^{-1}$), then the instrument switches to "tracking" mode, in which the external heaters are set to match the sample temperature so that as the sample undergoes self-heating due to the exothermic event, its surrounds will be maintained at the same temperature, hence creating an adiabatic environment and allowing the thermal event to progress under worst case, or no heat loss, conditions. These instruments will therefore give more meaningful onset data and maximum temperature and pressure rise rates which are useful for plant and equipment design. It is worth mentioning the phi factor at this stage as this is a key parameter for these types of tests. The phi factor (ϕ) is given by eqn (8.2):

$$\phi = 1 + \frac{(mC_p)_{\text{container}}}{(mC_p)_{\text{sample}}} \quad (8.2)$$

where m = mass and C_p = specific heat capacity.

It can be seen from the above that as the product of the mass and heat capacity of the sample increases relative to the product of the mass and heat capacity of the sample cell, then the closer phi becomes to a value of one, which is the situation that is approached in large-scale vessels (the closer the phi value is to one the more accurate the resulting data will be, and certainly for relief calculations a low phi factor test must be used, as it will have a dramatic influence on the resulting temperature and pressure rise rates).

8.4.3.2 *Characterisation of the Normal Reaction*

Section 8.4.1.1 discussed how it was possible to obtain estimates of heats of reaction from thermodynamic calculations. In some cases this approach may not be useful or indeed wise. For instance, it may be difficult to obtain reliable heat data, the chemistry may be complex or the reaction hazardous enough to warrant a more detailed study. There are instruments available to measure heats of reaction and they will often be used at this stage of development, if not before. There are simple instruments, such as the SuperCRC[15] and the Micro Reaction Calorimeter,[16] that can be used to measure basic heats of reactions, and these may often be used during the early development phases (reaction test volume is $\sim$5–15 mL) and route selection work to either supplement any calculated data or provide stand-alone information. However, this section focuses on a larger scale of reaction calorimeter often used during process development and optimisation activities to investigate reaction heats and reaction kinetics of the "desired" reaction.

Calorimeters of this type include RC-1[17] and Simular.[18] This chapter does not go into a detailed operation of these instruments, or the science behind their measurement techniques, but focuses on the type of information they can give and how it can be used. In a typical experiment, starting materials and solvent are charged to the reactor and calibrations performed to obtain the heat transfer coefficients (U) and the heat capacity of the reaction mass. The reaction is then initiated by adding another reagent, by heating, or perhaps by the addition of a catalyst, and the heat output profile is recorded. Further calibrations are then performed and with these data the heats of reaction are obtained. Also importantly, the instruments give a heat flow profile during the entire course of the reaction, enabling for example the reaction rate to be followed, induction periods recorded and crystallisation heats observed. The addition of gas flow meters to these calorimetry experiments enables gas volumes and flow rates to be obtained, which is essential safety information in the design of safe chemical reactions. Another specific piece of information resulting from these calorimetry experiments is the amount of reaction accumulation.

For ideal reaction control it is desirable to be able to control the rate of reaction by, for instance, the controlled addition of a reagent. This is described as a semi-batch process, whereas a process where all the reagents are added at the start of the reaction is called a batch process. In a batch process, control of the reaction once it proceeds may be difficult (for instance, cooling may be the only effective control measure, and if this were to be lost during the reaction then a runaway could ensue). Semi-batch operation is often the best control method and ideally it is desirable for the kinetics of the process to be sufficiently rapid that the feed material reacts as it is being added. In this way, all the heat is evolved during the addition and there is zero accumulated heat at the end of the addition period. One of the main advantages of this type of reaction is that if any operating problems arise, stopping the addition will immediately stop the heat production; this is the ideal semi-batch situation. In reality, many reactions will not behave in this ideal fashion and will require stir-out periods to effect complete reaction after the addition has finished. In this case, at the end of the addition there will be a proportion of unreacted materials in the vessel, and hence if a problem such as loss of cooling arose there would be no means of controlling any heat output and a runway scenario may result. Therefore the ability to be able to measure the amount of accumulation is key in defining a safe process, as steps can then be taken to minimise this accumulation and hence the possibility of self-heating issues. Two frequently used ways to reduce accumulation are:

- Employ higher process temperatures (to increase reaction rate, but this must be balanced against possible thermal stability issues).
- Extend feed durations (reduce accumulation).

8.4.3.3 Explosives

The initial screening and route selection activities would have highlighted the presence of high-energy functional groups and the process designed to avoid such materials if possible. However, in many cases these high-energy functional groups are extremely useful synthetic building blocks precisely because of their reactive nature. Early safety work and route selection will have focused heavily on these materials, to make sure that the process was not so dangerous to prohibit operation at any scale. However, as we move into the development and optimisation phase, further characterisation testing may be required. Prudent questions to ask are:

- Is it explosive? High-energy DSC results (>600 J g^{-1}) may indicate explosive properties and this information should already be available.
- How sensitive is it? Tests such as impact sensitivity, friction tests and burning tests may be required if DSC results indicate explosive-type behaviour.
- Detonation tests: specialised tests such as Koenen tube tests, the United Nations (UN) time/pressure test and UN gap tests may be required to fully characterise the explosive potential of the material in question.

The above tests are not described in detail here as they often require specialist testing facilities not normally found in most pharmaceutical companies. However, they are described in the UN transport of dangerous goods recommendations manual of tests and criteria.[19]

8.4.4   Scale-up

8.4.4.1   *Kilogram Laboratory/Pilot Plant*

The safety data on a process builds over time, growing from the initial screening work as progress is made through the development lifecycle. It is not uncommon for pharmaceutical companies to operate kilogram laboratory facilities with scales in the range 20–100 L. Pilot plants of scales typically up to 1500 L vessels may also be used. It is important as the process transfers into these types of facility that the potential hazards are carefully reviewed and documented. Typically at this size a full-scale HAZOP (hazard and operability study) will not be undertaken, but certainly process safety reviews, which may include the use of "check lists" and "what-if scenarios", should be used and recorded. The safety review team (usually consisting a project chemist, a process safety specialist and plant supervisor/engineer) will review the process to be operated and the available safety data and agree either to run the chemistry, to implement other safety controls or to request more safety data to make the required decisions and hence an informed basis for safe operation.

8.4.4.2   *Large Scale Production*

As the process moves into full-scale production facilities, most of the safety data required will already have been generated, but it must now be considered in the light of the plant and operational procedures to which it will now be subjected.

The previous pilot plant work will likely have been performed in a carefully controlled facility with scientifically trained operators and close supervision of the process. As the process moves into a different environment, it is important that the hazards are revisited. At this scale it is likely that a full-scale HAZOP of the process and the intended plant operation will take place, and this may require some upset scenarios to be further tested on a specific case-by-case basis. At these larger scales a well-defined basis of safe operation is essential and both process control and protective measures may be employed:

- Process control. This will include the use of sensors, alarms, trips and other control systems that either take automatic action or allow for manual intervention to prevent the conditions for uncontrolled reaction occurring. Addition control is a common process control measure, but specifying such measures does require a thorough understanding of the process involved, especially the limits of safe operation.
- Protective measures. These measures do not prevent a runaway scenario (as process control measures do), but reduce the consequences and impact should one occur. As these measures, which include relief venting, dumping, inhibition and crash cooling, only operate once a runaway scenario has commenced, it is important that a detailed knowledge of the reaction under runaway conditions is obtained. These type of data are normally obtained by the use of the low phi factor type adiabatic calorimeters discussed earlier.

8.5   Process Development Synergies

As we have seen, safety data can yield a lot of useful information about the process under investigation. Techniques such as reaction calorimetry are especially useful as heat output can be used as a measure of reaction rate, and hence can be used in kinetic modelling studies, as discussed in Chapter 7. The heat output profile also helps us understand the reaction progress, as we might observe induction periods, physical change events such as crystallisations or heat kicks due to different reactions occurring. The use of *in situ* techniques such as FTIR in association with heat calorimetry can help increase reaction understanding even further, as heat profiles can be matched to the appearance/disappearance of IR bands, thus enabling us to obtain a broader picture of the reaction mechanism and pathway.

8.6   Conclusion

Chemical processes require a thorough and rigorous assessment procedure as they are being developed, in order to ensure safety at all scales of operation. Many decisions which will impact on the inherent hazard of the process are made at early stages of development and hence chemists can play a fundamental part in helping safety professionals develop safer processes. Whilst it is important to have safety specialists to administer testing programmes and advise on potential issues, it is important that this process is integrated seamlessly into the development lifecycle. Training of chemists to understand the criticality of decisions they make on safety as they look to develop a process is an essential part of integrating process development and safety methodology. Having read this chapter, I would urge readers to take a look at the United States Chemical Safety Board report into the runaway reaction at Morton International in Paterson, New Jersey, which occurred in 1998.[20] This incident highlights multiple failings – runaway reaction, lack of data, lack of understanding of thermal issues, communication problems and scale-up issues – all of which this chapter will hopefully help chemists avoid in the future.

References

1. *Lees Loss Prevention in the Process Industries*, ed. S. Mannan, Butterworth Heinemann, Amsterdam, 3rd edn, 2004.
2. HSE Books, *HSG* 143, 2000.
3. R. Rodgers and S. Hallam, *Trans. Inst. Chem. Eng.*, 1991, **69**, 149.
4. *Bretherick's Handbook of Reactive Chemical Hazards*, ed. P. G. Urben, Academic Press, Burlington, MA, 7th edn, 2005.
5. National Fire Protection Association, *Manual of Hazardous Chemical Reactions*, ANSI/NFPA 491M, American National Standards Institute, Quincy, MA, 1997.
6. G. A. Weisenburger, R. W. Barnhart, J. D. Clark, D. J. Dale, M. Hawksworth, P. D. Higginson, Y. Kang, D. J. Knoechel, B. S. Moon, S. M. Shaw, G. P. Taber and D. L. Tickner, *Org. Process Res. Dev.*, 2007, **11**, 1112.
7. F. Stoessel, *Chem. Eng. Prog.*, 1993, **89**(10), 68.
8. P. Lambert and G. Amery, *Inst. Chem. Eng. Symp. Ser.*, 1988, **115**, 85–95.
9. T. Hofelich, presented at the International Symposium on Runaway Reactions, Boston, MA, 1989; CCPS/Institute of Chemical Engineers, New York, 1989, pp. 74–85.
10. Hazard Evaluation Laboratory, http://www.helgroup.co.uk (last accessed 14 December 2010).
11. A. Craven, *Inst. Chem. Eng. Sym. Ser.*, 1982, **102**, 97–111.
12. D. Townsend, *Thermochim. Acta*, 1980, **37**, 1.
13. Hazard Evaluation Laboratory, http://www.helgroup.co.uk (last accessed 14 December 2010).
14. Fauske and Associates, http://www.fauske.com (last accessed 14 December 2010).

15. Omnical, http://www.omnicaltech.com (last accessed 14 December 2010).
16. Thermal Hazard Technology, http://www.thermalhazardtechnology.com (last accessed 14 December 2010).
17. Mettler, http://www.metler.com (last accessed 14 December 2010).
18. Hazard Evaluation Laboratory, http://www.helgroup.co.uk.
19. Recommendations on the Transport of Dangerous Goods, *Manual of Tests and Criteria*, United Nations, Committee of Experts on the Transport of Dangerous Goods, New York, 3rd edn, 1999.
20. Chemical Safety Board, http://www.csb.gov (last accessed 14 December 2010).

CHAPTER 9

Physicochemical Data Requirements for the Design of Fine Chemical Processes: Acquisition and Application

JOHN H. ATHERTON

Department of Chemical and Biological Sciences, University of Huddersfield, Queensgate, Huddersfield, HD1 3DH, UK

9.1 Introduction

Process design is an essential precursor to the introduction of a pharmaceutical product into the market. Discovery chemists require a rapid synthesis of their target molecules; process efficiency is unimportant at that stage. It is rare that the synthesis process used in discovery is adequate for manufacture; even if it were, much additional information would be required before a product could be safely and economically manufactured. Once a new product has been selected for development, the next task for the process scientists is to select a route, as discussed in Chapter 5. This chapter seeks to illustrate briefly the physicochemical data requirements to support development of a chosen route into an optimised laboratory process and thence to a manufacturing process.

Process design is an interdisciplinary activity requiring knowledge of chemistry, including some aspects not normally taught in universities, as well as a number of other disciplines, as shown in Figure 9.1.

RSC Drug Discovery Series No. 9
Pharmaceutical Process Development: Current Chemical and Engineering Challenges
Edited by A. John Blacker and Mike T. Williams
© Royal Society of Chemistry 2011
Published by the Royal Society of Chemistry, www.rsc.org

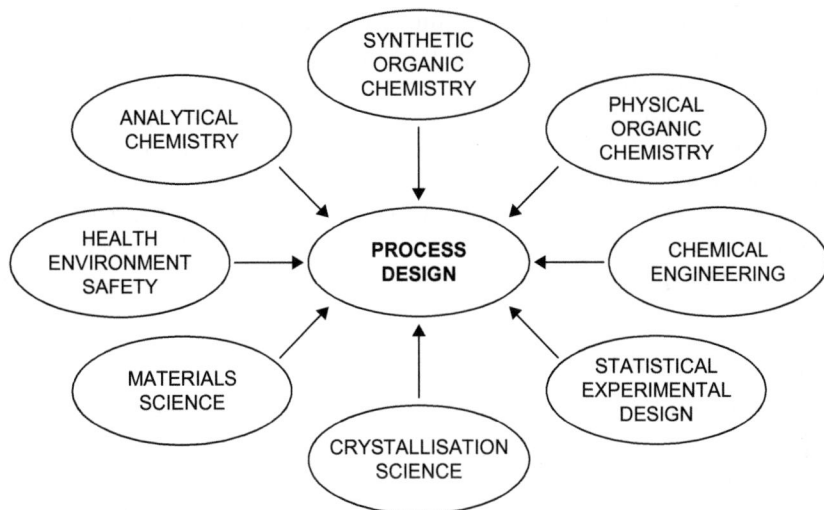

Figure 9.1 Disciplines contributing to process design.

Whilst it is not necessary that every scientist in a development team has an expert knowledge of all these areas, it is essential that they have an appreciation of the part that disciplines other than their own have to contribute to the total process. To quote J. H. Newman, "A man of well improved faculties has the command of another's knowledge. A man without them, has not the command of his own".[1]

9.2 Strategy for Identifying Data Requirements

The nature of the data requirements for process optimisation and scale-up depends on the complexity of the process. It is helpful to think about process complexity in three categories:

- Phase characteristics
- Chemical complexity
- Physical variables which interact with the chemistry.

9.2.1 Phase Characteristics

A majority of the processes used in the manufacture of pharmaceuticals are multiphase. A recent survey of processes used across the pharmaceutical and fine chemical industry[2] showed that over 60% of processes had at least one solid phase present during reaction, around 25% had two liquid phases present and 20% had a gas phase (usually hydrogen). Processes with three phases are not uncommon: a catalytic hydrogenation process with a heterogeneous catalyst is a common example, and if in such a process either a reactant or a product is out of solution, then there are four separate phases. Before reaction can occur, transfer of materials into (or onto) a common phase must occur (mass transfer), and the rate of the transfer processes will depend on physical parameters that are usually independent of the chemical rates. So the overall rate of a given chemical process will be a function of both the chemical rate and of the associated mass transfer parameters. If competing processes are differently coupled to mass transfer, then selectivities can change when the mass transfer conditions change. So for these reasons, knowledge of the phase characteristics throughout the process is a prerequisite to identifying other data requirements.

9.2.2 Chemical Complexity

Another essential part of process investigation is development of a reaction map, or picture, of the chemistry that is taking place. This should include all known reactions of the starting material, any intermediates and the product. Some side reactions may derive from impurities in the starting material. It is very helpful to display this as a picture, rather than trying to keep it in your head; this has the additional advantage that it is then more easily shared with co-workers. A generic example is shown in Scheme 9.1; this topic is discussed in more detail in Section 9.7.

  As part of the reaction map, it is helpful to include some indication of the chemical reaction rates. Very fast competing processes may lead to problems if they are faster than the rate of mixing of reactants. At the other extreme, slow chemical reaction rates may limit the reaction time. Thermochemistry must always be considered before scale-up, and can often be rate limiting at full scale. This topic is covered in Chapter 8. Exothermic processes are frequently constrained to fed-batch (or semi-batch) mode by the need to remove heat in a controlled way. Pre-reaction equilibria such as acid–base equilibria or ketone–enol tautomerism can have dramatic effects on reaction rates, and should also be added to the reaction map.

9.2.3 Physical Variables

In this section a summary is provided of the primary concerns. They are discussed in more detail in later sections.

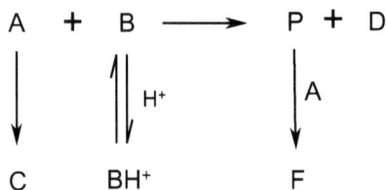

Scheme 9.1 Generic reaction picture showing reactant and product decomposition, and a reactant pre-equilibrium.

- **Time.** Many processes can be sensitive to the overall processing time; if problems are to be avoided, then relevant side reactions need to be identified so that they can be controlled. It is good practice to investigate the effect of variation in addition times and hold times on process performance.
- **Heat transfer.** Heating and cooling operations take longer on scale-up. As the scale increases, heat transfer becomes more difficult and can lead to the reaction time being extended. The need to remove heat from a reaction is a common reason for operating in "fed batch" mode. Distillation processes require heat input and will take longer as the scale increases.
- **Mixing.** Even in nominally homogeneous reactions, concentration transients exist during the addition of reagents; if there are fast competing reactions, then process selectivity may be unexpectedly low.
- **Rheology.** Low viscosity Newtonian fluids are the norm, but occasionally unusual rheologies are seen, generally when handling slurries or solutions of polymers. These can have a major impact on heat transfer and on mixing times.
- **Multiphase systems: solubility and mass transfer.** Solubility of a reactant in the reacting phase is often a limiting factor in controlling the rate of reaction. In particular, common reactive gases (hydrogen and oxygen) have solubilities in water and common solvents in the millimolar range, so reaction rates involving these gases are frequently controlled by the rate of transfer of gas into the reacting phase. Liquid–liquid systems are usually less problematic. For solid–liquid systems the particle size of the solid determines the surface area and hence the rate of dissolution, and the agitation system must be adequate to suspend the solid.

9.3 Literature

There is a wise saying that

A month in the lab can save a morning in the library!

A great deal of useful information can be gleaned from the literature; there is no point in expensively rediscovering known facts. Chemists in particular love to experiment. There is a feeling that you are not working unless a reaction is running, and this can result in poorly designed experiments. As a generalisation, not enough time is spent carefully designing experiments to maximise the data acquisition. This involves finding pre-existing information as well as ensuring that all useful information is extracted once the stirrer is turning.

There is a wealth of textbooks providing leads into the primary literature for reaction mechanisms and kinetics;[3] pK_a data can be obtained from paper[4] and from free online sources;[5,6] free on-line calculation packages are available,[7] as are more comprehensive commercial packages.[8] Physical property data can be more difficult to find, with the internet being the first port of call.[9] Multiphase reaction systems are well covered,[10] although much of this literature is unfamiliar to chemists. Patent searching can be difficult for those without access to costly commercial databases, but a free website is available.[11]

9.4 Efficient Data Acquisition: Reaction Profiling

The old fashioned way of carrying out process development was to guess the important process variables and then to vary them in a systematic way, possibly with the help of a statistical design package, and to carry out a "hill climbing" optimisation with successive experimental sets.[12] The output data from this approach were product yield and quality for a specified range of conditions. Whilst this procedure will provide an optimised set of conditions within the parameters chosen, it does not ensure that the appropriate variables are investigated, provides little or no understanding of the fundamental science determining process performance, and gives no information on the suitability of the process for scale-up.

The first objective of the data acquisition process should be to establish what the key process variables are. This involves constructing a reaction map, identifying the branch points and determining the factors that influence selectivity at each branch point.

Reaction profiling is now well established as best practice for efficiently gaining an understanding of the reaction map and for later investigating the effect of process variables.

9.4.1 Equipment

A good general purpose reactor design for laboratory use is shown in Figure 9.2.[13]

This vessel is suitable for most reaction types used in pharmaceutical research and development. For gas–liquid reactions it requires a feed tube which discharges the gas either below the agitator or close to the blade edges.

Reaction profiling involves sampling the reaction mass at intervals and analysing the composition *versus* time. It is useful for either batch or fed-batch reactions. The profile provides information on loss of starting material(s) and on the formation of intermediates, product and by-products. This greatly assists in the construction of the reaction map described earlier. The aim should be to obtain a mass balance throughout the reaction.

For a fed-batch reaction, some idea of the reaction rate can be obtained by quickly adding a small portion of reactant (say 5% of the total) and monitoring its disappearance rate. Relevant inorganic species should be monitored.

9.4.2 Profiling Methods

Sampling followed by chromatographic analysis by GC or HPLC is the most widely used method of obtaining reaction profiles. These techniques permit tracking of intermediates and by-products during the reaction, which is an

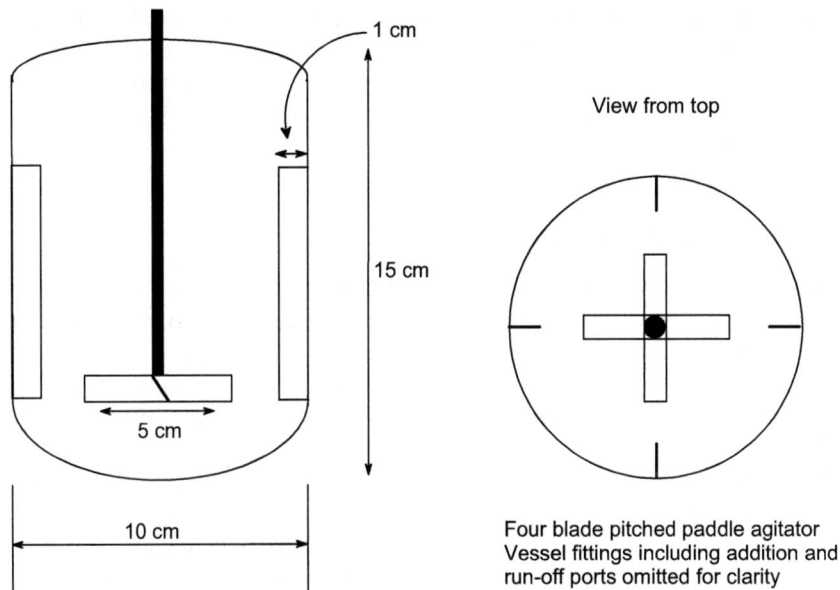

1 cm

15 cm

5 cm

10 cm

View from top

Four blade pitched paddle agitator
Vessel fittings including addition and
run-off ports omitted for clarity

Figure 9.2   General purpose reactor for process development.

essential part of understanding process performance. Care should be taken to ensure that an appropriate quench process is used prior to chromatography, such that a meaningful analysis is obtained. For example, salts of organic components require neutralisation before GC analysis. Where reactive products or by-products are involved, adventitious reactions during workup can cause difficulty in interpretation. For example, arylhydrazines oxidise rapidly to the des-hydrazine on contact with air. For gaseous reactants or products, measurement of their rate of consumption or production provides a good method of monitoring reaction progress.

Once a good understanding of intermediate and by-product formation has been obtained, less labour intensive methods can be used to gain an overall understanding of the process kinetics. Blackmond[14] has pointed out the advantages of a multi-technique approach to reaction profiling, using thermal methods together with spectroscopy and, where appropriate, measurement of gas evolution or consumption.

9.5   Reaction Kinetics

Real processes often have complex kinetics far removed from simple first- or second-order kinetics. The reasons include: multiphase processes (discussed later), where interphase transport may be rate controlling; catalytic processes, where the kinetics depend on the slow step in the kinetic cycle and may not show a simple dependence on the concentration of the reacting species; and processes involving ionic pre-reaction equilibria, where the rates are determined by solution conditions such as pH and not simply by the stoichiometric concentration of the reactants. Whilst understanding the main product forming reaction is important, it is also important to understand the nature and rates of impurity forming reactions, in particular the decomposition routes of reactants and products.

9.5.1 Zero-order Reactions

A zero-order reaction is one in which the reaction rate is substantially independent of the concentration of the reactants. There are two common cases: transport controlled reactions, where the rate-limiting step is transfer of a reactant from source phase into the phase where the reaction occurs, and catalysed reactions where the breakdown of an intermediate "resting state" is rate determining.

An example of the first case is heterogeneous catalytic hydrogenation of an alkene. For a given reaction condition the reaction profile is commonly linear with respect to time. The slope will show a linear dependence on applied pressure but a non-linear dependence on both catalyst concentration and hydrogen mass transfer rate constant.

Zero-order catalytic reactions are seen in organometallic catalysis and in enzyme-catalysed reactions. A recent example of the former is the reaction of benzophenone hydrazone with an aryl bromide using Pd(binap) as catalyst (Scheme 9.2).[15]

This reaction proceeds at a constant (zero-order) rate throughout most of its course, owing to accumulation of the intermediate complex (Scheme 9.3),

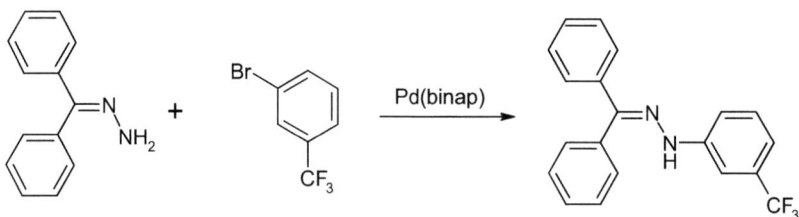

Scheme 9.2 Reaction of benzophenone hydrazone with an aryl bromide using Pd(binap) as catalyst.

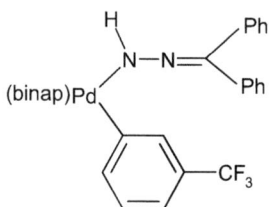

Scheme 9.3 Intermediate complex in the arylation of benzophenone hydrazone.

decomposition of which is rate determining. In contrast, reaction of the primary amine hexylamine shows first-order kinetics, with the rate-limiting step being oxidative addition of the aromatic substrate to the catalyst.

Zero-order kinetics are common in enzyme-catalysed reactions, because at high substrate concentrations the reaction rate is governed by the intrinsic turnover rate of the enzyme.

9.5.2   First-order Processes

Reactions showing first-order kinetics due to the rate being proportional to a single reactant concentration are sometimes but rarely seen in product formation; for example, decomposition of a diazonium salt[16] or solvolyses of triflates.[17] They are more common in reactions which are bimolecular but in which one reactant concentration is held constant. This could be because one component is in large excess, typically solvolysis processes, where a starting material or product reacts with the solvent, or reactions where one component is held at constant concentration. Examples are acid- or base-catalysed reactions where the pH is held constant either by a buffer or by manual intervention, and multiphase reactions where the concentration of a reactant is held constant by equilibration with a second phase. Acid-catalysed dehydrations may show first-order kinetics at a particular acidity (Scheme 9.4), but the rate will be proportional to the appropriate acidity function.[18] For a reaction of this type, measurement of the rate constant at one acidity enables extrapolation to other acidities.

9.5.3   Second-order Processes

Bimolecular second-order reactions can be studied as first-order process by maintaining a constant concentration of one component. This can be done either by using a large excess of one component, so that its concentration is effectively unchanged during the reaction, or by continuous adjustment of a reactant in order to maintain a constant concentration, as in the control of pH to keep $[H^+]$ or $[OH^-]$ constant. Second-order reactions between two different components, A and B, when run at equal reactant concentrations will slow down greatly as they approach completion. Use of a small excess of one component can dramatically reduce the reaction time (Table 9.1).

Doubling the concentrations will halve the time required for a given fraction to react. There is a tendency at very small scale to add solvent to make the materials more easily handleable or stirrable, but dilution beyond that

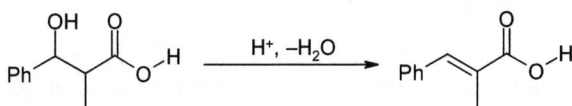

Scheme 9.4   Dehydration of a β-hydroxy acid.

Table 9.1   Effect of reactant excess on reaction time for a second-order reaction.

Amount of A reacted (%)[a]		Initial concentration of B (M)	
	1	1.02	1.1
		Time to react (min)	
95	19	15.6	10
99	99	54	23

[a]$k_2 = 1.0$ M^{-1} min^{-1}; initial [A] = 1.0 M.

necessary to bring the reactants into contact will always increase the overall reaction time.

9.5.4   More Complex Kinetics

Many modern synthetic procedures use metal catalysts and an understanding of the mechanism of catalysis can be a great help in optimising a process or in troubleshooting. All soluble metal catalysts function *via* redox mechanisms and a map of the catalytic cycle can be very helpful in identifying resting states and possible poisons. Simple rules of thumb for uncatalysed reactions, such as "to increase the rate of a second-order reaction, increase the reactant concentrations", do not usually apply to catalysed processes. For the commonly used synthetic reactions, for example alkene hydrogenation using Wilkinson's catalyst,[19] formation of biaryls by the Suzuki reaction, amination of aryl halides using the Buchwald–Hartwig procedure[20] or alkene metathesis using Grubbs's catalyst,[21] the catalytic cycles are reasonably well understood.

9.5.5   Application of Kinetic Information to Process Design

Understanding the reaction kinetics is useful for several reasons:

- It aids the understanding of process selectivity.
- It provides information on the time course of a reaction and enables reaction times to be predicted.
- It helps to specify safe hold times during the process.
- It permits an understanding of the effect of process concentrations on selectivity. For example, a desired second-order reaction competing with a first-order reaction will show poorer selectivity if the reactants are diluted.
- It can help to diagnose the rate-determining features of a multiphase process.
- The likely importance of mixing effects on the selectivity can be estimated.

9.6   Pre-reaction Equilibria

Ionic equilibria can have a big influence on overall reaction rates and selectivity, and can influence the distribution of components in multiphase reactions and separation processes. Such equilibria need to be understood and controlled. Common examples in aqueous solutions are shown in Scheme 9.5.

$$ArOH \rightleftharpoons ArO^- + H^+$$
$$+$$
$$RNH_3 \rightleftharpoons RNH_2 + H^+$$

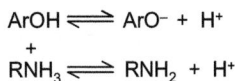

Scheme 9.5 Pre-reaction equilibria of amines and phenols.

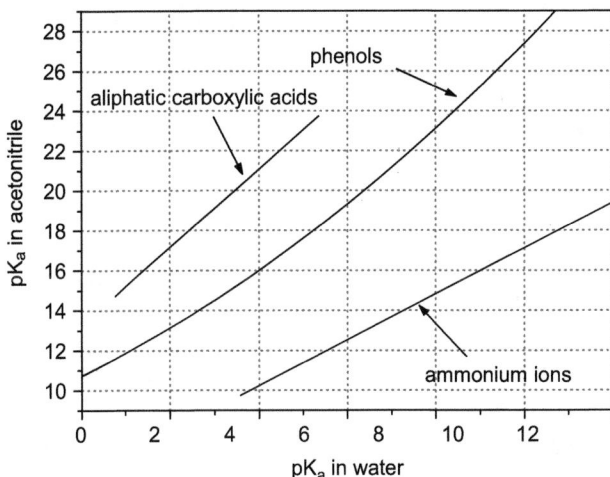

Figure 9.3 Relationship between pK_a in water and pK_a in acetonitrile for a selection of carboxylic acids, phenols and ammonium salts.

Reaction of amines and phenols with acylating agents, or with diazonium ions to give azo compounds, both proceed *via* the deprotonated component in the above equilibria. pK_a data can be used to calculate species availability using the Henderson–Hasselbach relationship.[22] King[23] has demonstrated an efficient methodology for optimisation of acyl transfer processes in aqueous media.

Non-aqueous systems are more complex. pK_a values are very different from those in water,[24] and ion-pairing and homoconjugation effects (hydrogen bonding of the uncharged acid species to the ionised species) dominate the equilibria.[25] Figure 9.3 shows the relationship between pK_a in water and pK_a in acetonitrile for a selection of phenols,[26] carboxylic acids[27] and amines.[28] Carboxylic acids are stronger acids than aliphatic ammonium salts in water, but the reverse is the case in acetonitrile. What this means in practice is that it is unwise to attempt to predict behaviour using aqueous pK_a data.

Nitration processes for unreactive substrates proceed via the nitronium ion (Scheme 9.6). The position of this equilibrium is extremely sensitive to acid strength, and nitration rates show a related sensitivity. Figure 9.4 shows how the nitration rate of 1-chloro-2-nitrobenzene correlates with the extent of conversion of nitric acid to the nitronium ion.[29] These data can be used to guide experimentation and avoid the need for extensive scouting work to approach useful reaction conditions. There is an extensive literature on acidity and reactivity in strongly acidic solutions.[30]

$$H^+ + HNO_3 \rightleftharpoons \overset{+}{N}O_2 + H_2O$$

Scheme 9.6   Equilibrium for nitronium ion formation.

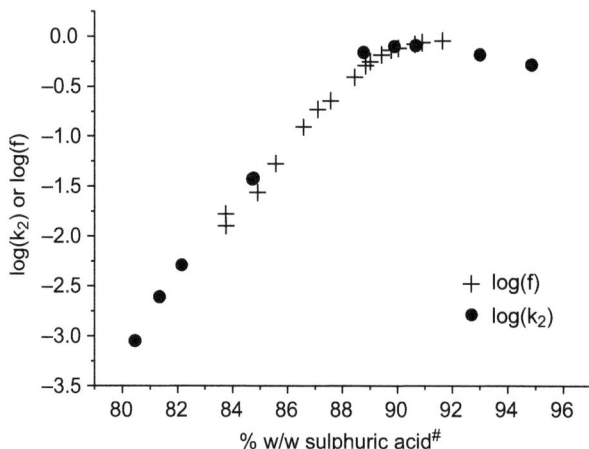

k_2 is the second order rate constant for nitration of o-dichlorobenzene at 40°C; f is the fraction of nitric acid present as nitronium ion.

expressed as (wt of sulphuric acid/(weight of sulphuric acid + weight water))

Figure 9.4   Correspondence between ionisation of nitric acid and nitration rate of 1-chloro-2-nitrobenzene.

9.7   Competing Reactions: Reaction Maps

Reaction maps derive from reaction profiling. The main reason for developing reaction maps is to identify branch points where the chemistry deviates from the required course, so that strategies can be developed to mitigate the effect of side reactions. Scheme 9.7 shows the reaction map for the reduction of a nitro compound where the desired product is the amine.[31]

The desired reduction requires three equivalents of hydrogen and goes through two intermediates, the nitroso compound and the hydroxylamine. The starting material can also react with the product amine. The hydroxylamine intermediate can, in some circumstances, disproportionate exothermically to the nitroso compound and amine, and the nitroso intermediate can capture product amine or hydroxylamine to generate azo by-products. A reaction map of this type can be used to identify branch points, where the reaction deviates from the desired course and thereby helps to develop strategies for minimising side reactions.

In the above case it is desirable to minimise the concentrations of starting material and to avoid conditions where hydroxylamine can accumulate under conditions where it may decompose. One approach to this would be to slowly

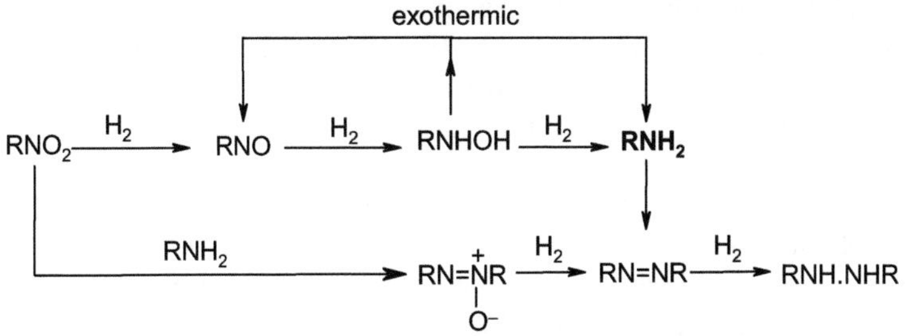

Scheme 9.7 Reaction map for the catalytic reduction of a nitro compound.

feed the nitro compound to the reaction solution whilst maintaining an adequate concentration of hydrogen, with the aim of minimising the concentrations of materials involved in side reactions. Knowledge of the temperature sensitivity of the side reactions is also helpful.

As part of the reaction map, it is helpful to include some indication of the chemical reaction rates. Very fast competing processes may lead to problems if they are faster than the rate of mixing of the reactants. At the other extreme, slow chemical reaction rates may limit the reaction time.

9.8   Mixing Effects in Pseudo-homogeneous Systems

Mixing effects occur because undesirable reactions occur faster than the time-scale of mixing.[32] A significant minority of pharmaceutical processes show some sensitivity to mixing. There are three common scenarios: consecutive, parallel and exothermic reactions. In the first case, when one reactant is added to another, there will always be a localised excess of the added component over the receiving component. If reaction is fast and there is a consecutive reaction forming a by-product, then selectivity will depend on the mixing rate. In the second case, exemplified by the neutralisation of a strong acid in a base-sensitive reactive component, the desired neutralisation is accompanied by a parallel reaction of the base-sensitive component. In the third case, which is common but has received little academic attention, localised heating caused by either an exothermic reaction or heat of mixing leads to a localised temperature rise and consequent side reactions.[33]

9.8.1   Diagnosis of Mixing Effects

A simple first question to ask is: would you have any concern if the reactants were mixed in the opposite order? If the answer is no, then it is unlikely that mixing will be a problem. An example of how this question could give rise to concern is the nitration of an organic compound using a mixture of concentrated nitric and sulfuric acids. This process would normally be carried out by slow addition of the nitrating mixture to the organic compound dissolved in a solvent. Slow addition of the organic compound to the nitrating mixture would be expected to lead to over-nitration. So in this case it would be prudent to investigate the possibility of mixing effects in the preferred contacting method. Three categories of mixing effect are recognised:

- Effects due to *slow bulk mixing*, on the timescale of circulation in the vessel, most commonly occur when one reactant is "hunting" another throughout the vessel. A common example is the neutralisation of a small amount of acid in an ester with a strong base, where the ester hydrolysis may not be especially quick but is significant on a time scale of tens of seconds.
- *Mesomixing* refers to problems associated with the dispersion of a reactive intermediate in the addition zone. In practice it is found that some reactions are sensitive to the feed rate, owing to local inhomogeneities causing a localised excess of the added reagent.
- *Micromixing* refers to mixing on the timescale (milliseconds) of the microeddy diffusion which leads to complete mixing on the molecular scale. Micromixing effects only cause problems with extremely fast reactions.

An experimental protocol to distinguish between these mixing types has been provided by Bourne[31] and is shown in Figure 9.5.

It requires the use of a well-designed laboratory reactor, such as that shown in Figure 9.2. With the exception of expt. 3, it requires the addition point to be in the region of the agitator suction or discharge. The logic for the first experiment is fairly obvious: if the reaction selectivity is good and unchanged over a wide range of agitator speed, for example 1–10 Hz, then there is not a mixing problem. The result of expt. 2 is designed to identify a micromixing

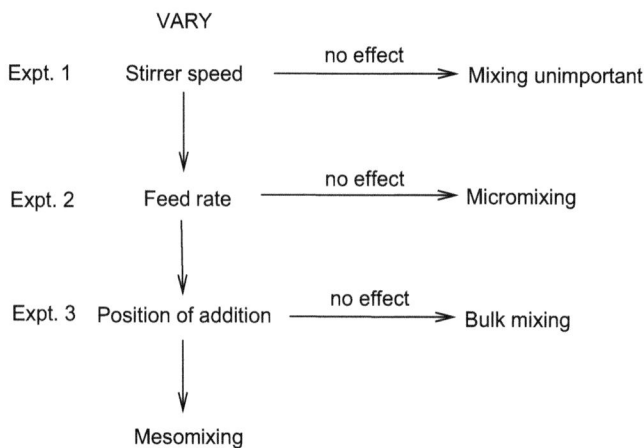

VARY

Expt. 1   Stirrer speed   ——— no effect ———→   Mixing unimportant

                              ↓

Expt. 2   Feed rate   ——— no effect ———→   Micromixing

                              ↓

Expt. 3   Position of addition ——— no effect ———→   Bulk mixing

                              ↓

                         Mesomixing

Figure 9.5   Procedure to diagnose mixing problems.

problem. For addition of a reactant to the suction zone of the agitator, only for a reaction that is controlled by micromixing is the selectivity insensitive to the feed rate. This is because, at a constant agitator speed, the energy dissipation rate is constant and micromixing is sensitive to the local energy dissipation rate.

Expt. 3 requires a comparison to be made between the selectivity for addition to the agitator suction zone and addition into a poorly stirred region. A process controlled by mesomixing will show extreme sensitivity to such a change in addition location; a macromixing process will not.

An alternative approach, used successfully by this author, is to determine the kinetics for the process of interest. This provides unambiguous data on the rates of the desired and undesired processes and enables the controlling mechanism to be deduced. Section 9.10 gives an example.

9.8.2   Solving Mixing Problems

Since mixing problems occur because the side reactions are occurring faster then the reactants are being intimately mixed, then solutions to mixing problems require either that the mixing be improved or that the rates of the chemical reactions are slowed down. Quite often reagents are added neat and their concentrations are quite high. For example, neat acetic anhydride is 10.6 M. Dilution of the reagent to be added is frequently very effective in mitigating mixing problems, by reducing the localised concentrations and hence the reaction rates. Localised exotherms are also reduced. Section 9.10 exemplifies this. Adjustment of pre-reaction equilibria also can be used to slow reaction rates relative to mixing.

The nature of useful improvements to the equipment used for mixing will depend on the type of problem. For bulk mixing problems, reducing the mixing time by using baffles and providing the correct agitator type will be necessary. Mesomixing problems require consideration of the rate of addition of reagents, which should be added in a region of high turbulence in the reactor. Micromixing problems can be addressed by the use of high-speed agitators such as those used for cell rupture.

9.9   Multiphase Systems

Multiphase processes show strong coupling of the physical characteristics of the system to the chemical kinetics. So the overall reaction rate will depend on the homogeneous kinetics, the solubility of the reactants in the reacting phase, and the rate of transfer of the reactant from the source phase to the reacting phase. The chemical kinetics are coupled differently to mass transfer, depending on the relative rates of the two processes. A brief introduction to "film theory" is necessary to explain this.

Most two-phase reactions occur after one reactant has transferred from a source phase into the phase where the reaction occurs. Very few reactions are actually interfacial. Figure 9.6 exemplifies this for cases where the reaction occurs in water or in the organic phase.

A_organic

A_aqueous

B

Product

or

Product

A_organic

+

Q⁺X⁻_(org)

Q⁺X⁻_(aq)

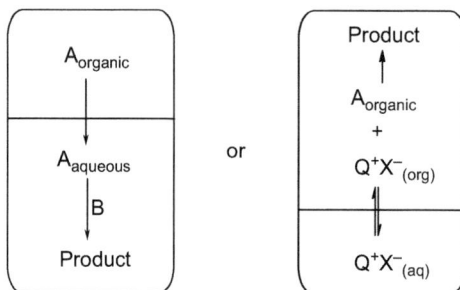

Figure 9.6   Two-phase reactions in aqueous–organic solvent systems.

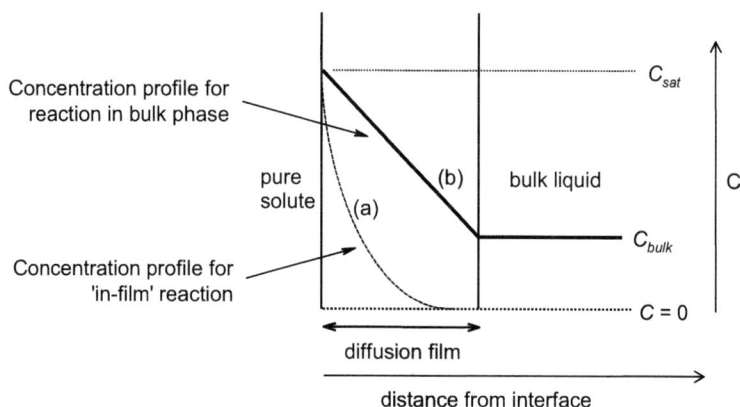

Concentration profile for
reaction in bulk phase

C_{sat}

pure
solute

(b)

bulk liquid

C

(a)

C_{bulk}

Concentration profile for
'in-film' reaction

$C = 0$

diffusion film

distance from interface

Figure 9.7   Concentration profiles across the diffusion film for "in-film" and bulk phase reactions.

Transport even from a neat source phase is not instantaneous: it is limited by diffusion across a thin unstirred layer close to the interface, which must either be crossed before convective mixing can occur in the bulk phase, or else reaction may occur within the diffusion film, as shown in Figure 9.7.

For liquid–liquid reactions, film transport times are typically a few seconds; for gas–liquid systems it is more difficult to generalise, but for stirred gas dispersions may be much less than this. A reaction that occurs on a timescale less than the transit time through the diffusion film, as shown by curve (a) in Figure 9.7, is referred to as an "in-film" reaction. The overall reaction rate is then dependent only on the interfacial area and is independent of the receiving phase volume. A reaction too slow to occur within the diffusion film will occur in the bulk receiving phase, and its standing concentration in that phase will depend on the relative rates of mass transfer and reaction, as shown by line (b) in Figure 9.7.

A simple protocol (Figure 9.8) is available to distinguish between these two cases.[34] A reaction that is slow relative to film transport times, such that it occurs in the bulk phase, will show a plateau in a plot of rate *versus* stirring

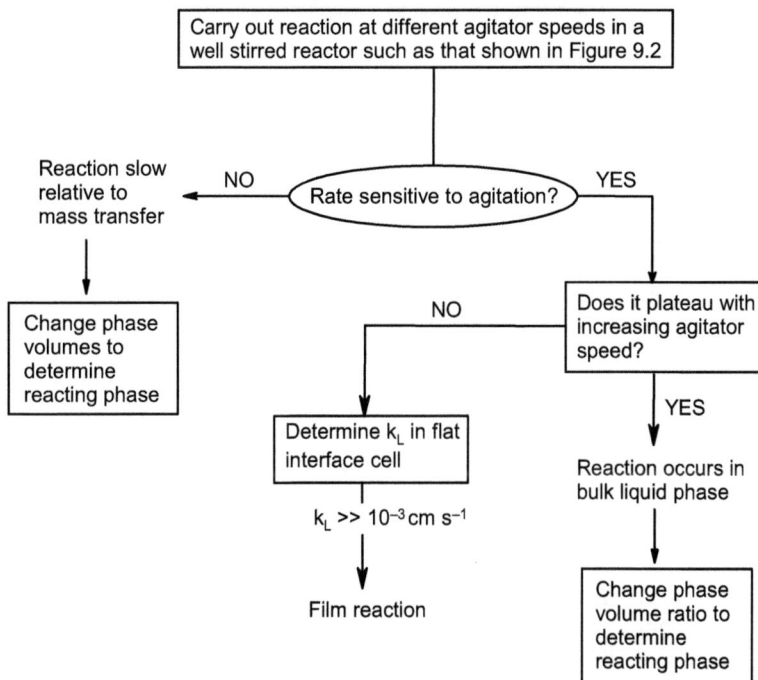

Figure 9.8 Protocol for distinguishing between bulk and "in-film" reactions.

speed, whereas a "film" process will not show such a plateau (except at very high agitator speeds, where droplet coalescence or agitator cavitation may limit further increase in interfacial area).

9.9.1 Reaction in a Bulk Phase

For a reaction between two components A and B that shows second-order kinetics in the receiving phase, and does show a rate plateau *versus* agitator speed, the rate expression is:

$$\frac{[A_{sat}]}{rate} = \frac{1}{k_2[B]} + \frac{1}{k_L a} \tag{9.1}$$

where $[A_{sat}]$ is the saturation concentration of the transferring reactant in the receiving phase, $[B]$ is the concentration of the B component, k_2 is the second-order rate constant and $k_L a$ is the mass transfer rate constant. The plateau rate occurs when mass transfer is much faster than the chemical rate, so that the rate is then determined only by the pseudo-equilibrium concentrations of reactants in the reacting phase:

$$rate = k_2[A_{sat}][B] \tag{9.2}$$

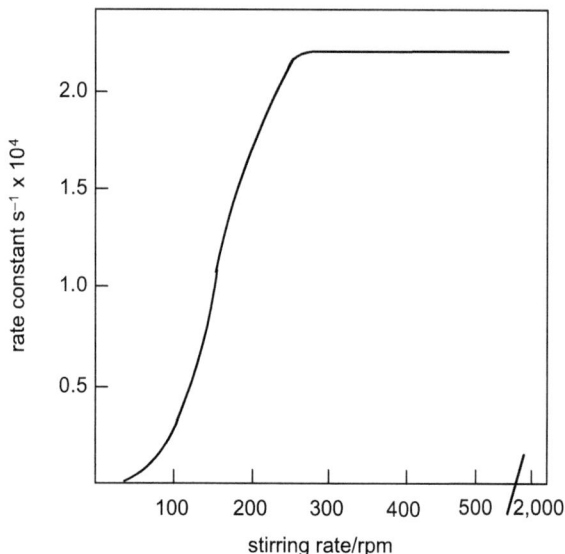

Figure 9.9   Effect of stirring rate on the reaction rate between 1-chlorooctane and aqueous sodium cyanide at 90 °C catalysed by $C_{16}H_{33}P^+Bu_3$ Br^- (reproduced with permission of the American Chemical Society).

where the rate is in units of $mol\ cm^{-3}\ s^{-1}$. A bulk phase reaction is thus usefully characterised by the rate constant for the reaction in the reacting phase and the distribution coefficients for the reactants between the two phases. Distribution coefficients are sensitive to both the solvent chosen and to the ionic strength of an aqueous solution and should be understood in order to facilitate optimisation. Increasing the ionic strength will have a different effect on the rate depending on whether the reaction occurs in the organic or the aqueous phase. Figure 9.9 shows the effect of agitation on the slow phase-transfer catalysed reaction between 1-chlorooctane and sodium cyanide.[35] In this example the reaction rate shows a plateau at a stirring speed of around 250–300 rpm.

9.9.2   Reaction in the Diffusion Film

Reactions of this type are not uncommon. The rate constant for solvolysis of benzoyl chloride in water is $1.54\ s^{-1}$ at 25 °C,[36] and will therefore be essentially complete "in-film". The second-order rate constants for the solvolysis of formate esters in water are close to $30\ M^{-1}\ s^{-1}$,[37] so their solvolysis by 1 M hydroxide ion is also "in-film". An "in-film" reaction will not show a rate plateau as the stirring speed is increased. Figure 9.10 exemplifies this: it shows how the hydrolysis rate of hexyl formate with 1 M aqueous sodium hydroxide increases with stirring speed.[38]

If mechanistic understanding of an "in-film" reaction is required, then, since the rate is dependent on interfacial area, it is necessary to control the interfacial

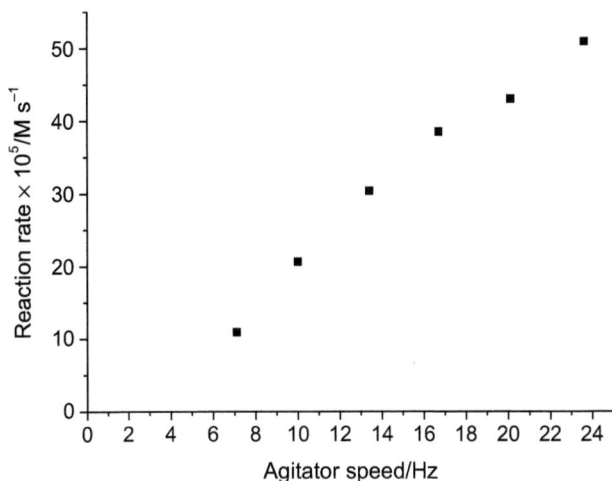

Figure 9.10 Effect of agitator speed on initial reaction rate of neat *n*-hexyl formate (0.23 L) with 1 M NaOH (0.46 L) in the 1 L vessel shown in Figure 9.2.

area. Sophisticated systems are available within academia for study of reaction kinetics in two-phase systems,[39] but a simple method which is often sufficient for development purposes is to use a Lewis cell,[40] in which a flat interface between two phases is stirred on each side at such a rate as to avoid rippling the interface. Reaction rates are typically of the order of nmol cm^{-2} s^{-1} and can be followed by analysis of the bulk phase, or, in the case of reaction of a gas, volumetrically. Where the reaction is "in-film", data treatment uses equation:

$$j = [A_{sat}]\sqrt{Dk_2[B]} \qquad (9.3)$$

where j is the reaction rate in mol cm^{-2} s^{-1}, $[A_{sat}]$ is the saturation concentration of A in the receiving phase, k_2 is the second-order rate constant, D is the diffusion coefficient of component A in the receiving phase, and $[B]$ is the concentration of reactant B in the receiving phase.[41] Several correlations are available to permit estimation of diffusion coefficients to within $\pm20\%$.[42]

9.9.3 Gas–Liquid Reactions: Catalytic Hydrogenation

Catalytic hydrogenation may use a homogeneous catalyst, in which case the methodology of the previous section is applicable to determining the controlling mechanism. Heterogeneous catalytic hydrogenation of alkenes is usually a fast reaction where the rate is limited by the rate of dissolution of hydrogen in the solution. Figure 9.11 shows the individual rate steps for an alkene hydrogenation using a solid supported catalyst, where product desorption from the catalyst is not rate limiting.

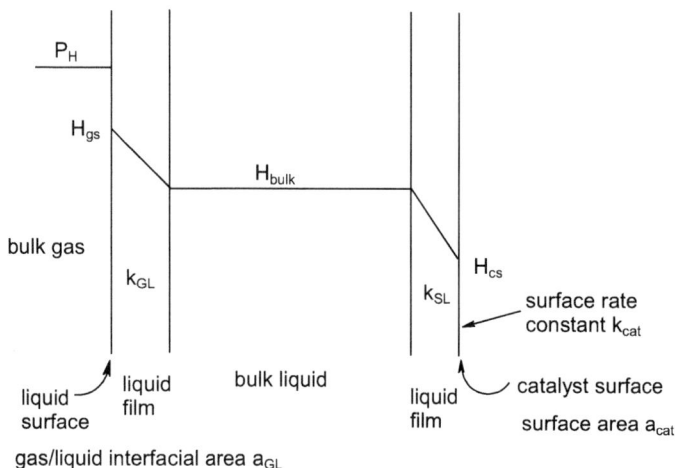

Figure 9.11   Two-film model for catalytic hydrogenation of an alkene.

The simultaneous equations for the two mass transfer rates and the reaction rate on the catalyst are:

$$\begin{aligned} \text{rate} &= k_{GL} \times a_{GL}(H_{gs} - H_{bulk}) \\ &= k_{SL} \times a_{cat}(H_{bulk} - H_{cs}) \\ &= k_{red} \times a_{cat} \times H_{cs} \end{aligned} \tag{9.4}$$

where the terms are as shown in Figure 9.11. It is assumed that the alkene is strongly adsorbed on the catalyst, and that the product is desorbed. Solving these gives:

$$\frac{H_{gs}}{\text{rate}} = \frac{1}{a_{GL}k_{GL}} + \frac{C}{a_{cat}} \tag{9.5}$$

where

$$C = \left(\frac{1}{k_{SL}} + \frac{1}{k_{red}}\right) \tag{9.6}$$

since both k_{SL} and k_{red} are constants.

Note that the rate is always proportional to the applied hydrogen pressure. In practice a key parameter is the gas–liquid mass transfer rate constant, $k_{GL}a_{GL}$. This parameter is strongly dependent on the equipment used, in particular the type of agitator, the use or otherwise of baffles and the agitator speed.

It should now be apparent that, in order to optimise (and scale-up) multi-phase processes, it is necessary to consider additional process parameters, in particular the mass transfer rate constant, k_La, and the solubility of the transferring reactant, and to understand how these are coupled to the process chemistry. Particularly in the case of gas–liquid reactions with poorly soluble gases, process performance can be critically dependent on achieving an appropriate mass transfer rate. A recent example from Merck[43] illustrates this. The relevant part of the synthesis is shown in Figure 9.12.

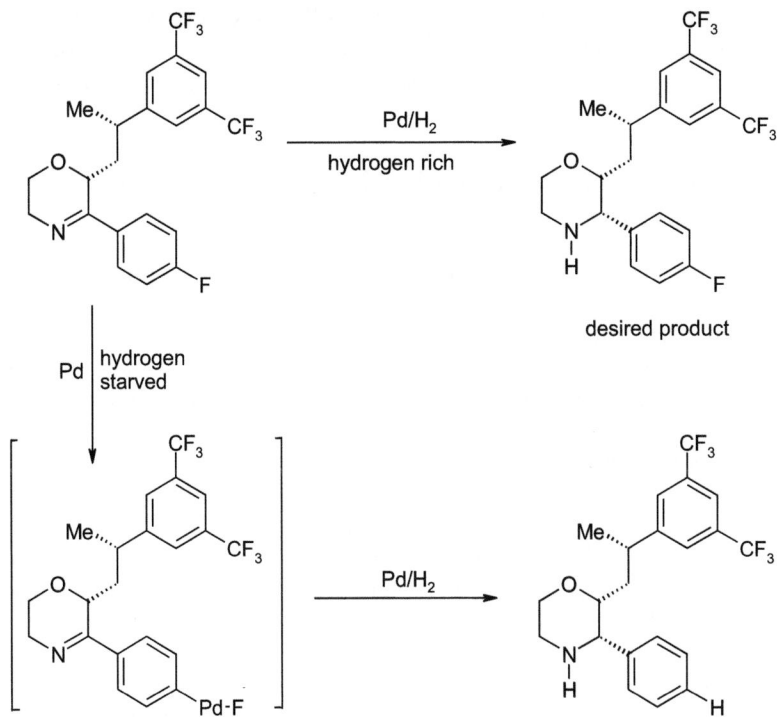

Figure 9.12   By-product formation in the reduction of an imine (reproduced with permission of the American Chemical Society).

Table 9.2   Effect of changing agitation speed and mass transfer rate constant (k_La) on impurity levels in the imine reduction.

Agitation speed (rpm)	k_La (s^{-1})	Defluoro compound yield (%)
500	0.03	23
1000	0.3	2

Under laboratory conditions shown to be hydrogen starved (that is, the chemical reaction rate was sufficient to deplete the hydrogen concentration in solution significantly below its saturation level), unacceptable levels of the desfluoro impurity were formed; increasing the agitation speed to produce a tenfold increase in k_La gave a major reduction in impurity level (Table 9.2). By reducing the catalyst loading, increasing the mass transfer rate constant and increasing the hydrogen pressure, the impurity level was reduced to a non-detectable level on scale-up.

Other examples include studies of the effect of hydrogenation conditions on regioselectivity[44] and enantioselectivity.[45]

Thus, carrying out a reaction involving a gaseous reagent, without knowing something about the mass transfer rate constant and the gas solubility, is akin to adding a critical reagent at an uncontrolled and unspecified rate. Good practice in gas–liquid reactions must involve obtaining a knowledge of these parameters and using them to choose appropriate reaction conditions.

9.9.4 Measurement of Gas-Liquid Mass Transfer Rate Constants

The measurement method used will depend on the equipment and the chemistry.[46]

The concentration of oxygen can be directly measured in water, so the mass transfer rate constant can be easily measured by measuring the rate of change of oxygen concentration against time in a solution (previously oxygen free) being sparged with oxygen.

For reactors where gas absorption is by entrainment through the surface, $k_L a$ can be measured by following the drop in pressure when the gas dissolves. The reactor is pressurised with the agitator off, and then the rate of approach of the pressure to its equilibrium value is measured. This method requires a highly sensitive pressure gauge, since the pressure drop for gases of low solubility may only be 1-2% of the initial pressure (depending on the ratio of the volume of gas to the volume of liquid).

In a system where a chemical reaction is consuming the dissolving gas, $k_L a$ may be obtained by measuring the rate of gas uptake or the rate of formation of a derived product in a situation where the rate is known to be transport controlled. A typical method involves measuring the reaction rate *versus* catalyst load (Figure 9.13) and extrapolating the rate to that where gas-liquid mass transfer is rate determining, that is, at infinite catalyst loading.

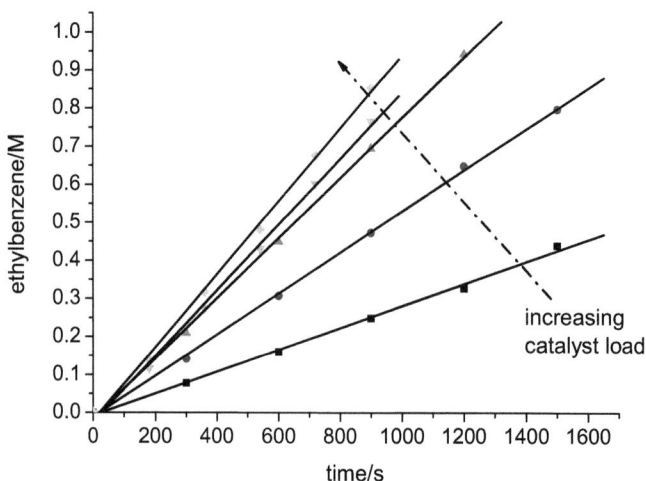

Figure 9.13 Plots of extent of reaction *versus* time for different catalyst loadings for the hydrogenation of styrene in a bubble column using 5% w/w Pd/C.

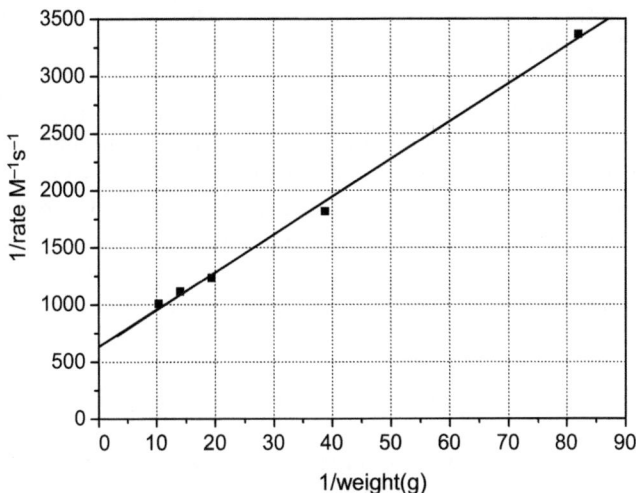

Figure 9.14 Double reciprocal plot for extraction of the mass transfer rate constant.

Extraction of the mass transfer rate constant is conventionally done *via* a double reciprocal plot of 1/(catalyst weight) *versus* 1/(reaction rate), as exemplified in Figure 9.14 for the characterisation of a small bubble column using as a test reaction the catalytic hydrogenation of styrene with a Pd/C catalyst.

From the intercept at 1/weight = 0, the transport limited rate is 1.72×10^{-3} M s^{-1}. The hydrogen solubility in the methanol solvent at the measurement temperature is 3.9×10^{-3} M, so:

$$k_{GL}a_{GL} = \frac{1.72 \times 10^{-3}}{3.9 \times 10^{-3}} = 0.44 \text{ s}^{-1} \qquad (9.7)$$

9.10 Scale-up

Once an adequate understanding of the chemistry and physics of the process has been obtained, then the scientist is much better placed to specify the full-scale process with reasonable confidence that it will perform as required. The key requirements for scale-up are that:

- The process will be safe to operate.
- The chemical selectivity (yield) will be maintained.
- The process cycle time will be as predicted.

These requirements apply both to the reaction and to the workup and product isolation stages. Typical problems that arise include:

- Side reactions of starting materials, intermediates or products.
- Incomplete or slow reactions.
- Fouling or blockages.

- Corrosion.
- Solids handling.

There are many examples of processes failing to maintain reaction selectivity on scale-up. Assuming that raw material quality and stoichiometric control are maintained, these problems are usually a result of changes in physical parameters associated with the process. A useful axiom,[47] which provides the intellectual basis for analysing the problem, is that

> *Chemical reaction rate constants are scale independent, whereas many related physical parameters are strongly dependent on scale.*

For reaction selectivity to be less than 100% there must be at least one "side reaction".

In order to identify potential selectivity problems it is essential to understand the nature of the side reactions. The position of the "branch points" – the points where the reaction deviates from the desired course – can then be clearly seen. Displaying the total reaction scheme, including intermediates, side reactions and branch points, is best done using a reaction picture that includes the physical variables associated with the reaction, the total scheme being usefully described as a "rich picture". Each of the branch points can then be interrogated to determine the response of the chemistry to changes in the system physical parameters.

The most common root causes of problems on scale-up are:

- Changes in processing time.
- Heat transfer.
- Mixing in pseudo-homogeneous reactions.
- Poor interphase mass transfer in liquid–gas and liquid–liquid systems.
- Poor dispersion of solids in solid–liquid processes.

9.10.1 Processing Time

Increases in processing time may be enforced by such factors as heat transfer, time taken to charge materials (it typically takes 2 hours to fill a 10 m^3 vessel with water *via* a 2.5 cm diameter pipe), manpower availability, phase separation times and filtration times. Thus it is essential to have an understanding of the impact of extended processing times on the process. An unpublished industrial example shown in Scheme 9.8 is the synthesis of an amino acid *via* a Strecker reaction, followed by hydrolysis of the aminonitrile.

Delays in initiating the hydrolysis step following the Strecker stage led to formaldehyde-catalysed hydrolysis of the nitrile to the amide, followed by another Strecker reaction on the amide. This could be circumvented by avoiding an excess of formaldehyde and by conducting the hydrolysis immediately the first-stage reaction was complete.

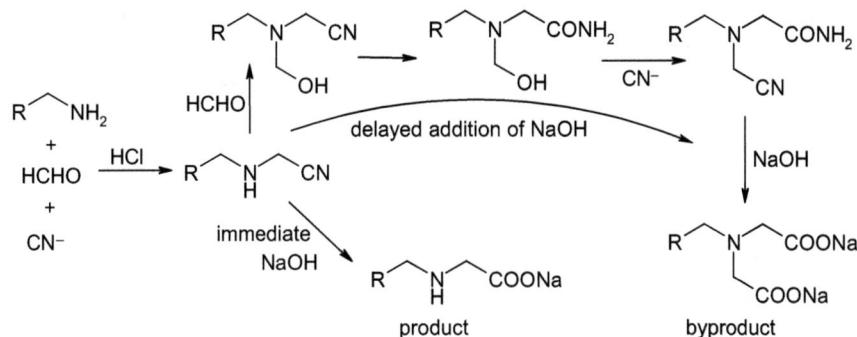

Scheme 9.8 Effect of processing delays on the outcome of a Strecker reaction.

Table 9.3 Change in surface area/volume ratio with increase in vessel size.

Vessel volume	10 L	100 L	$1\ m^3$	$5\ m^3$	$10\ m^3$
Reaction volume	Change in surface area/volume ratio				
1 L	0.464	0.215	0.100	0.058	0.046
10 L		0.464	0.215	0.126	0.100
100 L			0.464	0.271	0.215
$1\ m^3$				0.585	0.464
$5\ m^3$					0.794

9.10.2 Heat Transfer

Scale-up in batch equipment always results in a reduction in the surface area per unit volume of the reactor. Some examples are given in Table 9.3.

The most common impact of this is feedback to increased processing time, of which a common example is distillation time. A mixture of dimethylacetamide, xylene, a phenol and potassium hydroxide could be azeotropically dried during one hour on the 1 L scale, to provide a solution of the corresponding potassium phenolate. At full scale (10 m^3), where the heat input rate per unit volume is roughly one twentieth of that at the laboratory scale, the process took around 15 hours and led to significant solvolysis of the dimethylacetamide.

One secondary effect of poorer heat transfer per unit volume of reactants is the tendency to use a bigger temperature differential between the external heat transfer fluid and the reactor. Where a reaction is being heated this can lead to a higher heat transfer surface temperature, and possibly to increased thermal decomposition. For a process that is being cooled, an overcooled surface can result in higher viscosity near the surface and thus poorer heat transfer, or, in the case of a crystallisation, to a locally high desupersaturation rate, resulting in the production of "fines", or to crystallisation on the vessel surface.

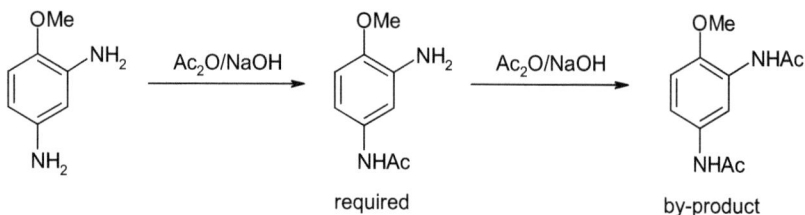

Scheme 9.9   Monoacylation of a diamine.

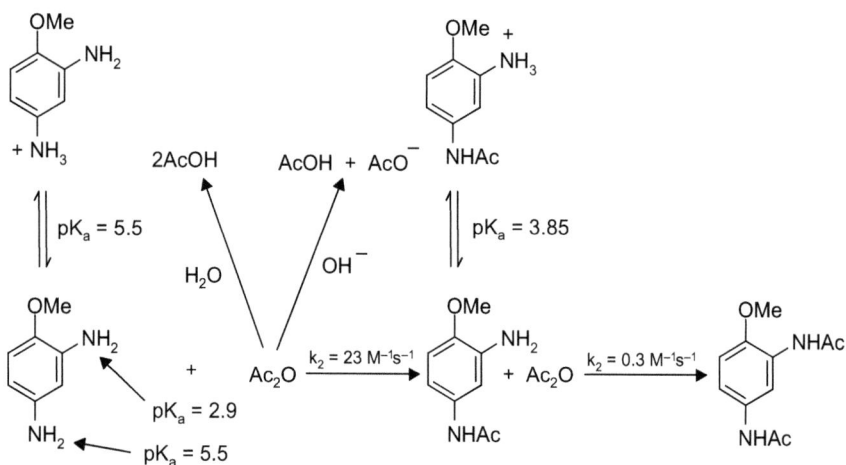

Scheme 9.10   Reaction picture for diamine acylation.

9.10.3   Maintaining Mixing Efficiency on Scale-up in Pseudo-homogeneous Systems

Scale-up of a mixing-sensitive process requires identification of the nature of the problem: whether bulk, meso- or micromixing. This can be carried out in the laboratory using the protocol described earlier. Alternatively, the reaction kinetics can be determined independently and used to diagnose the problem, as shown in the following example.

Monoacylation of a diamine was required (Scheme 9.9). The solvent was 3:1 v/v methanol/water. On the laboratory scale a yield of 93% was achieved, with 3.5% of unreacted diamine and 3.5% of the diacylated compound, but on scale-up to 20 m^3 the reaction yield was reduced to 75%. The acylation kinetics were determined by this author and are summarised in Scheme 9.10.

The process consisted of the addition of neat acetic anhydride over 4 hours to a solution of diaminoanisole (2 M) in 3:1 methanol/water with the simultaneous, radially opposed addition of sodium hydroxide solution to control the pH at 7.0 ± 0.2. A "rich picture" showing the key parameters is shown in Figure 9.15.

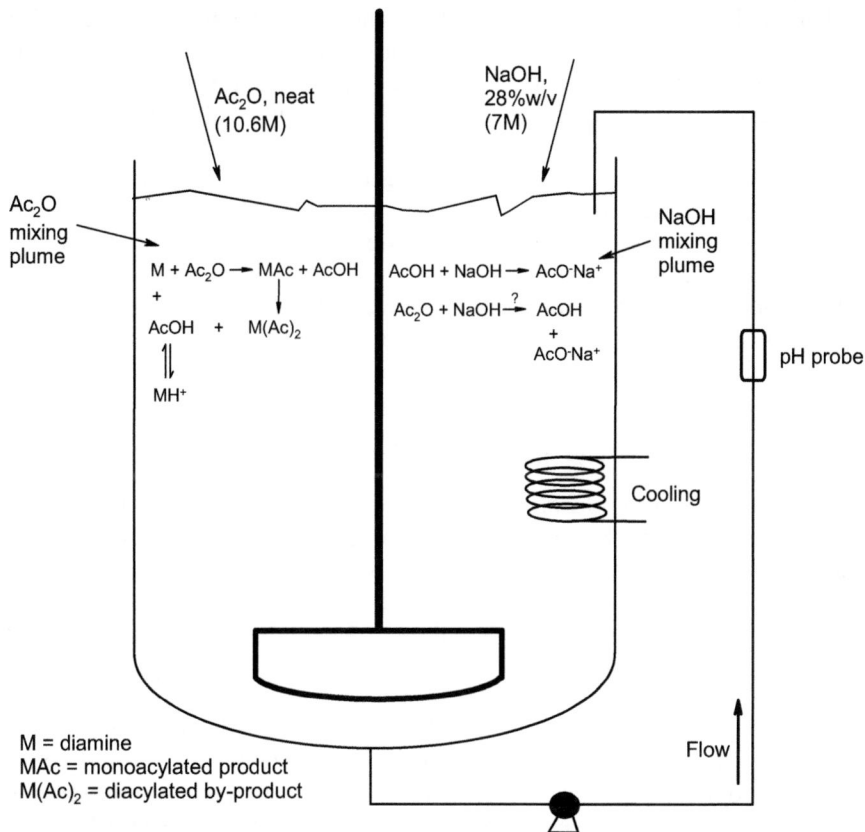

Figure 9.15 Example of the use of a "rich picture" to aid analysis of a scale-up problem.

Given the known reaction kinetics, it could readily be deduced that the selectivity would be sensitive to mixing parameters. Inspection of the rich picture reveals some further subtleties:

- Since acetic acid is formed in the acylation, the pH within the mixing plume will fall. The pK_a of the target amino group is 5.5, whereas that for the product is 3.85. Reducing the pH significantly below pH 6.5 will adjust the pre-equilibrium so as to disfavour the desired monoacetylation, and will therefore reduce the selectivity.
- A cursory inspection of the possible side reactions suggests that hydrolysis of the acetic anhydride by sodium hydroxide could be significant. In practice, analysis of the products showed that all the added acetic anhydride had reacted with the amines present, to form either the mono- or diacyl product.
- Since the pH is measured in the recirculation loop, which is fed from the bottom of the vessel, the measured pH may not represent that in the mixing zone at the top of the vessel.

This analysis permits a detailed and evidence-based analysis of the problem. The major problem was the concentration of the neat acetic anhydride. Following laboratory verification, a significant improvement in plant performance was achieved by in-line dilution of the acetic anhydride with methanol prior to its addition to the reaction mass.

Bulk mixing can be modelled by a geometrically scaled-down reactor, provided that care is taken that the flow is turbulent at the scale chosen. This typically requires a scale of around 5 L.[13] Mesomixing is more difficult to scale-up, but some detailed recommendations are available.[48] Scale-up of a process with a micromixing problem, where the rate cannot be reduced by manipulating chemical conditions, will require high-intensity mixing such as provided by a high-speed toothed rotor/stator device or a baffled tubular mixer.

9.10.4 Mass Transfer and Mass Transport

The key factor is an understanding of the mass transfer requirements that are coupled to the chemistry. For gas–liquid and gas–liquid–solid reactions, secure scale-up requires knowledge of the mass transfer rate constants at both scales. The mass transfer rate constant, $k_L a$, is a function of power per unit volume, but calculation is unreliable at a small scale. Laboratory work should include test reactions at the same $k_L a$ as the full scale.

Liquid–liquid reactions are generally less problematic, because it is easier to generate a good dispersion. Here the key is to obtain adequate dispersion of one phase in another. This can be critically dependent on the vessel configuration and the agitator speed, and in the laboratory is easily observable. Experimentation should be carried out at a range of agitator speeds to determine the sensitivity of the process to this parameter. Scale-up based on maintaining constant power per unit volume is generally satisfactory. The ease of phase separation after mixing can be dependent on the phase continuity, and this parameter should be examined.

A common problem with solid–liquid reactions is difficulty in properly dispersing the solid.[49] Again, inspection is easy in glass vessels. For scale-up, engineering correlations are available.

9.10.5 A Protocol to Check the Level of Understanding before Scale-up

It is commonplace during a project to design a new chemical plant to prepare engineering costs at the sanction stage that are accurate to ±10%. Yet there is no generally agreed procedure to interrogate the chemistry, and the interaction of the chemistry with the process equipment, commensurate with that level of accuracy. Whilst formal hazard assessments are carried out rigorously, they are designed to identify and eliminate hazard, and they do not address the questions of process efficiency or general operability. Project teams are required to make judgements as to scale-up predictability without a formal process for so

Develop a reaction picture for the chemistry. Include rates, pre-equilibria, and physical equilibria and rate processes	Draw the process equipment

List the main physical variables	and common problem areas:

• time • heat transfer • mass transfer • dispersion of solids • mixing	• side reactions • slow reactions • blockages • corrosion • solids handling

Interrogate the process step by step, questioning the relevance of physical variables and common problem areas at each stage

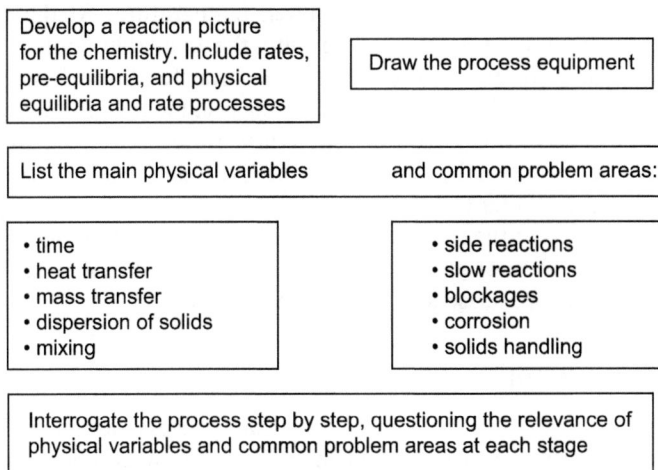

Figure 9.16   A protocol to aid secure scale-up.

doing. A simple protocol, operated in the same manner as a hazard assessment, but with a different set of questions, can significantly reduce the risk of unforeseen scale-up problems. A schematic of the protocol proposed here is shown in Figure 9.16. Muller and Latimer have proposed a similar but more formalised scheme.[50]

For this procedure to be useful a multidisciplinary group is required, including chemists, chemical engineers and someone with a detailed knowledge of the actual or proposed plant. This simple protocol puts to full use the information gathered during the process investigations described earlier, and invariably leads to new insights. Iteration is usually required to incorporate results from work identified as necessary to fill knowledge gaps.

Typically around three hours are required for the first iteration through this process, provided that adequate pre-work has been carried out. By comparison with the time or money that can be wasted by failing to identify critical factors, this is a very efficient use of resources. The procedure can also be useful before a process is transferred to pilot scale, as this vital step is often treated with too little respect. Note that, for example, scale-up from 1 L to 1 m^3 results in a reduction in the surface area to volume ratio by a factor of 10, whereas a subsequent scale increase to 10 m^3 only involves a factor of 2.2. The rate-limiting step may well change on quite modest scale-up, and this can quite easily be missed if care is not taken to analyse the situation. An example from real life illustrates this.

A catalytic hydrogenation process had been developed in the laboratory at the 300 mL scale. The $k_L a$ for the laboratory equipment was estimated at 0.17 s^{-1}, and the process was mass transfer limited, showing zero-order kinetics. Reaction time in the laboratory at a hydrogen pressure of 2 barg (~ 200 kPa) was 1 hour, with a maximum permitted temperature of 5 °C. The heat of reaction was found to be 360 kJ mol^{-1}, but no problem was found at the laboratory scale in maintaining the required temperature. At the 1 m^3 scale the heat transfer limited reaction time with the available cooling fluid was calculated to be 5 hours, whereas the mass transfer limited reaction time ($k_La = 0.07\,s^{-1}$) was 2.5 hours. So heat transfer became the limiting parameter, and it was necessary to change the contacting strategy to avoid overheating. This could be done either by controlling the hydrogen feed rate or by a controlled feed of the reactant.

Agitation systems at the small pilot scale (20–50 L) are often poorly designed and problems due to poor gas or solid dispersion are common. Care should be taken to make a proper assessment of these factors.

9.11   Conclusions

This chapter has sought to demonstrate that process design and scale-up requires consideration of a variety of interactions between chemistry and the physical environment in which the chemistry takes place, and has attempted to provide an overview of these factors. This approach to process development makes for more efficient processes and leads to less failures on scale-up.

Acknowledgements

I should like to thank the many colleagues I worked with in Avecia, Fujifilm, ICI, Syngenta and Zeneca, from whom I learnt a great deal.

References

1. J. H. Newman, *The Idea of a University*, 1852; Baronius Press, London, 2006, p. 141.
2. J. H. Atherton, J. M. Double and B. Gourlay, paper presented to World Congress of Chemical Engineering, Glasgow, 2005.
3. E. V. Anselm and D. A. Docherty, *Modern Physical Organic Chemistry*, University Science Books, Sausalito, CA, 2006; N. Isaacs, *Physical Organic Chemistry*, Longman, Harlow UK, 2nd edn, 1987; H. Maskill, *The Physical Basis of Organic Chemistry*, Oxford University Press, Oxford, 1984; T. H. Lowry and K. H. Richardson, *Mechanism and Theory in Organic Chemistry*, Harper & Row, New York, 2nd edn, 1981; W. P. Jencks, *Catalysis in Chemistry and Enzymology*, Dover, New York, 1969; E. S. Gould, *Mechanism and Structure in Organic Chemistry*, Holt, Rinehart and Winston, New York, 1959.
4. D. D. Perrin, B. Dempsey and E. P. Sergeant, *pK$_a$ Prediction for Organic Acids and Bases*, Chapman and Hall, London, 1981.
5. http://chemknowhow.com/forum/viewtopic.php?t = 25 (last accessed 5 November 2010).
6. http://en.wikipedia.org/wiki/Acid_dissociation_constant (last accessed 26 January 2011).

7. http://sparc.chem.uga.edu/sparc/ (last accessed 11 November 2010); http://aceorganic.pearsoncmg.com/epoch-plugin/public/pKa.jsp (last accessed 11 November 2010).
8. ACD/PhysChem Suite; see www.acdlabs.com (last accessed 11 November 2010).
9. www.engineeringtoolbox.com (last accessed 11 November 2010).
10. L. K. Doraiswamy and M. M. Sharma, *Heterogeneous Reactions: Analysis, Examples and Reactor Design*, Wiley, New York, 1984, vol. 2; E. L. Cussler, *Diffusion Mass Transfer in Fluid Systems*, Cambridge University Press, Cambridge, UK, 1984; J. M. Smith, *Chemical Engineering Kinetics*, McGraw-Hill, Singapore, 3rd edn, 1981; J. H. Atherton, in *Research in Chemical Kinetics*, ed. R. G. Compton and G. Hancock, Elsevier, Amsterdam, 1994, vol. 2, pp. 193–259.
11. www.FreePatentsOnline.com (last accessed 11 November 2010).
12. R. Carlson, T. Lundstedt and C. Albano, *Acta. Chem. Scand. B*, 1985, **39**, 79.
13. J. H. Atherton and K. J. Carpenter, *Process Development: Physicochemical Concepts*, Oxford University Press, Oxford, 1999.
14. D. G. Blackmond, *Angew. Chem. Int. Ed.*, 2005, **44**, 4302.
15. A. C. Ferretti, J. S. Mathew, I. Ashworth, M. Purdy, C. Brennan and D. Blackmond, *Adv. Synth. Catal.*, 2008, **350**, 1007.
16. M. L. Crossley, R. H. Kienle and C. H. Benbrook, *J. Am. Chem. Soc.*, 1940, **62**, 1400.
17. X. Creary and C. C. Gieger, *J. Am. Chem. Soc.*, 1982, **104**, 4151.
18. D. S. Noyce and S. K. Brauman, *J. Org. Chem.*, 1968, **33**, 843.
19. A. J. Birch and D. H. Williamson, *Org. React.*, 1976, **24**, 1.
20. S. Shekhar, P. Ryberg, J. F. Hartwig, J. S. Mathew, D. G. Blackmond, E. R. Strieter and S. L. Buchwald, *J. Am. Chem. Soc.*, 2006, **128**, 3584.
21. For a recent review, see G. C. Lloyd-Jones, in *The Investigation of Organic Reactions and their Mechanisms*, ed. H. Maskill, Blackwell, Oxford, 2006, ch. 12.
22. A. Albert and E. P. Sergeant, *Ionisation Constants of Acids and Bases, a Laboratory Manual*, Methuen, London, 1962.
23. J. F. King, R. Rathore, J. Y. L. Lam and D. F. Klassen, *J. Am. Chem. Soc.*, 1992, **114**, 3028; J. F. King, Z. R. Guo and D. F. Klassen, *J. Org. Chem.*, 1994, **59**, 1095; J. F. King, M. S. Gill and P. Ciubotaru, *Can. J. Chem.*, 2005, **83**, 1525.
24. K. Izutsu, *Acid-Base Dissociation Constants in Dipolar Aprotic Solvents*, IUPAC Chemical Data Series no. 53, Blackwell, Oxford, 1990.
25. I. M. Kolthoff and M. K. Chantooni Jr., *J. Am. Chem. Soc.*, 1969, **91**, 4621; F. G. Bordwell, R. J. McCallum and W. N. Olmstead, *J. Am. Chem. Soc.*, 1984, **49**, 1424.
26. I. M. Kolthoff and M. K. Chantooni, *J. Am. Chem. Soc.*, 1969, **91**, 4621.
27. I. M. Kolthoff and M. K. Chantooni, *J. Am. Chem. Soc.*, 1976, **98**, 5063.
28. J. F. Coetzee and G. R. Padmanabhan, *J. Am. Chem. Soc.*, 1965, **87**, 5005.

29. N. C. Marziano, A. Tomasin, C. Tortato and J. M. Zalvidar, *J. Chem. Soc.*, Perkin Trans. 2, 1998, 1973.
30. J. G. Hoggett, R. B. Moodie, J. R. Penton and K. Schofield, *Nitration and Aromatic Reactivity*, Cambridge University Press, Cambridge, 1971; H. Cerfontain, *Mechanistic Aspects in Aromatic Sulfonation and Desulfonation*, Interscience, New York, 1968; E. E. Gilbert, *Sulfonation and Related Reactions*, Interscience, New York, 1965.
31. T. H. Cranshaw, case study presented at Scientific Update meeting, Brunel University, 1991.
32. J. Baldyga and J. R. Bourne, *Turbulent Mixing and Chemical Reactions*, Wiley, New York, 1999.
33. J. Baldyga, J. R. Bourne and B. Walker, *Can. J. Chem. Eng.*, 2009, **76**, 641.
34. J. H. Atherton, *Trans. I. Chem. E.*, 1993, **71**(part A), 111.
35. C. M. Starks and R. M. Owens, *J. Am. Chem. Soc.*, 1973, **95**, 3613.
36. T. W. Bentley, G. E. Carter and H. C. Harris, *J. Chem. Soc.*, Perkin Trans. 2, 1985, 983.
37. A. K. Nanda and M. M. Sharma, *Chem. Eng. Sci.*, 1966, **21**, 707.
38. J. H. Atherton, unpublished work.
39. J. H. Atherton, in *The Investigation of Organic Reactions and their Mechanisms*, ed. H. Maskill, Blackwell, Oxford, 2006, ch. 5.
40. J. B. Lewis, *Chem. Eng. Sci.*, 1954, **3**, 248.
41. L. K. Doraiswamy and M. M. Sharma, *Heterogeneous Reactions: Analysis, Examples and Reactor Design*, Wiley, New York, 1984, vol. 2.
42. T. K. Sherwood, R. L. Pigford and C. R. Wilke, *Mass Transfer*, McGraw-Hill, New York, 1975.
43. K. M. J. Brands, S. W. Krska, T. Rosner, K. M. Conrad, E. G. Corley, M. Kaba, R. D. Larsen, R. A. Reamer, Y. Sun and F.-R. Tsay, *Org. Process Res. Dev.*, 2006, **10**, 109.
44. Y. Sun, J. Wang, C. LeBlond, R. A. Reamer, J. Laquidara, J. R. Sowa Jr. and D. G. Blackmond, *J. Organomet. Chem.*, 1997, **548**, 65.
45. Y. Sun, R. N. Landau, J. Wang, C. LeBlond and D. G. Blackmond, *J. Am. Chem. Soc.*, 1996, **118**, 1348.
46. M. Zlokarnik, *Stirring, Theory and Practice*, Wiley-VCH, Weinheim, 2001.
47. J. H. Atherton, in *Pilot Plants and Scale-Up of Chemical Processes II*, ed. W. Hoyle, Royal Society of Chemistry, Cambridge, 1999.
48. E. L. Paul, V. A. Atiemo-Obeng and S. M. Kresta, *Handbook of Industrial Mixing, Science and Practice*, Wiley-Interscience, Hoboken, NJ, 2004; J. R. Bourne, *Org. Process Res. Dev.*, 2003, **7**, 471.
49. J. D. Moseley, P. Bansal, S. A. Bowden, A. E. M. Couch, I. Hubacek and G. Weingartner, *Org. Process Res. Dev.*, 2006, **10**, 153.
50. F. L. Muller and J. M. Latimer, presented at the European Congress of Chemical Engineering (ECCE-6), Copenhagen, 2007.

CHAPTER 10

Liquid–Liquid Extraction for Process Development in the Pharmaceutical Industry

IAN F. MCCONVEY[a] AND PAUL NANCARROW[b]

[a] AstraZeneca, Pharmaceutical Development, Charterway, Silk Road Business Park, Macclesfield, Cheshire, SK10 2NA, UK; [b] School of Chemistry and Chemical Engineering, Queen's University Belfast, Stranmillis Road, Belfast, BT9 5AG, Northern Ireland, UK

10.1   Introduction

Relationships between chemists and chemical engineers are based on trust and understanding. Chemists do not trust chemical engineers, and chemical engineers do not understand chemists.

Currently, most pharmaceutical companies tend to manufacture an active pharmaceutical ingredient (API) batchwise. There are processes that are prevalently semi-continuous or continuous; however, these tend to be the exception due to the small manufacturing volumes typically encountered. The multi-stage batch pharmaceutical process can be further broken down into the following general steps: reaction, "drown-out"/extraction into a solvent, separation, isolation and drying. In some cases, not all the general steps will be present. This chapter covers the area of extraction and separation. Herein, we aim to establish process heuristics derived from process science, not to trivialise process

RSC Drug Discovery Series No. 9
Pharmaceutical Process Development: Current Chemical and Engineering Challenges
Edited by A. John Blacker and Mike T. Williams
© Royal Society of Chemistry 2011
Published by the Royal Society of Chemistry, www.rsc.org

development but to establish how process development can be done in a timely manner and with confidence.

When the process development team sets out to establish the best strategy to solve extraction and separation problems, a few key pieces of physical data information are usually available from very small-scale experiments or predicted *in silico*. From this base level of information, it is then possible to build up a process hierarchy as illustrated in Figure 10.1.

Early in process development, predictive science and engineering are used, especially for physical properties to track process performance. As scale-up continues, high-throughput screening for chemical routes and for analytical methods tends to take precedence. In the background, the pharmaceutical engineer will be focusing on bioavailability to ensure the appropriate delivery platforms are being utilised. As cost-effective scale-up continues, then more emphasis is placed on the basic unit operations necessary to make the primary and secondary pharmaceutical material, and testing performance in the patient becomes more important. If the new drug survives through the clinical attrition period, then all the science and technology previously developed is pulled together to derive the overall state space analysis,[1] more commonly referred to as design space, and the rules for the lifecycle management of the final product.

Many pharmaceutical chemical synthesis processes can be characterised by reacting two or more components in the presence of a base or acid, catalyst and solvent. The solvent chosen is normally one that will not interfere with the reaction and will allow the reaction to proceed homogeneously in solution until completion. In some reactions it is possible to take advantage of a telescoped process (one where a number of reactions can be carried out in series without isolating or

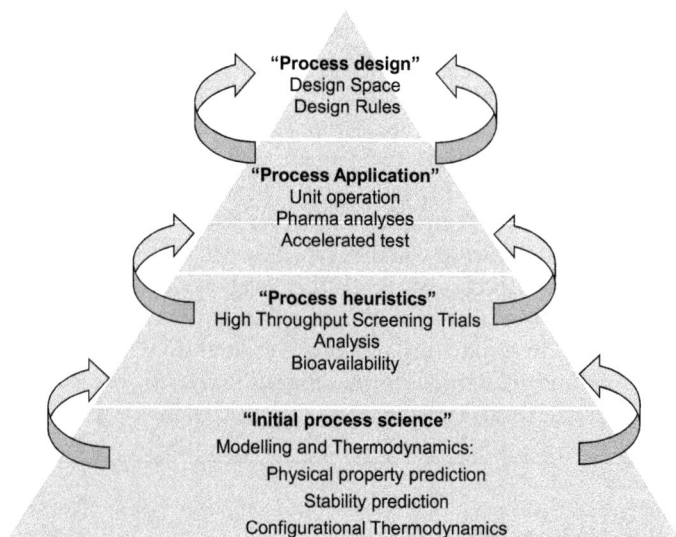

Figure 10.1   Typical hierarchy of process development in the pharmaceutical industry.

purifying the product of the reaction) if the solubility of the feed materials and the product vary widely enough to undertake a reaction where the product then supersaturates in the solvent, and either crystallises, or precipitates out of solution. For those reactions that are homogenous, the end of reaction mass is contacted with an anti-solvent to either extract the product or the reaction by-products and impurities. Normally this will include a salt of some description from the acid or base reaction and then a series of other impurities related to the reaction.

In reality, most new chemical entities are first encountered as part of a discovery processes; by the time they arrive in process development, they have already undergone some kind of selection process based on their physical properties. For example, in the pharmaceutical industry the pharmacist may be using a heuristic rule such as Lipinski's "rule of five" (Ro5)[2] or the application of log D. Ro5 adjustment as proposed by Bhal[3] is a way of reducing the complexity involved in predicting bioavailability. The key factors in the Ro5 are: molecular weight < 500, log $P < 5$ and/or log $D < 5$, hydrogen bond donor atoms < 5 and a sum of nitrogen and oxygen atoms < 10. Box and Comer[4] have taken the previous ideas one step further to relate log P to drug solubility and Biopharmaceuticals Classification Scheme (BCS) class. Class I are considered to be highly soluble and permeable, class II have limited solubility but are permeable, class III are soluble but poorly permeable and class IV have low solubility and low permeability.

The nirvana of the process developer is to be able to predict the behaviour of the various materials in the system using the minimum amount of information across a range of operating conditions. Increasingly, theoreticians are using quantum chemistry to derive new methods and one such approach is a conductor-like screening model for real solvents (COSMO-RS), which we shall return to later in the chapter.

10.1.1 What is Liquid–Liquid Extraction?

In liquid–liquid extraction, components are separated by their distribution between two immiscible liquids. The feed solution containing the solute is contacted with the extracting solvent and the solute is partitioned between the two solvents, called the raffinate and the extract, respectively. Liquid–liquid extraction is usually considered to be an indirect method of separation because it involves introducing new material to effect a separation, whereas direct methods, such as distillation, do not introduce any new material. From a process mass efficiency point of view, adding additional materials should be considered carefully, as well as the consequences for "green operation". However, there are several cases where separation via the introduction of a new component is the only viable option, such as:

(a) separation of close boiling liquids
(b) separation of liquids with relatively poor volatility
(c) recovery of high boiling point materials
(d) reduction in the cost of evaporation

(e) replacement of fractional crystallisation
(f) separation of heat sensitive materials
(g) separation of mixtures that form azeotropes
(h) separation according to chemical type
(i) replacement of more expensive separation techniques
(j) assisting chemical reactions.

Manufacture of new chemical entities (NCEs) in the pharmaceutical industry usually involves quite complex chemistry, where functionality is added to molecules and the reactions themselves are rarely quantitative in terms of yield. Therefore many of the scenarios in (a) to (h) above are commonly encountered. Frequently, the number of synthesis steps varies from six to twenty for NCEs and extraction is necessary to manufacture relatively pure materials for the next step. For efficient synthesis with the different process chemistries involved, several extractions to either purify the intermediate, remove impurities or avoid interference from interfering solvents are also often necessary.

10.1.2   ICH Guidelines for Solvent Classification and Usage

The European Medicines Agency (EMEA) guideline on residual solvents recommends the acceptable amounts for residual solvents in pharmaceuticals for the safety of the patient. The guideline recommends use of less toxic solvents and describes levels considered to be toxicologically acceptable for some residual solvents. The classification system for residual solvents is listed in Table 10.1.

The octanol–water distribution can be characterized by the log P value (eqn 10.1), which is particularly useful in the pharmaceutical industry to give a likely distribution of drugs in the body:

$$\log P_{ow} = \log\left(\frac{\text{solute concentration in octanol}}{\text{solute concentration in un-ionised water}}\right) \qquad (10.1)$$

If a compound has a log $P < 0$, then it tends to be hydrophilic, whilst log $P > 0$ indicates that the compound is lipophilic.

Table 10.1   ICH classification guide for solvents.[a]

Class	Usage	Comments
1	Solvents to be avoided	Known human carcinogens, strongly suspected human carcinogens, and environmental hazards.
2	Solvents to be limited	Non-genotoxic animal carcinogens or possible causative agents of other irreversible toxicity such as neurotoxicity or teratogenicity. Solvents suspected of other significant but reversible toxicities.
3	Solvents with low toxic potential	Solvents with low toxic potential to man; no health-based exposure limit is needed. Class 3 solvents have permitted daily exposures of 50 mg or more per day.

[a]Note for Guidance on Impurities: Residual (CPMP/ICH/283/95) (March 1998).

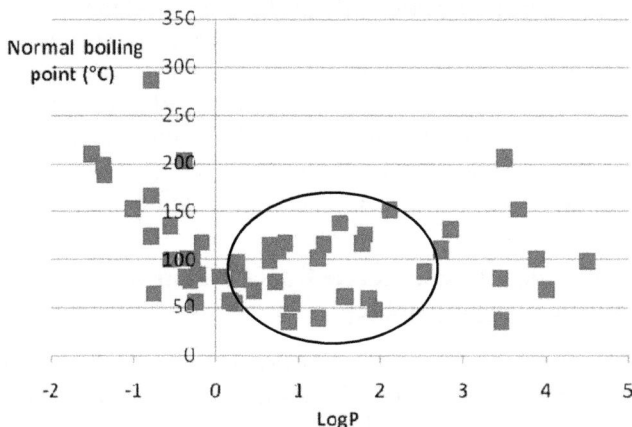

Figure 10.2   Normal boiling point (°C) *versus* log *P* for ICH class 2 and 3 solvents.

Table 10.2   List of some major commercial drugs and their log *P* values.

Generic substance	Therapy area	log P
Atorvastatin	Cholesterol regulator	5.7
Clopidogrel	Platelet aggregation inhibitor	2.5
Esomeprazole	Antiulcerant	0.6
Fluticasone + salmeterol	B2 stimulant + corticoids	3.4/4.2
Quietiapine	Antipsychotic	2.8
Olanzapine	Antipsychotic	2
Montelukast	Antileukotriene antiasthmatic	7.9
Lansoprazole	Antiulcerant	1.9
Venlafaxine	Antidepressant	2.8
Rosuvastatin	Cholesterol and triglyceride regulator	2.4

Plotting the log *P* against the normal boiling point for ICH listed class 2 and 3 solvents as shown in Figure 10.2 reveals that the operating zone for the most popular reaction types and solvents can be identified. Within the normal operating conditions for standard batch manufacturing plants $(-20\,°C < T_{batch} < 160\,°C$ and $0\,bar < P_{batch} < 3\,bar)$ there is a concentration of solvents shown within the elliptical area that are commonly used by the pharmaceutical industry for a wide range of different reactions. The physical properties of these solvents are also widely available from standard, open and easily accessible sources.

The usefulness of log *P* can also be extended to drugs themselves; ten of the top selling commercial pharmaceutical drugs and their log *P* values are listed in Table 10.2.

10.1.3   Impurities

The ICH guidelines relate acceptable impurity levels to the dose per day, the period over which the drug may be taken and the method of administration.

Normally any impurities formed early in the drug substance pathway may be removed by attrition on further processing. Typical impurity levels early in development are <0.5% w/w (phase 1) and usually reduce to <0.2% w/w (phase 2, 3 and commercial), unless there is evidence of genotoxic activity, when the levels would be set substantially lower. The ultimate level set on the drug substance target specification reflects the best view of what is acceptable for human safety and this should always be borne in mind.

10.1.4 Commonly Encountered Reactions in the Pharmaceutical Industry

To the non-chemist, process chemistry can seem like a series of case histories with a large database of publication history. Chemical engineers tend to take the reactions and write them in the form of reaction sets with the following nomenclature:

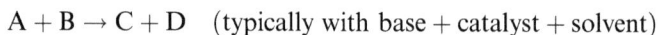

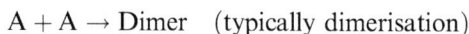

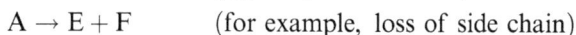

$$A + B \rightarrow C + D \quad \text{(typically with base + catalyst + solvent)}$$
$$A + A \rightarrow \text{Dimer} \quad \text{(typically dimerisation)}$$
$$A \rightarrow E + F \quad \text{(for example, loss of side chain)}$$

Whilst this approach is useful for understanding reaction kinetics and can help to reduce complexity by a lumped parameter approach, unfortunately a lot of key learning and understanding may be lost by not comprehending fully the chemical mechanism. If we instead look at the problem from the chemist's point of view, we may be able to gain knowledge by reviewing the more common reactions.

Groups of chemists have surveyed the types of reactions that they most often use,[5,6] and these surveys have also been used to establish historically the most popular process synthesis chemistry. The top five reaction categories are listed here with a number of important sub-categories identified:

- Heteroatom alkylation and arylation (*N*-substitution): reductive amination
- Acylation: *N*-acylation of an amide
- Deprotection of an amino group
- Functional group interconversion: alcohol to halide
- C–C bond forming.

An example of C–C bond forming is Suzuki coupling; this is considered as a case study later in Section 10.4.3.

10.2 Theoretical Considerations for Liquid–Liquid Extraction

Whilst the design and implementation of a liquid–liquid extraction step in pharmaceutical processing is very much reliant on an heuristic approach, which will be described later, a sound understanding of the underlying theoretical principles is essential to reduce the search space and maximise success. Liquid–liquid extraction is a mass transfer process that involves transferring a component from a liquid in which it has a relatively low solubility into another liquid in which it has a relatively high solubility. The theoretical extent to which the desired component can transfer from one phase into another is determined by the thermodynamics of phase equilibria, which is therefore the most fundamental theoretical consideration for successful liquid–liquid extraction. However, the time taken for two phases to reach this equilibrium is also important for efficient process design, and so understanding the factors affecting the rate of interphase mass transfer is also a key consideration. Furthermore, successful liquid–liquid extraction depends upon efficient dispersion of one phase into another to maximise the contact area between phases, whilst also enabling phases to be subsequently separated without the formation of a stable emulsion. The science underpinning liquid–liquid extraction is vast and, therefore, the aim of this section is to provide a brief introduction to the most important theoretical considerations and direct the reader, where appropriate, to further sources of information.

10.2.1   Phase Equilibria

The thermodynamic properties of a system determine how many liquid phases will be formed and how the components distribute between these phases. The separation of a liquid mixture of a given composition into two liquid phases in thermodynamic equilibrium can be explained by the variation of the energy of the system as a function of composition. Equilibrium is achieved when the composition of the system is such that the total energy is minimised. The total energy of the system may be represented by the Gibbs free energy (G), defined in eqn (10.2):

$$G = H - TS \tag{10.2}$$

where H = enthalpy, T = temperature and S = entropy. The change in molar Gibbs free energy (G^m) for one mole of mixture from n pure components at a specific temperature and pressure can be represented by eqn (10.3):

$$G^m/RT = \sum (x_i \ln a_i) \tag{10.3}$$

where, for component i, x_i = mol fraction and a_i = activity = $x_i \gamma_i$ (for an ideal system, the activity coefficient $\gamma = 1$).

It is therefore possible to divide G^m into two components, the free energy of the ideal system, G^{ideal}, and the excess free energy of the non-ideal system, G^{ex} (eqns 10.4–10.6):

$$G^m = G^{ideal} + G^{ex} \tag{10.4}$$

where

$$G^{ideal}/RT = \sum (x_i \ln x_i) \tag{10.5}$$

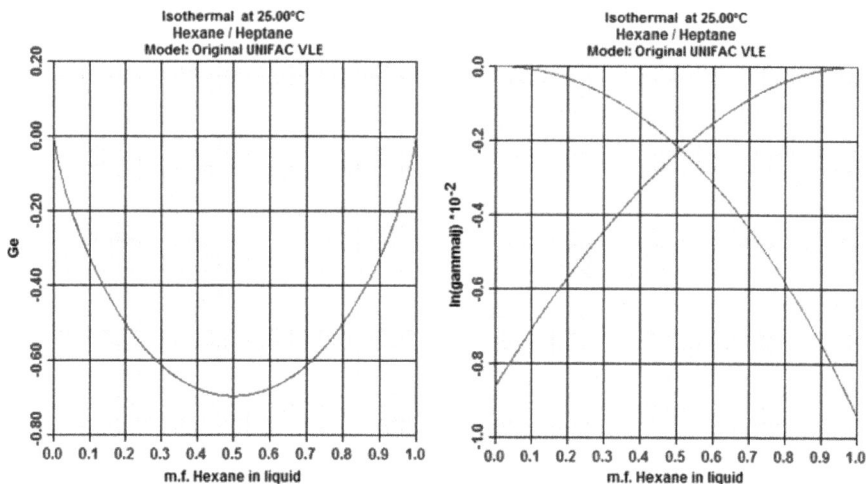

Figure 10.3   Variation of thermodynamic properties of the hexane/heptane system
as a function of composition: (*left*) Gibbs energy and (*right*) activity
coefficient (γ).

and

$$G^{ex}/RT = \sum (x_i \ln \gamma_i) \tag{10.6}$$

For an ideal system, such as the hexane/heptane system shown in Figure
10.3, it is not possible to find a pair of liquid phases with a molar Gibbs free
energy that is less than any single liquid phase and, therefore, we tend to have a
standard concave-shaped graph when the standard state energy is plotted as a
function of composition.

However, for a non-ideal system it can be seen, on the left of Figure 10.4, that
the excess free energy curve has two minima. Generally, in moderately non-ideal
systems, the symmetry is destroyed but at no point does the second
derivative of the free energy change sign. The system is stable or metastable when
$d^2G/dx^2 > 0$ and the boundary of instability is $d^2G/dx^2 = 0$. To demonstrate this
effect, consider the hexane/acetonitrile system as shown in Figure 10.4. Minima
in Gibbs energy are observed at 0.05 and 0.97 mole fraction (m.f.) hexane at
25 °C. Any liquid mixture with a composition falling between these two points
will tend to split into two liquid phases: a hexane-rich phase with 0.97 m.f. hexane
and an acetonitrile-rich phase with 0.05 m.f. hexane. Mixtures with less than
0.05 or greater than 0.97 m.f. hexane will form a single liquid phase. Non-ideality
tends to decrease as temperature increases, whereas pressure has little effect
on liquid–liquid equilibria (LLE) and is usually ignored.

One of the most commonly encountered binary systems in organic synthesis
is toluene/water; the miscibility limits for this system are shown in Figure 10.5.
Toluene reaction systems are quite often drowned out into water to separate
organic from inorganic material. Another commonly encountered application

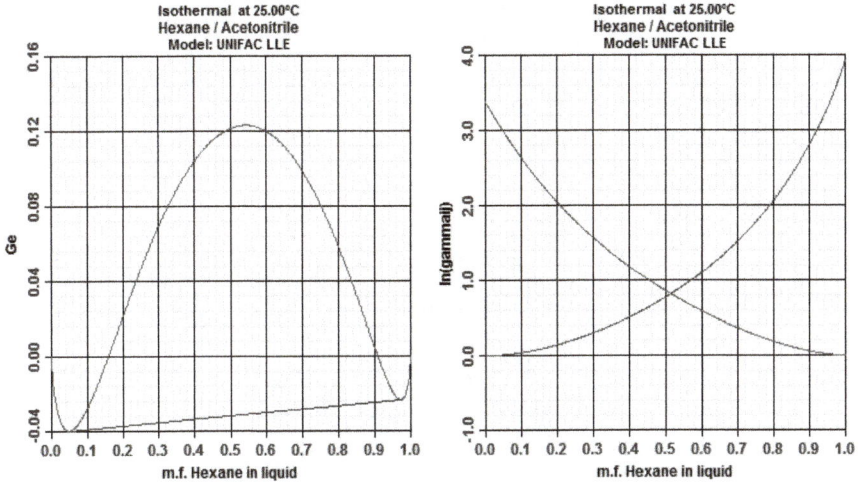

Figure 10.4   Varation of thermodynamic properties of the hexane/acetonitrile system as a function of composition: (*left*) Gibbs energy and (*right*) activity coefficient (γ).

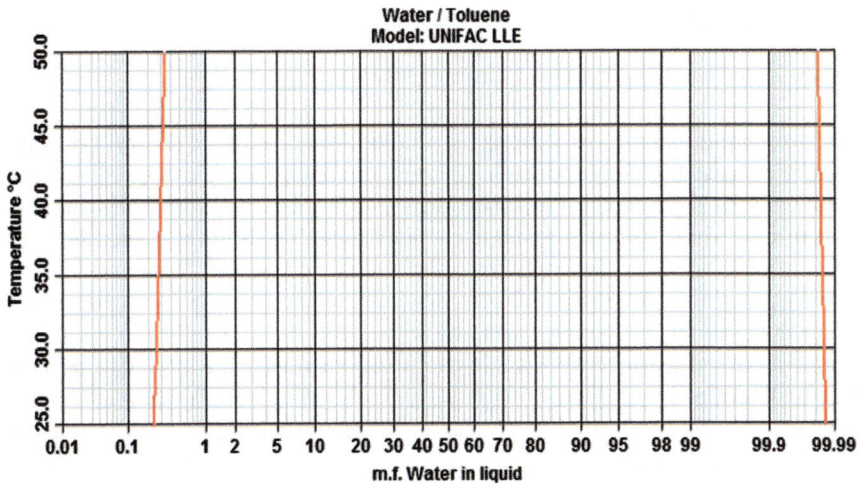

Figure 10.5   Liquid–liquid phase equilibria for the toluene/water system as a function of temperature.

of the toluene/water system is to remove water azeotropically from solution and separate the water by condensation from the vapour.

10.2.2   Multicomponent Systems

Simple binary systems are useful for explaining the theoretical thermodynamic basis for phase equilibria in liquid–liquid systems. However, in

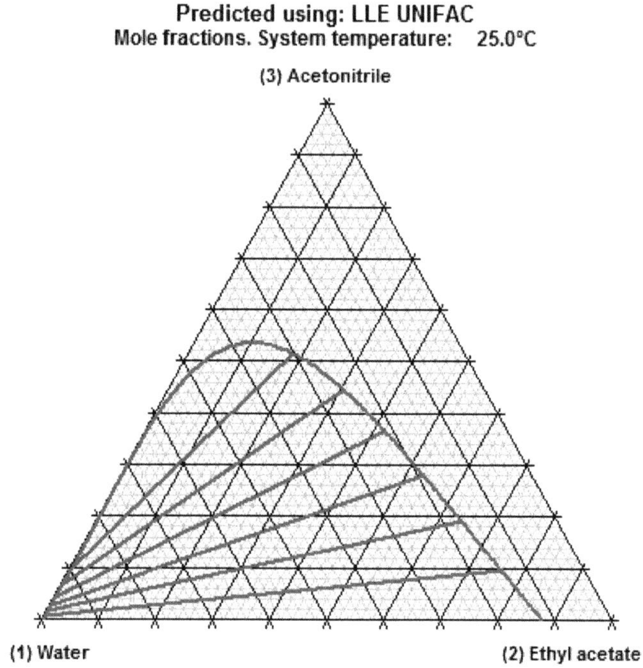

Predicted using: LLE UNIFAC
Mole fractions. System temperature: 25.0°C

(3) Acetonitrile

(1) Water   (2) Ethyl acetate

Figure 10.6   Example of a type 1 ternary system.

practice, liquid–liquid extractions in the pharmaceutical industry typically involve at least three components: (i) the component being extracted, (ii) the initial solvent from which the component is being extracted and (iii) the new solvent into which the component is being extracted. If all three components in a ternary system are completely miscible, a single phase will form and liquid–liquid extraction is not possible. Therefore, it is necessary to employ an extractant which has limited miscibility with the initial solvent. To fully represent the equilibria over the range of possible compositions for such a system, a ternary phase diagram is necessary. Figure 10.6 shows a typical ternary phase diagram for the water/ethyl acetate/acetonitrile system. Equilateral triangular coordinates are used to allow the complete range of compositions for a three-component mixture to be represented graphically. At any point within an equilateral triangle, the sum of the perpendicular distances to the three sides is equal to the height of the triangle. Therefore each apex represents 100% of a single component, whilst each axis represents a binary mixture of the two components at the adjoining apexes. Any point within the triangle represents a ternary mixture containing each of the three components, with the perpendicular distances to each side representing the percentage or fraction of each component. The binodal curve (see Figure 10.6) represents the limit of miscibility for the ternary system. Any mixture with a composition outside the curve will exist as a single phase, whereas any mixture with a composition within the curve will tend to split into two liquid phases whose compositions lie at points on opposite sides of the binodal curve. These equilibrium compositions are connected by an infinite number of tie lines, which converge at a point on the binodal curve known as the plait point.

For those ternary systems which demonstrate limited miscibility, a range of scenarios is possible, which can be classified as follows:

Type 1: forms a pair of partially miscible components (and two miscible pairs)

Type 2: forms two pairs of partially miscible components (and one miscible pair)

Type 3: forms three pairs of partially miscible components

Type 4: solid phases can be formed.

A typical example of type 1 is the acetonitrile/ethyl acetate/water system, as shown in Figure 10.6. This system has one partially miscible pair (water/ethyl acetate) and two miscible pairs (water/acetonitrile and acetonitrile/ethyl acetate) of liquids. This type of system is most common in liquid extraction.

For type 1 systems the closer the liquid phase separation line is to the axes, then the more efficient is the separation in terms of the number of components. On the left-hand side of the diagram one liquid phase will contain principally water and acetonitrile, whereas on the right-hand side of the diagram the phase will contain principally ethyl acetate and acetonitrile; however, significant quantities of water will be present as contaminant.

Two examples of type 2 systems are shown in Figure 10.7. The ternary phase equilibria for the aniline/heptane/methylcyclohexane system, shown on the left of Figure 10.7, illustrate that both the aniline/heptane and aniline/methylcyclohexane binary pairs have regions of immiscibility, whereas the methylcyclohexane/heptane pair shows complete miscibility.

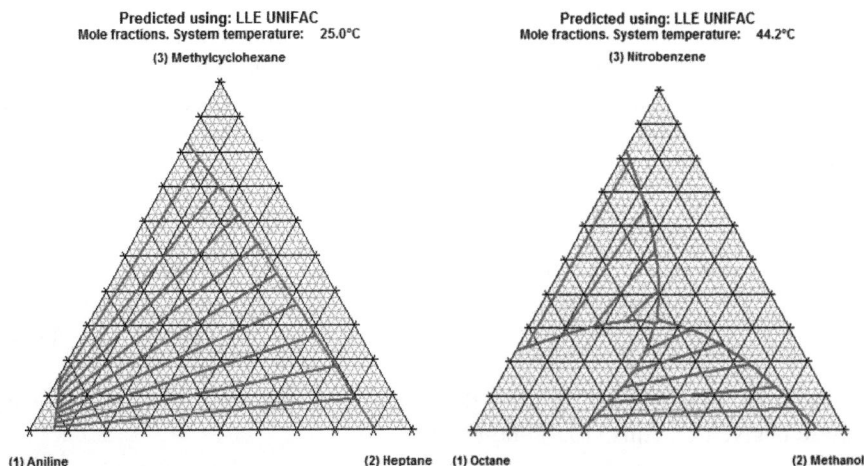

Figure 10.7 Two examples of type 2 ternary behaviour.

Other more complex behaviour may also be observed in ternary systems, such as the formation of three liquid phases, but such systems are not desirable for liquid–liquid extraction and will not be discussed here. In the pharmaceutical industry, liquid extraction of dissolved solids is a commonly encountered problem. In this case, in addition to the liquid phase regions the ternary phase diagram may show composition regions in which solid phases are formed. For further details on these more complex systems, refer to Treybal.[7]

10.2.3 Interphase Mass Transfer

The rate of mass transfer of a component from one liquid phase to another is an important consideration in the design of any extraction process. The driving force for any mass transfer process is the concentration gradient; the higher the concentration gradient of a component between two regions, the higher the rate of mass transfer. In addition, interphase mass transfer is dependent upon the interfacial area between the two phases and the mass transfer coefficient, which is a function of the geometry of the system and the physical properties of the phases. Therefore any liquid–liquid extraction process can be optimised by maximising the concentration gradient, interfacial area and mass transfer coefficient.

Two-film theory is one of a number of mechanisms postulated to represent the conditions present in the boundary region between phases having an influence on steady-state mass transfer. Figure 10.8 illustrates this mechanism. Consider the mass transfer of component A from liquid phase 1 to liquid phase 2. The concentration of component A, $C_{A,1}$, is constant throughout the bulk of phase 1. However, owing to resistance to mass transfer in an imaginary laminar film on the phase 1 side of the interface, the concentration decreases as the interface is approached until it reaches the interfacial concentration, $C_{Ai,1}$. Similarly, owing to resistance to mass transfer in phase 2 imaginary film, the concentration falls from an interfacial concentration of $C_{Ai,2}$ to the bulk concentration, $C_{A,2}$. It is assumed that resistance to mass transfer across the interface is negligible; therefore, the equilibrium concentrations of component A in phases 1 and 2 are the interface concentrations, $C_{Ai,1}$ and $C_{Ai,2}$, respectively. Over the range of possible concentrations, the equilibrium relationship is given by eqn (10.7):

$$C_{Ai,2} = f(C_{Ai,1}) \qquad (10.7)$$

which, in relatively dilute systems, can be simplified to eqn (10.8):

$$C_{Ai,2} = mC_{Ai,1} \qquad (10.8)$$

For steady-state mass transfer, the rate of mass transfer of A, N_A, in each phase is given by eqns (10.9) and (10.10):

$$N_A = k_1(C_{A,1} - C_{Ai,1}) \qquad (10.9)$$

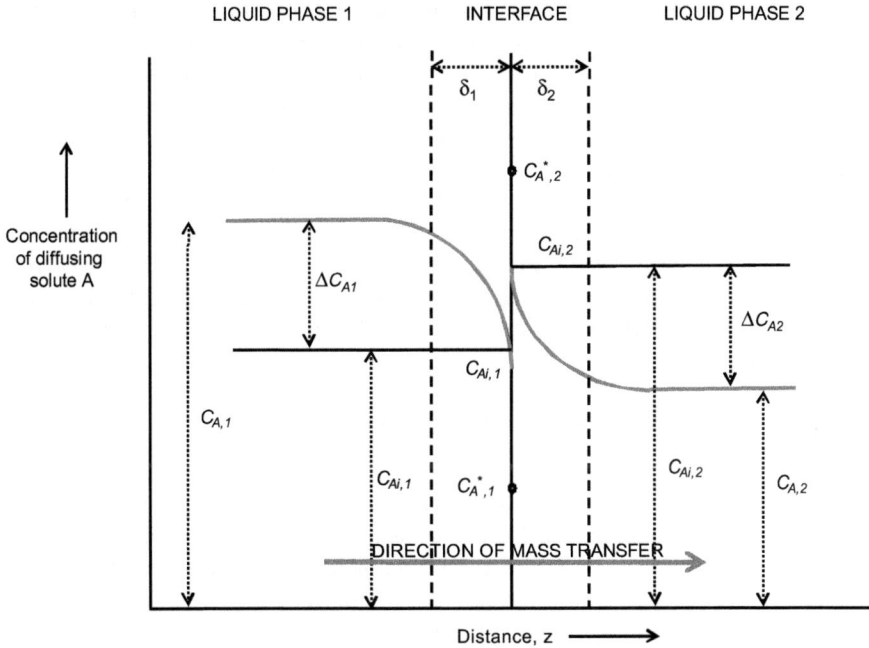

Figure 10.8   Schematic driving force diagram for interphase mass transfer in a typical liquid–liquid system.

$$N_A = k_2(C_{Ai,2} - C_{A,2}) \qquad (10.10)$$

for liquid phases 1 and 2, respectively, where k_1 and k_2 are the mass transfer coefficients in phases 1 and 2, respectively. Equating and rearranging gives eqn (10.11):

$$N_A = k_1(C_{A,1} - C_{Ai,1}) = -k_2(C_{A,2} - C_{Ai,2}) \qquad (10.11)$$

or eqn (10.12):

$$N_A = k_1 \Delta C_{A1} = -k_2 \Delta C_{A2} \qquad (10.12)$$

where ΔC_{A1} and ΔC_{A2} are the concentration driving forces, as illustrated in Figure 10.8 above. On this basis, the mass transfer coefficients are defined as the ratio of the molal mass flux, N_A, to the concentration driving force; they typically have units of kmol s^{-1} m^{-2} (kmol m^{-3})$^{-1}$ or kmol s^{-1} m^{-2} (mole fraction)$^{-1}$ in the case where concentration is given in terms of mole fraction. In all but the most simple of cases, mass transfer coefficients cannot be calculated from first principles and must be determined experimentally or, more commonly, estimated using correlations. A wide range of correlations exists, depending on the type and geometry of the mass transfer system; many of these can be found in the literature.[8] Such correlations relate the mass transfer coefficients to the physical properties of the liquid phases involved (such as diffusivities, viscosities and densities) and geometric parameters (for example, droplet diameters). In general, high diffusivities and low viscosities are conducive to high mass transfer coefficients. In addition, it is desirable to achieve a high interfacial area between phases to maximise the rate of mass transfer; this can be achieved by efficient dispersion, as described in the next section. Since liquid–liquid extraction in the pharmaceutical industry is often carried out batchwise, owing to the low volumes of liquids involved in the separation, unsteady state mass transfer is also an important consideration. In such cases, mass transfer is also dependent on liquid–liquid contact time and, therefore, adequate design of the extraction system can significantly reduce processing time. Batch extractions are typically carried out using a single-stage mixer-settler approach.

10.2.4   Phase Dispersion and Separation

Mixing of one liquid phase with another immiscible liquid phase normally leads to a dispersion of one liquid phase within the other; this is necessary in liquid–liquid extraction to achieve a high interfacial area and, thus, achieve efficient mass transfer. The liquid that is dispersed is normally referred to as the dispersed phase and the other phase is referred to as the continuous phase. Dispersions are generally unstable and when mixing or agitation is stopped, then the mixture will normally settle into two distinct phases. However, on occasions a stable emulsion may be formed where, once the agitation is removed, the phases do not settle out into two distinct phases within a reasonable time period. Unfortunately, when emulsions are encountered in processing they consume a lot of development time in terms of understanding and resolving the problem. Let us consider some useful heuristics that help to avoid or overcome emulsions in liquid–liquid extractions; the following may be useful tips for the reader:

- Low volumetric ratios of one of the phases can cause emulsions.
- The presence of emulsifying or surface active agents will influence the performance of the system and may cause the phase with the dissolved surface active agent to be the continuous phase.
- The higher viscosity phase may also tend to be the continuous phase.
- Selective wetting of the extraction vessel or its internals by one of the phases may cause this phase to be continuous.
- Any preferentially wetted solids present may cause one of the liquids to be become continuous.

10.3   Liquid–Liquid Process Development Considerations

10.3.1   Process Chemistry

For substitution reactions such as S_NAR, S_N1 and S_N2, Delhaye[9] has taken a pragmatic experimental approach of establishing the ability to remove dipolar aprotic solvents (DAS) from reaction and isolation mixtures. They have tabulated the performance of 10 DAS with five extracting solvents and also considered the interaction with water and a 10% w/w saline solution.

10.3.2 Process Heuristics

For selection of a suitable extraction solvent, a set of process heuristics has been developed to give some guidance to the developer. The steps that will help you identify a particular solvent are as follows:

1. Identify the feed components and the solute to be extracted.
2. Use miscibility data to identify solvents that are immiscible with the feed but miscible with the solute.[10]
3. Check the ICH classification of the solvent to ensure that it is acceptable.
   (a) Lowest possible toxicity is desirable. Environmentally acceptable.
   (b) The solvent should be stable at the process conditions in order to minimise losses by degradation and generation of further impurities.
4. For suitable solvents, evaluate or predict the selectivity.
   (a) A high selectivity is preferred. This factor will determine the number of extraction stages. The distribution or partition coefficient will affect the selectivity and the amount of solvent phase required.[11]
   (b) The greater the density difference, the easier it will be to obtain phase separation.
   (c) A low viscosity is desirable (<10 cP) and this will help material transfer and phase separation.
   (d) Interfacial tension influences settling, coalescence and the system material transfer coefficient. Coalescence and settling are normally aided by high interfacial surface tension whilst material transfer is hindered.
   (e) A solvent with low volatility is desirable due to the ease of removal by further process such as evaporation, distillation or drying.
   (f) Corrosivity will cause problems with equipment and process compatibility, and corrosive solvents should be avoided.
   (g) Some solvents suffer from high cost and at times low availability, such as acetonitrile and propionitrile, and these should also be avoided.
5. Predict the complete phase diagram for the remaining solvents to identify those likely to be adequate in respect of both selectivity and capacity.
6. Obtain experimental data to verify the previous step. Even small amounts of data are quite useful to check that your extraction strategy is correct.
7. If multiple extractions are required, carry out stagewise calculations and check for efficiency and any yield loss.

There is generally no solvent that meets all the criteria and a compromise is therefore necessary. Reduction in the overall process complexity should be high on the list of attributes as this helps with process robustness and supply chain efficiency.

A more detailed predictive approach to solvent selection is proposed by Gani[12] using a combination of the ICAS software, an estimation of the solvent's environmental acceptability and a selection chart for the likely reaction mechanism. If the ICAS software is available to the user, then a wider range of solvents can be identified and for simplicity a common solvent set has been identified that reflects those solvents regularly used in industry.

10.3.3 Predictive Screening of Solvents

The use of correlative and predictive tools to screen solvents for liquid extraction has become increasingly important in recent years. Experimental screening is both expensive and time consuming and, therefore, the use of computer simulations to reduce the search space can significantly reduce costs associated with the preliminary design of extraction processes. Typical pharmaceutical extraction systems involve non-ideal liquid mixtures for which activity coefficient models are best suited. These can be categorised as correlative models, which rely on the availability of a certain amount of experimental data, and predictive models, which are used in the absence of experimental data. Predictive approaches can be further sub-divided into semi-empirical methods, such as the universal functional activity coefficient (UNIFAC) model, and quantum-based methods, such as the conductor-like screening model for real solvents (COSMO-RS).

10.3.3.1 Correlative Activity Coefficient Models

Various correlative models have been developed to enable the estimation of ternary phase equilibria from binary data, such as the Wilson equation, NRTL (non-random two liquid) model and UNIQUAC (universal quasi-chemical) model. These are typically used in process simulation packages such as Aspen Plus for the calculation of thermodynamic data at different processing conditions. When sufficient data are available, such correlative models can be useful at the screening stage for solvent selection. However, since pharmaceutical mixtures often contain "exotic" components for which experimental thermodynamic data are limited, correlative approaches are typically only employed at the unit operation design stage after solvents have been selected and the necessary preliminary experimental data have been obtained. Detailed descriptions of these methods can be found elsewhere[13] and only a brief description of the UNIQUAC model is given here. In UNIQUAC, the activity coefficient of the ith component in a binary mixture is represented by combinatorial (denoted C) and residual (denoted R) terms in eqn (10.13):

$$\ln \gamma_i = \ln \gamma_i^C + \ln \gamma_i^R \qquad (10.13)$$

The combinatorial term accounts for differences in shapes between molecules and the resulting effects on deviation from ideality in solution, and is calculated using the van der Waals radii and surface areas of the pure chemicals. The residual term accounts for the changes in interacting forces between different molecules upon mixing and it depends on the determination of empirical interaction parameters by fitting against experimental data. However, with adequate binary phase equilibria data for the components involved, the UNIQUAC model allows the determination of ternary phase equilibria data over a range of conditions and is used extensively in detailed process design.

10.3.3.2 UNIFAC

At the early stages of pharmaceutical extraction process development, particularly at the solvent selection stage, often very little experimental thermodynamic data are known for the range of possible binary or ternary systems. Therefore, correlative models are of little use and the process development chemists and engineers must rely on predictive tools. The most commonly used predictive activity coefficient model is UNIFAC,[14] which is similar to the UNIQUAC model described previously except that the volume, surface area and interaction parameters for the molecules involved are determined from previously determined group contribution parameters rather than experimental data. This group contribution method (GCM) assumes that the interactions between the molecules present in solution can be adequately represented by the interactions between the functional groups which make up those molecules. Therefore, only the interaction parameters between the different functional groups need be known. Many of these group interaction parameters have been determined previously by fitting vast quantities of phase equilibria data and are available in published tables and process simulation software. This allows the phase behaviour of mixtures involving a wide array of components to be predicted without any direct experimental data and, therefore, is very useful in screening a wide range of solvents for an extraction process. A number of case studies will be investigated later in the chapter to illustrate the usefulness of UNIFAC in developing a process design strategy. Whilst it is somewhat less accurate than the correlative approaches and would not be considered suitable for detailed process design, such a predictive screening approach allows the search space to be significantly reduced and can generate a small list of potential solvents which can be further investigated experimentally. The main disadvantage of the UNIFAC approach in pharmaceutical process development is that group interaction parameters for some of the less common functional groups present in APIs are not available.

10.3.3.3 COSMO-RS

More recently, the predictive thermodynamic model COSMO-RS,[15] based on the quantum COSMO continuum solvation model,[16] has been increasingly used in the prediction of the phase behaviour of liquid mixtures in the pharmaceutical industry. An excellent overview of the COSMO-RS method is given by Klamt and Eckert,[17] so a relatively brief explanation is given here. The method consists of two main parts: the COSMO quantum calculation, which is performed separately for each individual species, and the COSMO-RS statistical thermodynamics calculation, which determines the electrostatic interactions between the individual components, and hence, the phase behaviour of the mixture. The COSMO quantum model is essentially a continuum solvation model (CSM), which describes the quantum chemistry of a molecule within the dielectric continuum of a surrounding ideal conductor with dielectric strength $\varepsilon = \infty$. This approximation significantly improves the efficiency of computation, while maintaining an accuracy of within 0.5% for solvents with strong dielectrics, such as water ($\varepsilon = 80$), and within 10% for very weak dielectrics solvents, such as hexane ($\varepsilon = 2$). While the COSMO model itself is a useful tool in understanding the behaviour of pure components, it does not consider the effects of temperature, nor does it consider the behaviour of mixtures. The extension of this model, COSMO-RS, is based upon the concept of interaction energy between neighbouring molecules in a mixture. For this model, the virtual conductor environment described in COSMO is considered to be a reference state for molecules in solution. A liquid is considered to consist of closely packed ideally screened molecules, with each piece of surface directly contacting another piece of surface on a neighbouring molecule. A major benefit of the COSMO-RS thermodynamic model over GCMs is its consideration of longer range molecular interactions such as hydrogen bonding.

One of the main advantages of COSMO-RS is that it can be used to predict phase equilibria in systems where the UNIFAC interaction parameters are not available. One such system is the Suzuki–Miyauri reaction, discussed later in Section 10.4.3. This system involves boron-containing species, such as 4-carboxyphenylboronic acid, for which no UNIFAC interaction parameters exist. In Table 10.4 (Section 10.4.3), the miscibilities of the various components in the system were predicted using COSMO-RS. As illustrated in Figure 10.9, COSMO-RS is a very useful qualitative tool in predicting the liquid–liquid or solid–liquid equilibria of API/solvent systems. The ternary LLE predicted by COSMO-RS for the very commonly encountered system of acetonitrile/toluene/water is shown. Importantly, the trends in the order of solubility of different solvents and the shapes of the solubility–temperature curves can be adequately predicted for starting materials, reagents and products and various examples are reported in the literature.[18]

10.3.4 Unit Operation Design and Laboratory Testing

From two simple tests it is possible to gain an effective understanding of the liquid–liquid performance of the system. These tests can highlight the agitation speed to operate the system, which system to be dispersed and how long phase separation should take.

Start with solvent A and add solvent B; increase the agitation speed until there appears to be no second layer remaining on top of the first layer, that is,

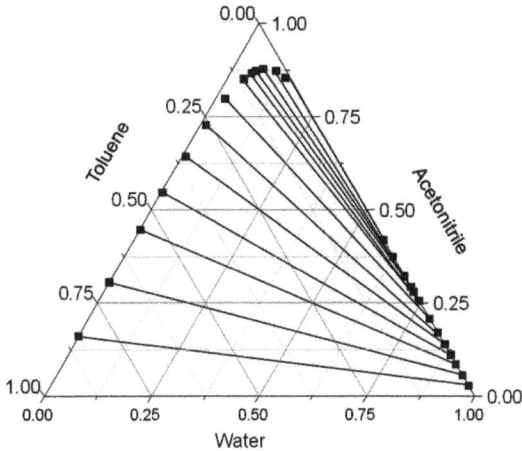

Figure 10.9 Ternary phase equilibria for the toluene/acetonitrile/water system predicted using the COSMO-RS model.

B is fully dispersed in A. Repeat the previous step but reversing the addition. There can be major differences in the agitator speed required to fully disperse the material. Also it is worth noting if one of the systems gave smaller dispersed droplets. The agitation speed can be scaled up on power per unit volume to predict the speed required in future operations.

Measurement of the phase separation time for both the previous tests will also give an indication of the best addition strategy. The separation time can be crudely scaled up on droplet settling velocity to predict the likely separation time in future operations.

The same test may be used to check the time to reach an acceptable end point or even equilibrium by sampling and analysing at frequent intervals, possibly every 5 or 10 minutes.

10.3.4.1 Stage Contacting of Liquid–Liquid Extraction Systems

For continuous extraction of a relatively dilute solute of two immiscible phases and where F is the feed and G is the extracting phase, then an estimate of the number of contacting stages can be made quite simply by the following process.[19]

Material balance on any stage in the process:

$$Gx_1 + Fy_1 = Gx_0 + Fy_2 \tag{10.14}$$

From eqn (10.18) (below), the equilibrium relationship can be written in its simplest form as:

$$y_n = mx_n \tag{10.15}$$

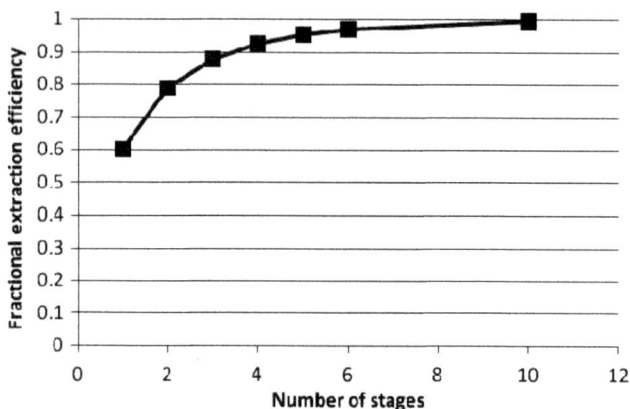

Figure 10.10   Typical fractional extraction efficiency *versus* number of stages plot for the recovery of an active material.

Combining the two previous equations and in the case where $x_0 = 0$ gives eqn (10.16):

$$y_2 = (1 + G/mF)y_1 \tag{10.16}$$

Defining an extraction factor, E, as equal to mF/G, then eqn (10.16) can be written as:

$$y_2 = (1 + 1/E)y_1 \tag{10.17}$$

For N stages the result can be extended to eqn (10.18):

$$y_{N+1} = (1/E^{N+1} - 1)y_1/(1/E - 1) \tag{10.18}$$

This result allows an estimate of the number of equilibrium stages required to achieve a desired separation. The equations, however, do assume that the efficiency of extraction is close to 100%.

Equation (10.18) is not very intuitive for process chemists and a suggested simpler approach when presenting these types of data for a system where $G = 12F$ and $m = 0.125$ is demonstrated in Figure 10.10. These values are typical of what might be encountered for benzoic acid from toluene to water, which will be examined in more detail in a later case study.

10.4   Case Studies in Liquid–Liquid Extraction

10.4.1   Liquid–Liquid Extraction and Process Safety

Process safety personnel are often challenged with having to use liquid–liquid extraction to test the stability of various intermediates or final products or their residues. The challenge is then to produce concentrated solutions for safety testing without causing the materials of interest to degrade by concentrating the process. An example of how this can be a problem if the phase equilibria are not fully understood is covered in the next section.

10.4.1.1 Distillation Residue Stability: 2-Methyltetrahydrofuran/ Water/Methanol Mixture

Distillation processes can be potentially hazardous, as over-distillation can leave residues prone to exothermic decomposition at the still service temperature. Generally, an investigation of the thermal stability of the residues is undertaken. A sample is generated at low pressure and temperature to eliminate exposure to elevated temperatures and thus reduce the risk of thermal decomposition prior to thermal stability testing. Unfortunately, in doing this the vapour–liquid equilibria (VLE) boundaries/separatrices can change and it is possible that the residue could have a significantly different composition to that of an atmospheric distillation, as was the case in this particular instance.

A limited amount of physical properties data is available in the literature[20] and these are summarised in Table 10.3.

The approximate composition of the reaction liquid phase, not including the reaction product(s), prior to the start of distillation was 27.8% w/w methanol, 61.9% w/w 2MeTHF and 10.3% w/w water.

The initial reaction sample was distilled down to a limited volume under reduced pressure (5 Torr) and resulted in a two-phase mixture (organic and aqueous phases). This phase behaviour was not encountered when carrying out the distillation under atmospheric pressure, whereby water was easily removed. It was obvious that the VLE boundary had changed under the reduced pressure conditions to favour concentration of water rather than removal.

SMSWin[21] and ProPred[22] were used to predictively model the ternary system using the UNIFAC method. The ternary diagram for the system including predicted LLE is shown in Figure 10.11. For successful removal of water, it was shown to be necessary to add an appropriate amount of anhydrous 2MeTHF to move the composition into the region of the diagram where the 2MeTHF was the stable node. This would ensure the effective removal of water from the system to below the specified target.

The two-phase LLE region is also shown on the ternary diagram (Figure 10.11). There was a lower level of confidence in the accuracy of this LLE region and if necessary a more detailed model would have had to be generated.

Table 10.3 Physical properties of the 2MeTHF/water/methanol system.

	Normal boiling point (°C)	Liquid density (g mL⁻¹ @ 20 °C)
Water	100	1
Methanol	65	0.791
2MeTHF	79	0.855

Azeotropic composition of 2MeTHF/water = 89.4/10.6 wt%
Azeotropic boiling point of 2MeTHF/water = 71.7 °C
2MeTHF solubility in water = 14% w/w at 20 °C
Water solubility in 2MeTHF = 4.4% w/w at 20 °C

Predicted using: VLE UNIFAC
Mole fractions. System pressure: 760.0 mmHg

(3) Methanol (64.94°C)

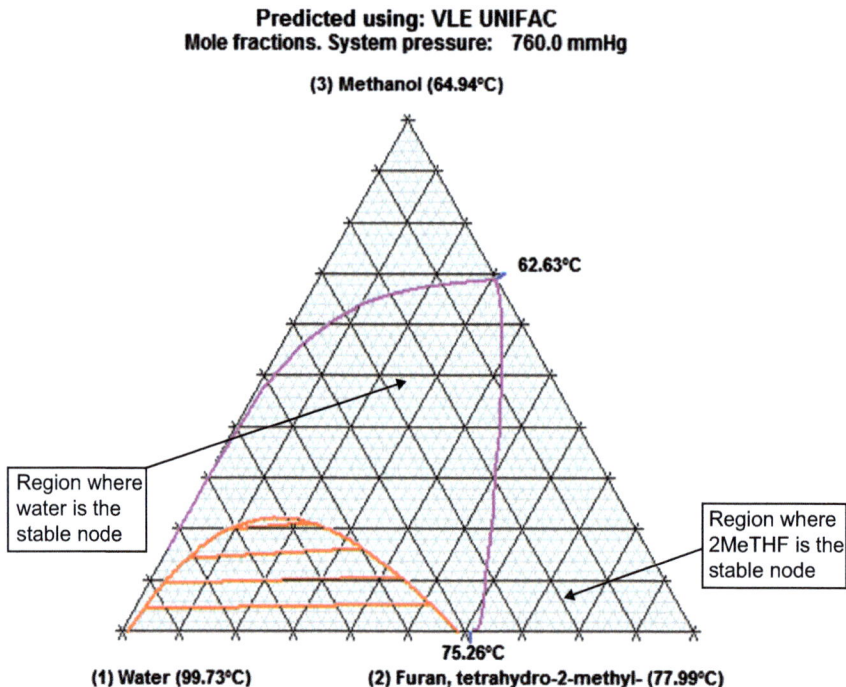

62.63°C

Region where water is the stable node

Region where 2MeTHF is the stable node

75.26°C

(1) Water (99.73°C)   **(2) Furan, tetrahydro-2-methyl- (77.99°C)**

Figure 10.11   Predicted phase equilibria for water/2MeTHF/methanol.

However, the reduced distillation was attempted with the additional 2MeTHF added and was shown to work by analysis of residual water in the residue.

The pink lines shown are the separatrices (a separatix is an equation to determine the borders of a system). Once within a region shown on the diagram, it is not possible to cross these lines except by adding material from an external source.

Understanding the VLE/LLE equilibria when carrying out low-temperature, low-pressure distillations to simulate the residues that would be encountered during atmospheric is therefore an important consideration. It may require appropriate changes to the initial composition prior to the distillation to ensure the correct final composition is achieved in the final sample.

10.4.2   Toluene/Water Separation during the Manufacture of Benzoic Acid

This example will be used to exemplify some of the principles outlined earlier. Benzoic acid can be manufactured by the aerial oxidation of toluene using a metal catalyst such as cobalt acetate at elevated temperatures and pressures, typically 136–160 °C and 2–7 bar. The reaction scheme is given below (Scheme 10.1) and typical incomplete and parasitic reactions occur, to give a mixture with the following components: benzyl alcohol, benzaldehyde, benzyl

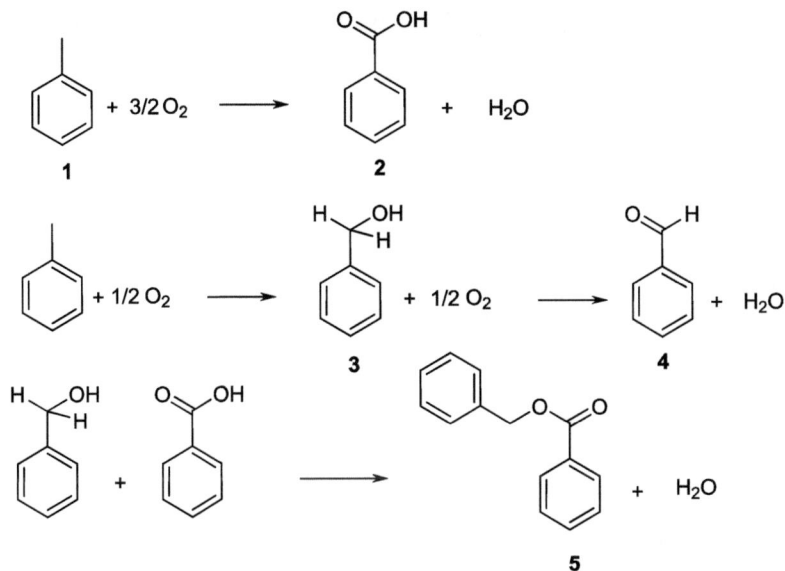

Scheme 10.1 Overall reaction scheme for the manufacture of benzoic acid and the stepwise reaction and formation of by-products.

benzoate, acetic acid, formic acid and other organics in smaller quantities. Water is produced as a by-product and is immiscible with the toluene phase. Reaction concentrations of benzoic acid can be typically in the region of 10–60% w/w. Some of the main reactions are shown in Scheme 10.1.

The water may be removed from the system by distillation of the water/ toluene azeotrope and subsequent condensation of the vapour to give toluene-rich and water-rich phases. The toluene can be returned to the reaction to allow the further manufacture of benzoic acid, whilst the water phase can be sent either for further purification or for safe disposal. After removal of water it is normal to continue with the distillation process to remove the impurities. For pharmaceutical grade benzoic acid, a recrystallisation is normally undertaken from methanol.

A preliminary distillation step will remove water, toluene, acetic acid and formic acid as indicated in Table 10.4. Further impurities such as acetic acid and formic acid may be removed by extraction using water, leaving a toluene solution rich in benzoic acid, benzyl alcohol, benzaldehyde and benzyl benzoate.

To manufacture USP grade benzoic acid, extraction using water could be utilised. However, typical solubilities of benzoic acid at 25 °C are as follows: in water 3.4 g/L and in toluene 85.5 g/L.[23] Most of the benzoic acid will be present in the toluene phase. The phase diagram in Figure 10.12 gives a prediction of the behaviour of the system. If the prediction is assumed to be correct, and this should always be checked, it is possible to carry out a preliminary design. A multistage extraction would be required and the likely efficiency and number of stages are shown in Figure 10.12.

Table 10.4   Predicted binary azeotrope temperatures (°C) for the manufacture of benzoic acid during the stripping phase of the reaction at a pressure of 1 bar using UNIFAC, with log P values at 25 °C in parentheses.

Component	Toluene						
Water	84.1	Water					
Benzoic acid (1.9)	None	99.8	Benzoic acid				
Benzyl alcohol (1.1)	None	99.3	None	Benzyl alcohol			
Benzaldehyde (1.48)	None	97.7	None	None	Benz- aldehyde		
Benzyl benzoate (4.0)	None	99.7	None	None	None	Benzyl benzoate	
Acetic acid $(-0.17)^{a}$	103.2	98.4	None	None	None	None	Acetic acid
Formic acid $(-0.41)^{b}$	85.0	98.7	None	None	None	None	None Formic acid

^{a}Acetic acid/toluene/water ternary at 84.0 °C.
^{b}Formic acid/water/toluene ternary at 83.4 °C.

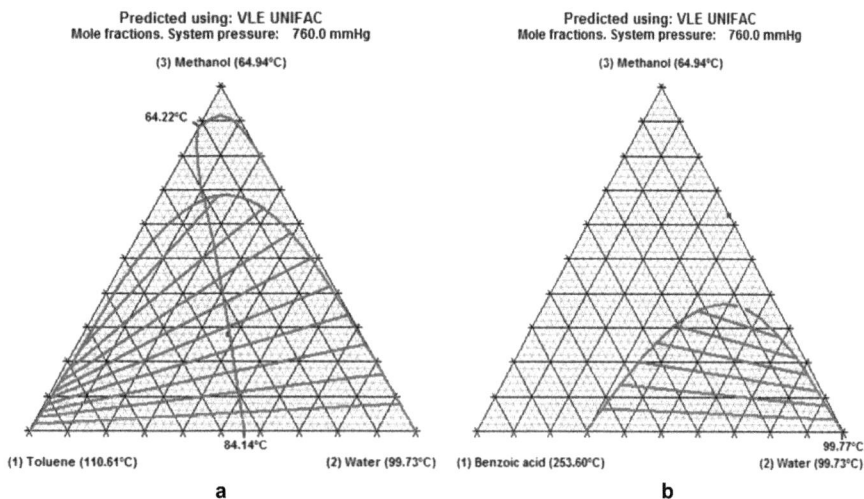

Figure 10.12   Ternary phase equilibria for (a) methanol/toluene/water and (b) methanol/water/benzoic acid.

Adjusting the water extracting phase to improve the efficiency of extraction whilst maintaining an acceptable two-phase system can be achieved using methanol. The distribution of methanol between the water and toluene phases is shown in Figure 10.12(a) The solubility of benzoic acid in methanol is 71.5 g per 100 g at 23 °C. Figure 10.12(b) shows the predicted distribution that might be expected for the water/methanol/benzoic acid system and indicates that there is only a small area where two-phase activity may be encountered.

This example demonstrates that with a restricted amount of physical property data it is possible to use a simple process analysis system to understand how a preliminary design may be carried out to work up the active ingredient from a reaction mixture.

10.4.3 Analysis of the Suzuki–Miyauri Reaction

The Suzuki–Miyauri reaction (Scheme 10.2) it is often carried out with a small excess of the haloaromatic compound as this is normally the less expensive material. If the base is a carbonate, then carbon dioxide is liberated from the system. The metal catalyst, typically palladium, may require scavengers for its removal if this step is late in the synthesis. The impurities formed are normally as shown in Scheme 10.3 and result from dimerisation (typically homocoupling) and loss of the side chain (typically deborination, dehalogenation).

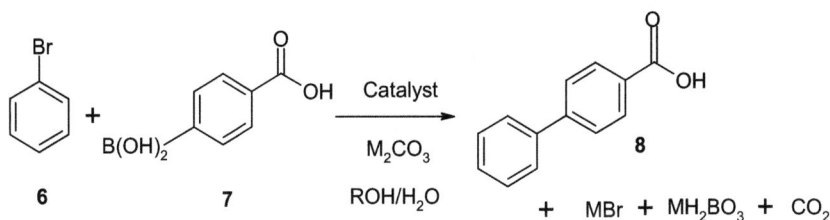

Scheme 10.2   The Suzuki–Miyauri reaction.

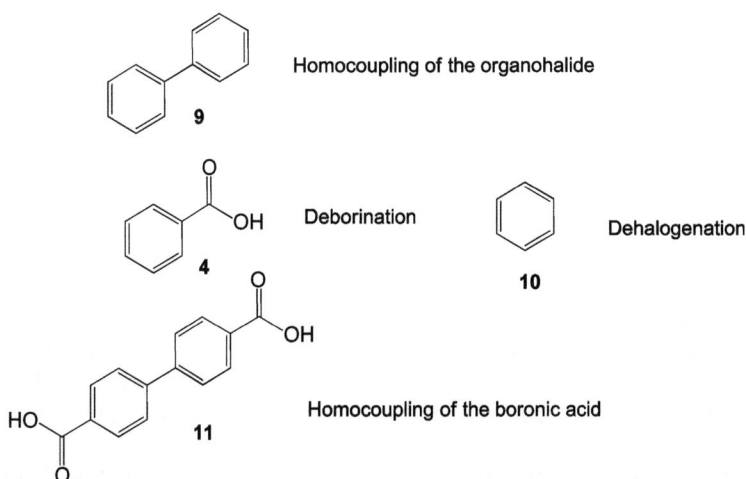

Scheme 10.3   Typical impurities formed in the Suzuki–Miyauri reaction.

Table 10.5 Miscibilities of the major components and impurities in the Suzuki–Miyauri reaction predicted using COSMO-RS.

Component	EtOH							
H_2O	yes	H_2O						
10	yes	no	10					
9	low	no	yes	9				
4	yes	no	no	n/a[a]	4			
8	yes	no	no	n/a	n/a	8		
11	yes	no	no	n/a	n/a	n/a	11	
6	yes	no	yes	yes	no	no	no	6
7	yes	no	no	n/a	n/a	n/a	n/a	no

[a]n/a = both materials are solid at room temperature.

Table 10.6 Azeotropes of the major components and impurities in the Suzuki–Miyauri reaction at atmospheric pressure (in °C).[a]

Component	EtOH							
H_2O	78.24	H_2O						
10	67.82	69.01	10					
9	80.8	99.45	none	9				
4	none	99.8	none	246.9	4			
8	none	none	n/a	n/a	n/a	8		
11	none	none	none	n/a	n/a	n/a	11	
6	80.77	95.31	none	n/a	none	n/a	n/a	6
7	n/a	n/a	n/a	n/a	n/a	n/a	n/a	n/a	7

[a]Note: there are also ternary azeotropes predicted as follows: benzene/water/ethanol at 64 °C and bromobenzene/water/ethanol at 79.0 °C.

Other by-products and wastes include inorganic compounds such as sodium bromide, sodium borate and carbon dioxide.

A simple method to start the process of solvent selection is to examine the incompatibilities of solvents. Tables 10.5 and 10.6 can be drawn up to show some of the miscibilities and form the start of a phase behaviour picture.

Note that various azeotropes are predicted with the carboxylic acids; however, the compositions in terms of carboxylic acids are very low and therefore for all intents and purposes these can be ignored. If the reaction is carried out batchwise, then it would be possible to remove benzene from the reaction mass by distillation and this would remove the potential for contamination with a genotoxic component. Using a toluene/water mixture for extraction leads to the phase behaviour shown in Figure 10.13. This type of system will allow separation of the polar components into the water phase and non-polar components into the toluene phase.

There is a ternary azeotrope predicted with a bubble point of 73.6 °C (Figure 10.14).

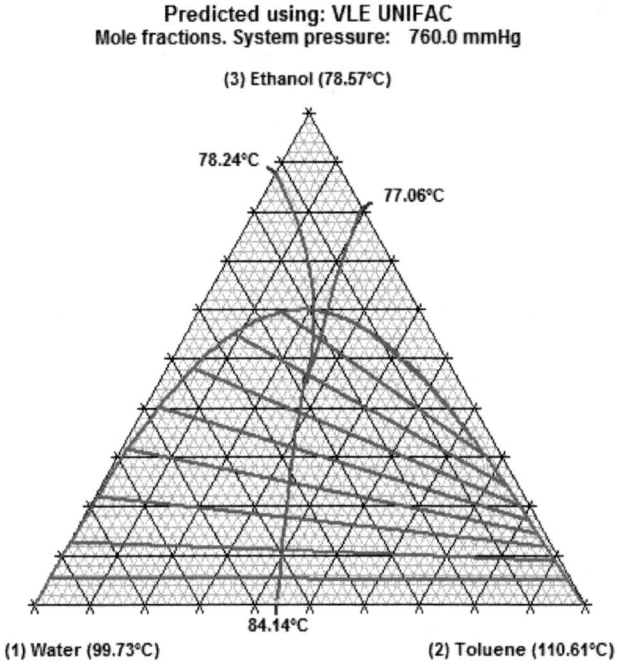

Figure 10.13 Ternary phase equilibria for the toluene/water/ethanol system.

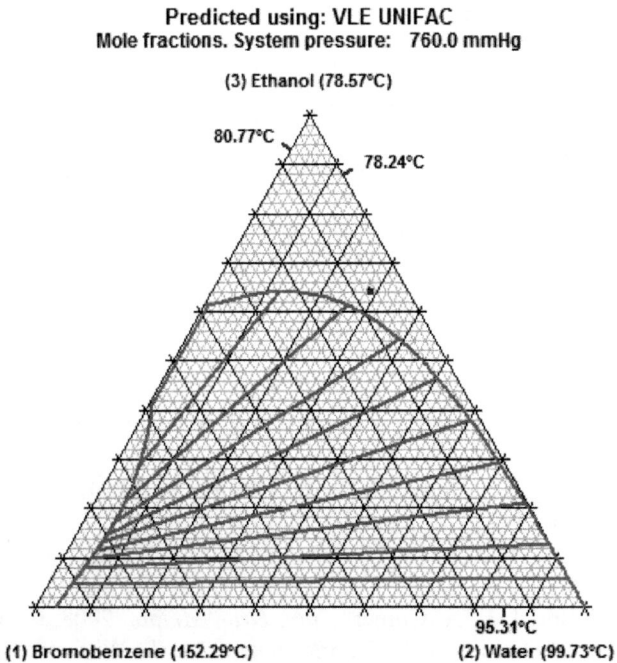

Figure 10.14 Ternary phase equilibria for bromobenzene/ethanol/water.

10.5   Summary

This chapter has been written to cover pragmatically the most common aspects of liquid–liquid extraction that a process developer may encounter on a day-to-day basis. More detailed information is available on the more demanding areas of liquid–liquid extraction elsewhere and the reader can develop their own expertise by further research and practice in this area. By using some simple physical property data such as log P, it has been demonstrated how process design strategies may be developed for work-up post the reaction phase. For more complex systems the use of property methods such as UNIFAC for predictive liquid–liquid extraction performance has been used in some worked examples. More modern physical property methods such as COSMO-RS are introduced and this method will undoubtedly become more common in the future.

   To close the gap between chemists and chemical engineers, the chapter has referred to the most common reactions that, historically, are experienced in process development in the fine and pharmaceutical industry and their likely by-products. An introduction into binary and ternary liquid–liquid systems with a potential work flow approach using regulated solvents may be chosen to ensure an efficient and effective separation. Some simple calculation methods are recommended to examine the effect of multistage extraction. Finally, some information is given on how to evaluate liquid–liquid extractions practically in the laboratory.

Acknowledgements

Professor R. Gani (DTU) and his team, who gave his permission to use the output from SMSWin, as this allowed the authors to concentrate on the main learning points associated with use of liquid–liquid extraction in process development. The talent and insight of Jim Morrison, who constructed SMSWin and the attendant philosophy, which has left a major footprint in chemical engineering history. Dean Woods (QUB) and Dr Gavin Walker (QUB), for the useful feedback and discussion provided on various aspects of the chapter.

References

1. J. Van de Vegte, *Feedback Control Systems*, Prentice Hall, Upper Saddle River, NJ, 1986, pp. 316–339.
2. C. A. Lipinski, F. Lombardo, B. W. Dominy and P. J. Feeney, *Adv. Drug Delivery Rev.*, 1997, **23**, 3.
3. S. K. Bhal, K. Kassam, I. G. Peirson and G. M. Pearl, *Mol. Pharmacol.*, 2007, **4**, 556.
4. K. J. Box and J. E. A. Comer, *Curr. Drug Metab.*, 2008, **9**, 869.
5. J. S. Carey, D. Laffan, C. Thomson and M. T. Williams, *Org. Biomol. Chem.*, 2006, **4**, 2337.

6. R. W. Dugger, J. A. Ragan and D. H. Brown Ripin, *Org. Process Res. Dev.*, 2005, **9**, 253.
7. R. E. Treybal, *Mass-Transfer Operations*, McGraw-Hill, New York, 1980.
8. D. W. Green and R. H. Perry, *Perry's Chemical Engineers' Handbook*, McGraw-Hill, New York, 8th edn, 2007.
9. L. Delhaye, A. Ceccato, P. Jacobs, C. Kottgen and A. Merschaert, *Org. Process Res. Dev.*, 2007, **11**, 160.
10. N. B. Godfrey, *Chem. Technol.*, 1972, **2**, 359.
11. C. Hansch, J. E. Quinlan and G. L. Lawrence, *J. Org. Chem.*, 1968, **33**, 347.
12. R. Gani, C. Jimenez-Gonzalez, A. ten Kate, P. A. Crafts, M. Jones, L. Powell, J. H. Atherton and J. L. Cordiner, *Chem. Eng.*, 2006, **113**(3), 30.
13. (a) H. Renon and J. M. Prausnitz, *AIChE J.*, 1968, **14**, 135; (b) D. S. Abrams and J. M. Prausnitz, *AIChE J.*, 1975, **27**, 116.
14. J. G. Gmehling, T. F. Anderson and J. M. Prausnitz, *Ind. Eng. Chem. Fundam.*, 1978, **17**, 269.
15. A. Klamt, *J. Phys. Chem.*, 1995, **99**, 2224.
16. A. Klamt and G. J. Schuurmann, *J. Chem. Soc., Perkin Trans. 2*, 1993, 799.
17. F. Eckert and A. Klamt, *AIChE J.*, 2002, **48**, 369.
18. A. Klamt, F. Eckert, M. Hornig, M. E. Beck and T. Burger, *J. Comput. Chem.*, 2002, **23**, 275.
19. E. L. Cussler, *Diffusion-Mass Transfer in Fluid Systems*, Cambridge University Press, Cambridge, 2nd edn, 1997.
20. Methyltetrahydrofuran, PENN Speciality Chemicals, 2004, www.pschem. com (last accessed December 2010).
21. SMSWin, CAPEC, www.capec.kt.dtu.dk (last accessed December 2010).
22. ProPred, CAPEC, www.capec.kt.dtu.dk (last accessed December 2010).
23. (a) M. J. Chertkoff and A. N. Martin, *J. Am. Pharm. Assoc.*, 1960, **49**, 444; (b) R. F. Pires and M. F. Franco Jr., *J. Chem. Eng. Data*, 2008, **53**, 2704.

CHAPTER 11

Development Enabling Technologies

MIKE J. MONTEITH AND MARK B. MITCHELL

GlaxoSmithKline, Five Moore Drive, Research Triangle Park, NC 27709, USA

11.1 Introduction

Through the last 30 years of process development in the pharmaceutical industry the widespread development and implementation of enabling technologies has become commonplace, and no longer restricted to the realms of "early adopter" specialist groups, finding routine application in all phases of the development cycle. A number of factors may lie behind this increased uptake; for example, increasing pressure to deal with more drug candidates from technologically advanced discovery organisations and the increasing need to get products to market in a shorter cycle time whilst still having well-developed, cost-efficient, sustainable manufacturing processes. Additionally, there has been a general, well-documented, need to increase productivity across the industry, resulting in higher workloads with less resource as the number of new chemical entities (NCEs) being approved has fallen commensurate with the decrease in R&D productivity.[1]

During the last five years, the regulatory landscape has also undergone a major change with the implementation of the U.S. Food and Drug Administration's (FDA's) process analytical technology (PAT) initiative. This has affected the way companies view the development process, with an increasing focus on the application of technologies to ensure that processes are robust, well controlled and sources of variability are well defined and managed.[2]

RSC Drug Discovery Series No. 9
Pharmaceutical Process Development: Current Chemical and Engineering Challenges
Edited by A. John Blacker and Mike T. Williams
© Royal Society of Chemistry 2011
Published by the Royal Society of Chemistry, www.rsc.org

That is, the move from optimising all individual parameters to fundamentally understanding interactions, minimising variations and allowing some flexibility in inputs. As part of this, a renewed focus on technology has been one response from the industry to address the concepts being described by the FDA and worldwide regulatory agencies.[3]

11.1.1 Process Enabling Technologies

Process enabling technologies encompass a gamut of methodologies (hardware, software and workflow related) and, as the various factors listed above have impacted the industry, the commercial availability of equipment with more user-friendly interfaces has increased. Technology enables reactions to be explored to a greater level of detail and allows scientists to rapidly investigate a wider range of synthetic transformations and/or different processing conditions whilst simultaneously using predictive tools to increase diversity within studies. The result has been increased understanding and control of the chemical processes under development.

A key component to leveraging the potential of various technologies is the application of statistical experimental design. This is now well established as a tool in the arsenal of techniques applied to pharmaceutical process development in both the active pharmaceutical ingredient (API) and the drug product side of the industry. As such, this represents a dramatic shift in the industry's way of working over the last decade. At the start of the 21st century, experimental design was little more than a mathematical oddity, mostly ignored by the bench scientist,[4] whereas today it is routinely applied to all aspects of the API development process: reaction screening, optimisation, critical parameter identification and process response surface modelling. Moreover, experimental design is increasingly part of a holistic approach to attain fundamental process knowledge; the real power is in the synergy that stems from use in parallel experimentation workflows in combination with other techniques, including kinetic analysis, mechanistic interpretation and application of visualisation and multivariate statistical tools. The methodology behind statistical experimental design has been well reviewed and it is not our intent to repeat this here, other than to indicate where it has been applied.[5,6]

In dealing with technologies applied to development we have attempted to provide some historical perspective, given that several of the key concepts and developments from around the year 2000 are still very relevant. From there we have segmented technology into three development phases, namely synthetic route investigation, process optimisation and finally process verification, before progressing into process intensification.

11.2 Technology Applied to Parallel Experimentation: Pre-2000

Initial developments had their foundation in equipment commonly used in high-throughput discovery activities in the mid 1980s, such as Zymark or variants.[7] These early attempts focused on automation of the synthetic process but had significant opportunities to pursue further automation. For example, the early platforms were solution dispensing centric and not geared toward handling solids.[8,9] Some authors, in 2000, had realised this limitation and were actually progressing some initial equipment development.[10] However, significant development, detailed later, has occurred in this field in the last 10 years.

In early publications, groups at Glaxo Wellcome and SmithKline Beecham recognised the opportunity to automatically prepare samples, analyse reactions and develop equipment to maintain inert atmospheres, which resulted in the development of the SK233 Reactarray platforms.[11,12] The commercial availability of these relatively small-scale kits opened a significant opportunity in process development laboratories. The ability to automate experiment set-up, reactions and analysis lent itself extremely well to the idea of statistical design of experiments (DoE) applied to pharmaceutical process development. The Glaxo Wellcome group has authored notable contributions to the field.[13,14] Since these early publications, statistical DoE is now routinely applied to all aspects of the development process, whether this involves automation or not.[5,6] At around the same time, some groups were adapting workflows to slightly larger equipment more designed as a means to evaluate process scale-up with better parameter control, such as Automate (Hazard Evaluation Laboratory), Multimax (Mettler Toledo) and Flexylab (Systag).[15] Updated reviews for this phase highlight the diversity and range of equipment commercially available or used in industrial laboratories.[16,17] On a larger scale, some researchers had automated larger equipment to more closely mimic production conditions.[10,18]

In all of these activities it was clear that researchers were utilising technology to help answer questions relevant to their phase of development, whether the questions be synthetic route discovery, process development or later stage process verification. In all cases the equipment was designed to operate several concurrent reactions with, or without, associated HPLC analysis, thereby offering the opportunity to accelerate the development process by generating extra data, examining different alternatives for new reactions, or evaluating a wider array of processing conditions. Around that time, the analytical and data manipulation load was perhaps starting to be identified as a potential gap and significant advances have occurred in this field in the last 10 years.[19]

Since approximately 2000, notable advances have occurred in the development and application of technology to other aspects of the pharmaceutical development process, such as version and form screening.[20,21] Technology implementation to small-scale reaction screening has also seen marked progress in a similar time frame, such that integrated parallel reactions are possible on a much smaller scale.[22] Later stage development has seen marked developments in the application of technology, in part in response to increased regulatory expectations and, in part, due to new and improved technologies.[3,23]

11.3 Developments Since 2000

11.3.1 Enabling Technologies at Route Scouting/Screening Stage of API Development

The start of the pharmaceutical development process is typically reaction screening/route scouting and it is in this area that much progress has been made in recent years. Pfizer was amongst the first to implement high-throughput experimentation incorporating automated solid and liquid handling to evaluate typically 100 reactions at a time at 1 mL scale in crimp-sealed HPLC vials. Their workflow incorporates a mix of in-house and commercial software tools to create designs, implement the experiments and then rapidly extract raw HPLC data into the visualisation package Spotfire (Figure 11.1). Using this technique, conditions to effect the Horner–Wadsworth–Emmons (HWE) reaction with good regiocontrol were readily identified (Scheme 11.1).[24]

A similar workflow was established by Avantium for parallel hydrogenation screening, combining commercial solid and liquid handling robotics with in-house hydrogenation tools, the QuickScreen (QS) and High Pressure Unit (HPU) blocks (Figure 11.2).[25] Using this equipment, rapid and routine screening

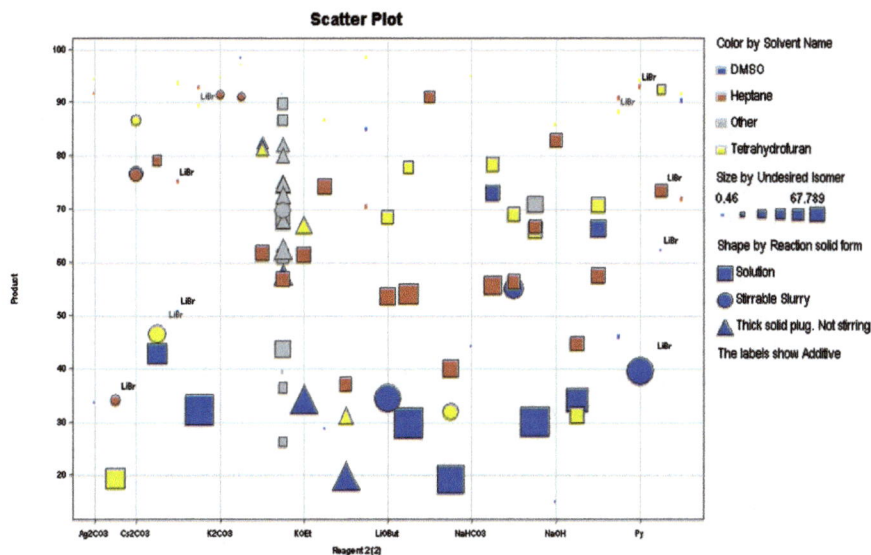

Figure 11.1 Spotfire visualisation of data for screening of conditions for the HWE reaction (reproduced with American Chemical Society permission).

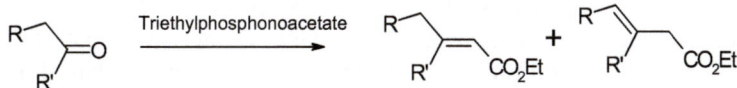

Scheme 11.1 The HWE reaction.

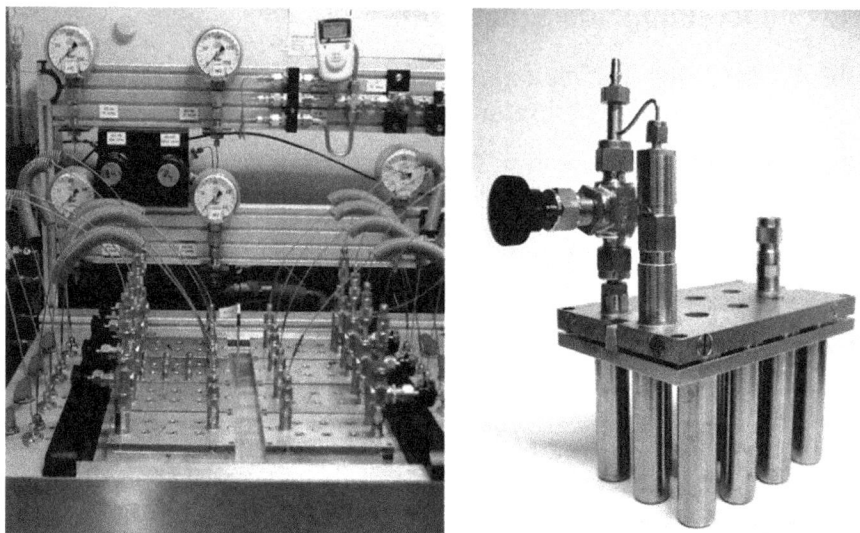

Figure 11.2   The Avantium QS workstation (reproduced with American Chemical Society permission).

of hydrogenation conditions with up to 192 reactions at a time was demonstrated and offered as a commercial service. This was applied to evaluate challenging chemoselective nitro group reduction for three different Pfizer intermediates, allowing one route to be immediately discarded and effectively identifying conditions (solvent, catalyst, additive, temperature, pressure) for the others.[26]

More recently, Symyx has offered an enterprise solution for large organisations to effect screening workflows based upon its Library Studio, Renaissance and Intellichem software suite. Experiments are typically performed in microtitre format glass vials utilising an Autodose Powdernium solid handling robot and Cavro liquid handling/reaction workstation (Figure 11.3).

Merck, for instance, has utilised this technology in unison with statistical design and kinetic profiling to identify effective conditions for a challenging Suzuki coupling of a 5,7-dichloro-1,6-naphthyridin-2-one with 2,4-difluorophenylboronic acid, obtaining a product with excellent conversion (97%) and regiocontrol (>97%) under their optimised conditions (Scheme 11.2).[27]

The workflow can also incorporate high pressure (HiP) reactor blocks, and Merck has reported catalyst screening for a number of hydrogenation reactions using this approach, generating key intermediates for sitagliptin, laropiprant and taranabant, for example.[28–30] The catalyst screening approach enabled the identification of mild conditions for the hydrogenation of unprotected enamino esters and amides using commercially available ligands, demonstrating that there was no requirement for *N*-acyl protection as previously believed. Indeed, hydrogenation of the intermediate enamino amide (Scheme 11.3) with t-Bu-Josiphos as ligand is now employed for the commercial manufacture of sitagliptin by Merck.

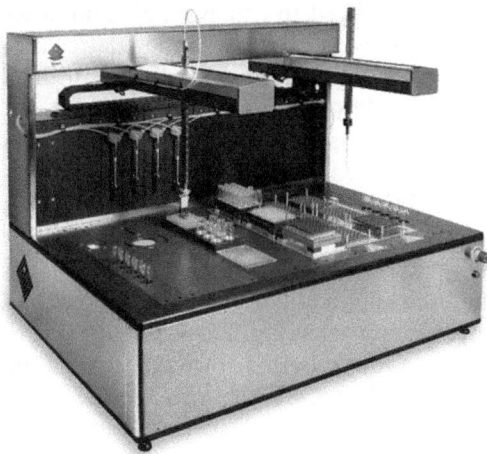

Figure 11.3 The Symyx Cavro workstation (reproduced with permission from Elsevier).

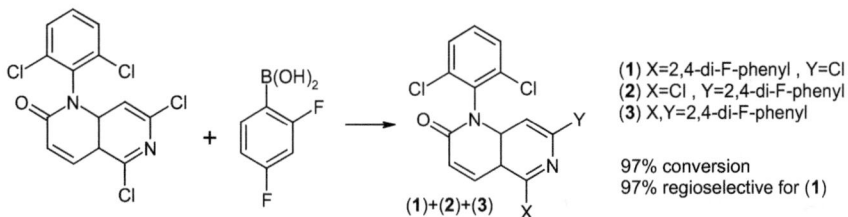

(1) X=2,4-di-F-phenyl , Y=Cl
(2) X=Cl , Y=2,4-di-F-phenyl
(3) X,Y=2,4-di-F-phenyl

97% conversion
97% regioselective for (1)

Scheme 11.2 Regioselective Suzuki coupling (reproduced with American Chemical Society permission).

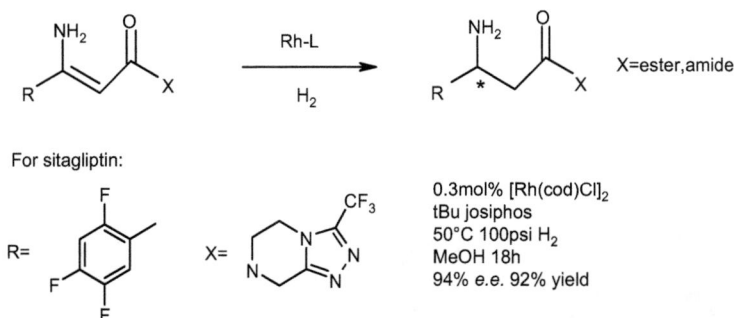

X=ester,amide

For sitagliptin:

0.3mol% [Rh(cod)Cl]$_2$
tBu josiphos
50°C 100psi H$_2$
MeOH 18h
94% *e.e.* 92% yield

Scheme 11.3 Asymmetric hydrogenation for synthesis of sitagliptin (reproduced with American Chemical Society permission).

For the case of laropiprant, the screening workflow was used to assess a variety of second-generation Noyori catalysts, but in this case the most notable finding was the identification of a significant pressure dependence upon the selectivity of the reaction. It was found that use of low hydrogen pressure [<30 psi (<1550 Torr)] was required to achieve high enantioselectivity, a detailed mechanistic study ultimately tracing this to the relative kinetics of asymmetric hydrogenation and ruthenium-mediated "walking" of the double bond in the starting material (Scheme 11.4).[29]

The automation platform has also been applied to biocatalysis screening, for instance the ketoreductase reduction of diaryl ketones.[31]

At the other end of the spectrum, cheaper tools for reaction and hydrogenation screening are now available, notably iChemExplorer and HEL Cat96. iChemExplorer (Figure 11.4) incorporates a heater/stirrer module into an Agilent LC system, allowing reactions to be performed directly in the HPLC vial and with facile reaction sampling. Examples of using this tool for catalyst screening and reaction optimisation through kinetic profiling have been reported by GlaxoSmithKline (GSK).[32,33] The power of this tool is also linked to the associated software that enables facile collation, visualisation and export of the experimental data. Cat96 allows reactions to be performed at up to 35 bar pressure ($\sim 26 \times 10^3$ Torr), again using HPLC or microtitre vials as the reaction vessel, and has been applied to catalyst selection in combination with a

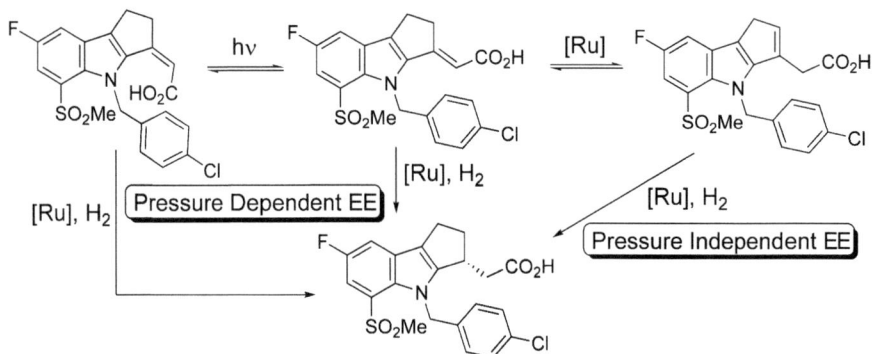

Scheme 11.4   Mechanistic understanding of laropiprant asymmetric hydrogenation (reproduced with American Chemical Society permission).

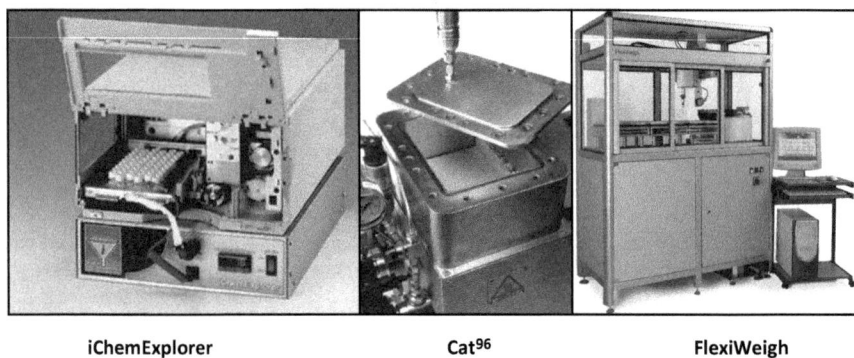

iChemExplorer   Cat⁹⁶   FlexiWeigh

Figure 11.4   Automated screening tools.

FlexiWeigh solids dispensing robot for both hydrogenation and carbonylation screening (Figure 11.4). The development of Rapid Resolution High Throughput (RRHT) HPLC columns based upon use of 1.8 μm silica should also be noted,[34] which has allowed the development of super-fast HPLC methods (sometimes as low as 1 minute in duration) for analysis of reaction enabling tools such as iChemExplorer to profile many reactions in parallel within a reasonable time frame. Even more recently, poroshell columns have been introduced which have comparable efficiency to the RRHT column but at lower back pressure, paving the way for more routine use of 1–2 min methods especially in combination with ultra-high-pressure pumps [600 bar ($\sim 450 \times 10^3$ Torr) or higher].[35]

Microchannel flow reactors for the screening of reaction conditions is an emerging area, still largely restricted to the academic domain.[36] Various chemistries have been demonstrated using MEMS (microelectric mechanical systems) chips, and there are reports of catalyst screening with this technique;[37,38] however, this is still highly specialised and far from mainstream in the industry.

The heat evolved from a reacting system is a direct measure of the overall reaction kinetics, and the application of reaction calorimetry in the optimisation of processes is well utilised. It is not generally applied to route scouting; however, there is an interesting report by Blackmond relating to comprehensive screening of catalysts using reaction calorimetry.[39,40] In ideal cases she reports that the efficacy of different catalysts can be evaluated by examination of the heat flow signal upon sequential injection of various catalysts to a reaction mixture. Each catalyst gives rise to a step change in the heat flow signal, the size of the change being proportional to the catalyst efficacy (Figure 11.5). This methodology assumes that the catalysts themselves do not interact, a condition that may not be true in many instances.

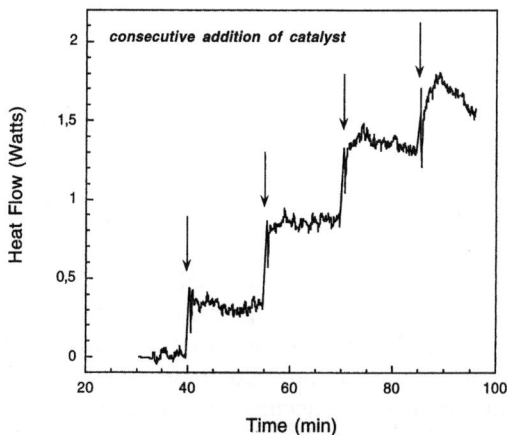

Figure 11.5 Calorimetric investigation of catalysts for a Heck reaction (reproduced with American Chemical Society permission).

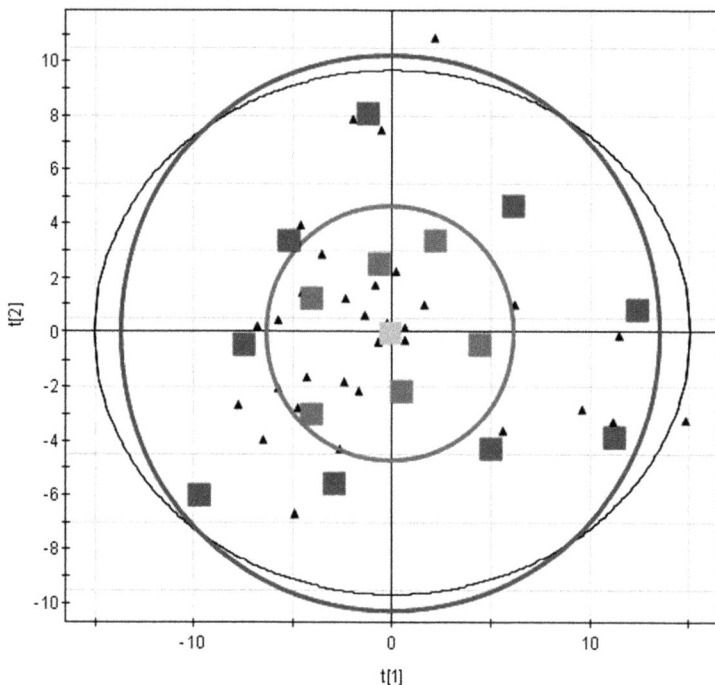

Figure 11.6   Concentric layers of a D-Onion design (reproduced with permission from AstraZeneca).

At the screening stage, use of statistical design to ensure both a good coverage of reaction space and to understand the trends in the experimental data is clearly essential. Typically, traditional fractional factorial designs or computer generated D-Optimal designs have been employed to delineate the main process effects. More recently, the application of a modified D-Optimal design, the so-called D-Onion design, has been introduced by Umetrics, which offers better spread of reaction space for screening applications.[41,42] D-Optimal designs tend to favour the edges of the design space at the expense of the centre; however, by creating a series of concentric D-Optimal designs and combining into a new "D-Onion" design, this problem is minimised (Figure 11.6). In an alternative approach, Carlson has recently reported the computer generation of candidate sets leading to highly efficient near orthogonal designs.[43] This methodology looks promising to the application of screening and optimisation designs, but is not yet available in commercial software packages.

A crucial aspect in creating screening arrays is the diversity of the reagents selected for the design. The use of principal component analysis (PCA) to classify diversity of solvents is now commonplace in the industry, but its application to other classes of reagents is less widespread. This is particularly true when it comes to ligands, where a lack of consistent physicochemical data in the literature has rendered generation of these models impossible. One approach

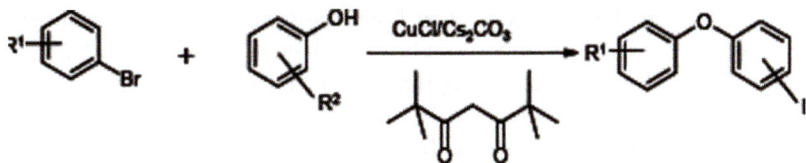

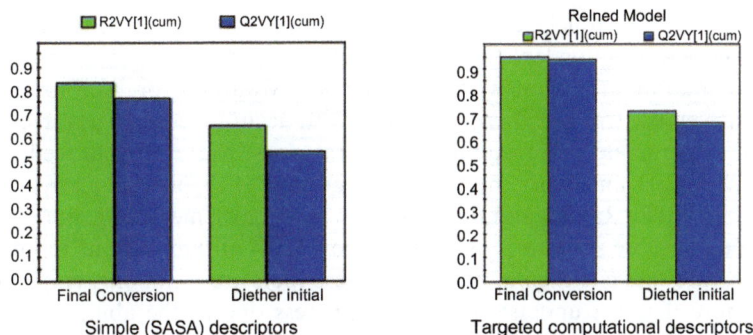

Figure 11.7 Targeted *versus* simple descriptors (reproduced with permission from AstraZeneca and the American Chemical Society).

to address this is to create numerical descriptions for the reagent set (ligands, bases, and so on) *in silico,* then perform a PCA dimensionality reduction upon this data. This has been performed by use of either computational modelling, use of QSSR/QSAR type descriptors or a combination of both. Avantium has utilised the descriptor approach to model ligands for the Corey CBS reaction, and AstraZeneca has applied computational models to classify the diversity of a set of bidentate phosphine ligands.[44,45] Interestingly, AstraZeneca has also compared simple descriptors with more complex computational models for describing diketone promoters of the Ullmann coupling.[46] Although computational models were indeed better, it was found that use of solvent accessible surface area (SASA) descriptors gave a respectable fit (green-R^2) and predictability (blue-Q^2) in the statistical models (Figure 11.7).

11.3.2 Enabling Technologies at Route Optimisation Stage of API Development

Technology applied to route optimisation is now somewhat mature, has been well reviewed[17,18] and for the most part has not evolved greatly in recent years. At this stage, reactions are typically performed at the 10–100 mL scale in systems typically comprising between 4 and 12 reactors.

Use of Peltier-driven RS-10 STEM blocks has become commonplace in the pharmaceutical industry for studying up to 10 reactions at a time, largely due to their ease of use, independent temperature/stir control and working range of −35 to 150 °C. In consequence, these have been a workhorse for performing statistical experimental designs to follow up on leads from the route scouting stage. The systems have been used in both stand-alone mode and integrated into robotic optimisation tools, notably the Anachem SK233 and 215 platforms.[21] Application to reaction optimisation in conjunction with statistical design has been thoroughly reviewed in the literature and will not be repeated here.[13,14] The Integrity-12 has recently been introduced to replace the RS-10; however, this is an evolutionary rather than revolutionary change. This system boasts 12 reactors, touch screen control and improved reliability, yet in essence it is functionally equivalent to its predecessor.

The common "four-pot systems" such as Multimax or Automate offer recipe-driven control of experimentation, with accurate control and recording of process parameters (for example, temperature, stir rate, reagent feed, T_r–T_j, reaction power) and when necessary integration with process analytic tools such as ReactIR, Raman and focused beam reflectance measurement (FBRM). The value of these systems is clear from the wealth of publications reported in the literature that cite their application. These tools are commonly employed to glean more detailed understanding of a process due to the ability to "scaledown" process conditions. This includes both geometric similarity of the vessel/ agitators to larger scale equipment, and the ability to employ the accurate temperature control of the systems to emulate heat transfer characteristics of larger process vessels. Obtaining kinetic information from a reaction is typically started at this scale, which at its simplest is monitoring the difference in reactor and jacket temperature, T_r–T_j. This often provides vital information about the reaction since it is directly proportional to reaction power, which itself describes the overall rate of reaction. Through use of calibration or compensation heaters the true reaction power may also be recorded, and once available a wealth of kinetic information becomes accessible such as information on induction periods and reagent accumulation. This knowledge, usually in synergy with design aspects, leads to improved understanding of the process, and has been well reviewed in the literature.[47] By integrating these tools with either on-line sampling (for example, Mettler MultiMax ART) or with process analytic tools (such as ReactIR, Multimax IR, iC10, NIR) the kinetics of individual species may be recorded.[48] This information, either alone or in combination with the overall rate, has then been used to model reaction kinetics *in silico* with tools such as BatchCad or DynoChem.[49–51] Use of the Omnical micro-calorimeter has occurred in both academia[52,53] and industry[54] to study catalytic kinetics, typically for palladium-mediated cross-couplings. The differential data obtained in the form of the power trace is used to understand the catalytic cycle by fitting to a fundamental model. This has been the workhorse for the reaction progress analysis method devised by Blackmond,[55] and is described in more detail in Chapter 7. Complementary to calorimetric methods, iChemExplorer in conjunction with fast analytics has been used to collect extensive integral data for reaction progress analysis, leading to process understanding for a Suzuki coupling.[56]

There are now a number of vendors (Mettler, HEL, Systag, Biotage, Chemspeed, to name a few) offering systems of varying price and complexity, but once again there is little difference in functionality to the pioneering systems offered by Mettler and HEL some 10 years ago. On this note, it is interesting to see that despite all the commercial systems now available, some organisations still choose to employ in-house set-ups, such as the recently reported Oscar Wilde at AstraZeneca.[57] It is then quite sobering to re-read Pollard's review on process development automation from 2001, where it is clear that not much has changed when you compare the "home-built" systems described in these two articles.[18] In the final analysis, four-pot systems at the 50–100 mL scale are a mature technology which you can choose to buy complete from a vendor, or put together yourself. They provide highly valued process information, but operationally have not really evolved too far over the last decade.

11.3.3 Design Space and Enabling Technologies at Process Validation Stage of API Development

Collectively, the concept of a design space and the determination of normal operating ranges (NOR), proven acceptable ranges (PAR), criticality and control strategy is relatively new.[58,59] That said, the fundamentals are well established and detailed in several reviews.[13,60] Furthermore, increasing regulatory focus has driven new efforts to better understand and define how industry responds to these concepts and their ultimate application to commercial processes.[61,62]

As a result, there is clearly a link between the delineation of design space and use of several of the tools and methodologies discussed earlier in this text. The concept of a multivariate design space lends itself well to the idea of statistical DoE. The key experimental aspects may be a significant undertaking with a need for very precise parameter control, which is well addressed by some of the equipment and workflows available.

Accordingly, reactions at this stage are typically performed at 1 L scale or larger and, as with process optimisation technology, equipment in this arena is somewhat mature. The so-called Automated Lab Reactor systems (ALRs) in common use include Mettler Toledo RC1, LabMax, HEL Auto-Lab, Systag Flexylab, *etc*. These tools have been well reviewed[17] and no effort will be made to repeat this here. In addition to these ALR systems, it has become increasingly common for fully fledged mini-plants to be incorporated into the research and development setting. These are typically self-contained units, sitting on their own skid, which offer all the capabilities of a typical pilot plant (for example, load cell equipped batch reactors, reflux and distillation capability, automated reagent charging, PC SCADA control) with a degree of modularity and flexibility so that they can be applied to various synthetic sequences. The commercial AP mini-plant system, for instance, is widely used within GSK (Figure 11.8). Similar systems are deployed in Proctor and Gamble.[63] Use of process analytic probes is commonplace with these tools, especially the crystallisation probes particle vision and measurement (PVM), FBRM and Raman, and *in situ* reaction monitoring tools (ReactIR, near-IR).

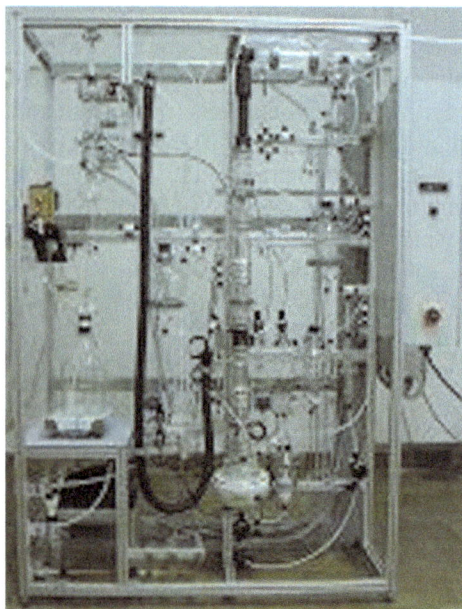

Figure 11.8   An AP mini-plant.

Several recent publications have exemplified advances in practical applications in this field; Cimarosti and co-workers at GSK have elaborated some of these themes in the development of a genotoxin control strategy.[64] In the drug product arena, Huang and co-workers at Wyeth have neatly demonstrated the combined power of DoE in combination with PCA and partial least-squares regression techniques to enable better understanding of a pharmaceutical product process development and to support the establishment of a design space for the product.[65] HPLC analysis can also benefit from similar methodology being applied to how reactions are monitored and developed. Some interesting early work in this area, from Delaney and co-workers, had highlighted the coupling of parallel reaction methodology with parallel HPLC analytical techniques, namely four columns and six mobile phases being applied to every reaction sample.[19] In a more recent development, Molnar and co-workers have translated into practice the definition of design space as applied to analytical methods, highlighting the importance of both optimisation and robustness as HPLC analytical methods are defined.[66]

To date, there are limited published works applying these technologies to drug substance process definition and manufacture. One notable exception to this is the recent work by Bravo and co-workers at GSK. Several of the concepts pertinent to design space have been elegantly described in the development of a manufacturing process to casopitant mesylate.[59]

11.3.4 Tools for Optimisation of Hydrogenation/ Carbonylation: Across Scales

We overviewed tools used for screening of hydrogenation/carbonylation reactions earlier and in this section shall highlight some of the tools typically used for the optimisation and validation of these processes. Typically, hits obtained from plate screening have been further characterized at the 5–15 mL scale using either the Biotage Endeavor or HEL HP-ChemScan instruments. Both of these tools are in widespread use within the pharmaceutical industry and also with catalyst vendors, including both Johnson Matthey and Solvias. Both systems have eight individual reactors and allow for individual pressure control, independent or zoned temperature control and measurement of gas uptake to allow reaction progress to be followed. At the 100 mL scale, leading tools include HEL's HP-AutoMate and Systag's Pressure Flexylab, and like their low-pressure counterparts allow for full control and monitoring of reaction parameters. In addition to gas uptake, calorimetry is commonly used with HP-AutoMate to follow the reaction kinetics, and ReactIR probes are routinely included to follow hydrogenations and observe intermediate hydrogenation species, for example the accumulation of hydroxylamine or nitroso intermediates during nitro group reduction.[67] At process validation, 1–2 L pressure vessels are often employed, using for example HP-AutoLab, LabMax or traditional Parr vessels. Once again, little has changed over the last decade in this respect. However, recent advances have been made with flow hydrogenation equipment, in particular the H-cube reactor which we shall now address within the context of process intensification.

11.4 Process Intensification

11.4.1 Continuous Flow Processing

Process intensification is used to describe methodology that can maximise overall efficiency of a process through reduction of processing cost, time, waste, equipment size and land use. The application of flow reactor systems and continuous processing is the central methodology to achieving this goal, and although common in the petrochemical industry it is still not widely used, as yet, in the drug substance side of the pharmaceutical industry. That said, there is an increased interest in use of these tools within the pharmaceutical industry. For example, at GSK the Innovative Manufacturing Initiative has been in place for several years,[68] which has demonstrated continuous processing as a cost effective alternative to the traditional batch approach for several projects at pilot scale. A typical workflow is depicted in Figure 11.9, where the API is prepared by a cascade between functional processing blocks, primarily continuous tubular reactors, continuous liquid–liquid extractors, continuous distillation and crystallisation units. The use of PAT tools[69,70] (typically ReactIR) is heavily employed to ensure processes remain within specification, which for the case of continuous flow equates to maintenance of steady-state conditions established for the process. Statistical tools are typically employed in real-time

A Typical Continuous Process

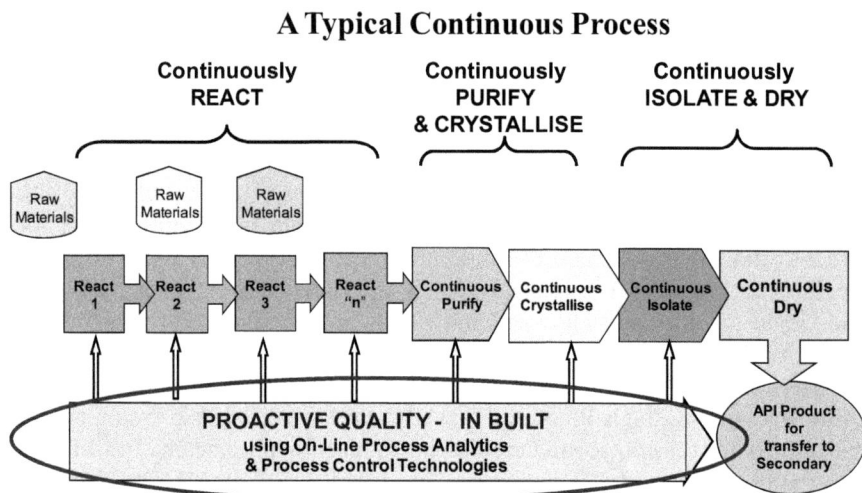

Figure 11.9   Typical continuous process workstream.

to monitor the process analytic data and provide early detection of deviation from the desired specification.

Combination of continuous flow to manage certain issues, for example reactive intermediates,[71] and batch operation are also prevalent in the industry.

In order to develop commercial flow processes it is clearly important to have laboratory based tools to screen and optimise reaction conditions; two systems which are not infrequently encountered in the industry are VapourTec and Accendo Conjure. The latter has been extensively used by Pfizer in combination with experimental design tools to screen and optimise conditions for a number of reactions under flow conditions, including nucleophilic aromatic substitutions, Diels–Alder cycloadditions, Michael additions, cross-coupling decarboxylations and "click reaction" triazole synthesis.[72,73]

Hydrogenation may also be performed in a continuous manner, as has been common in the fine chemical industry for many years. The H-cube flow reactor system from ThalesNano is a compact flow hydrogenator that allows for rapid and safe development of continuous hydrogenation processes using a proprietary CatCart catalyst cartridge system for support of the active catalyst (Figure 11.10). This tool has so far largely been applied to route scouting and discovery activities by companies including Eli Lilly[74] and Abbott;[75] however, it is anticipated more use will be made for development of continuous manufacture in the coming years.

11.4.2   Microwave Heating

The application of microwave heating is an intensification methodology since it has been demonstrated that, in comparison to conventional heating, microwaves may lead to a significant rate enhancement, improved conversion and

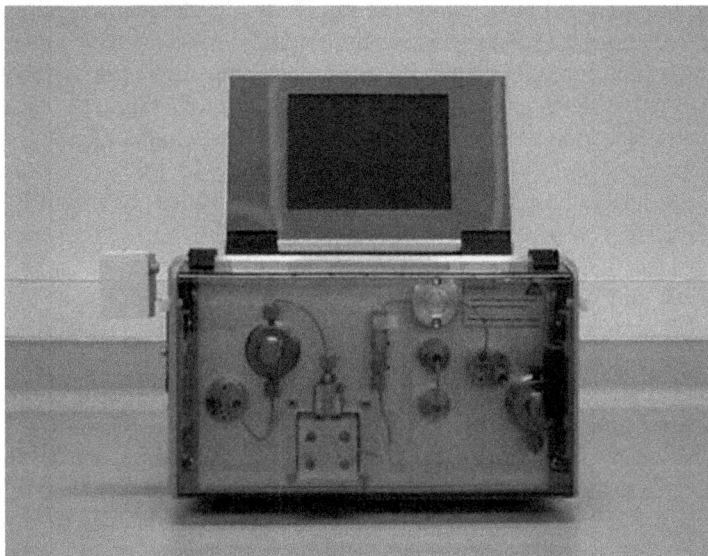

Figure 11.10 H-cube continuous hydrogenator.

reproducibility.[76] Microwave irradiation also requires less energy when compared to a conventional heating approach. The use of laboratory scale tools such as the Personal Synthesizer has demonstrated the significant time reduction that can be achieved for route scouting/optimisation activities. However, for pilot/ production scale, many issues have to be conquered to use these tools effectively, least of all the ability of microwaves to penetrate only a few cm into a sample. A recent review by AstraZeneca[77] compared seven commercially available micro-wave systems (both batch and continuous) and concluded that no single com-mercially available system is capable of meeting all of the needs of the pharmaceutical industry at >1 kg scale, further stating that "for true process development and pilot scale there is presently no commercial microwave scale-up solution". Researchers at the University of Connecticut have recently reported good linear scaling from small-scale microwave heating to a developmental 12 L batch microwave reactor,[78] suggesting this tool may prove valuable for general kg scale microwave synthesis. Overall, however, the use of microwave heating under flow conditions seems to be the most practical method at scale.[79] AstraZeneca has been particularly active in this area, having employed a flow microwave reactor to the classical high-temperature Newman–Kwart rearrangement[80] and a stop-flow microwave system for scaling out nucleophilic aromatic substitutions.[81]

11.4.3 Process Analytical Technology: Application to Process Development

Process analytical technology (PAT), as a process intensification tool (for example, ReactIR, UV, HPLC, Raman, Lasentec), referred to earlier in several sections, has seen a dramatic increase in process development applications. It is now commonplace across most development phases and activities. As technology has developed, in large part becoming more user friendly, and the industry has developed a greater appreciation of chemometric techniques, PAT has been widely applied to areas of process understanding, process control, process safety and sustainability. The ability to access additional data, analyse on-line coupled with chemometrics has added significantly to the overall efficiency in process development.

A very elegant application of PAT to process development has been detailed by Werner and co-workers at Lilly in the development of the PPAR-alpha agonist LY518674 (Figure 11.11).[82] The authors detail how a wide range of on-line techniques (for example, ReactIR, MS, FBRM and PVM) has added significant understanding to help them develop and importantly simplify and control their process. The result is a very well characterised process, in large part, enabled by the application of PAT to both the laboratory development and pilot plant manufacture of the compound.

In a similar vein, Sistare and co-workers at Pfizer have detailed their reaction control in removing most of the excess bromine following the bromination of 6-aminopenicillanic acid (Scheme 11.5). An on-line electrical potential measurement was used to determine the end point where, following reduction of excess bromine, the reaction mixture had a suitably low bromine concentration for product stability during onward processing.[83]

Application of PAT methodology to augment understanding and control of processes has also been applied to continuous flow processing applications. In a recent example, Ley and co-workers detail several examples of the use of a mid-IR in-line methodology to enable rapid process development. In an interesting

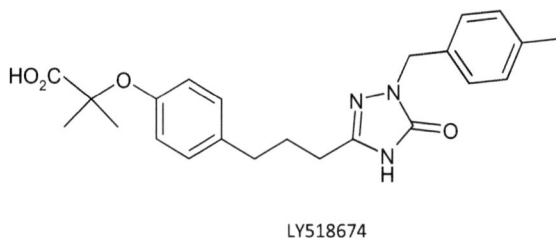

LY518674

Figure 11.11 PPAR-alpha agonist LY518674.

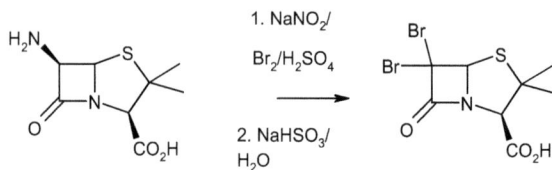

Scheme 11.5 Bromination of 6-aminopenicillanic acid (reproduced with American Chemical Society permission).

Scheme 11.6 DAST fluorination (reproduced with American Chemical Society permission).

observation, the fluorinated product (Scheme 11.6) was shown to elute from the flow cell slower than predicted by simple flow rate, as a result of dispersion, with obvious implications for onward processing.[69]

Application of LASENTEC monitoring to crystallisation development and process control is employed in practically all pharmaceutical crystallisations, as the desire to understand the interaction between physical attributes of drug substance and drug product performance has gained increasing focus since 2004. On-line monitoring of the physical attributes has allowed process development scientists to better control this key processing step.

Several authors have detailed the use of PAT to enable drying end-point detection to render processes more sustainable and reducing manufacturing cycle times. Tewari and co-workers have detailed their studies toward applying mid-IR spectroscopy, coupled with chemometrics, as a potential on-line drying end-point determination.[84] Wiss and Burgbacher highlight the use of near-IR spectroscopy on-line coupled with several industrial dryers.[85] The widespread use of on-line spectroscopy to detect drying end points in drug product manufacture is perhaps significantly more established than in drug substance manufacture.[86]

In conclusion, the variety of examples of PAT being applied to enhance process understanding across a range of unit operations throughout the pharmaceutical process development arena clearly indicates the transformation of this technology into a routine technology widely employed by process development scientists.

11.5 Conclusions

Over the last decade there have been some significant changes and improvements to enabling technologies in use in the pharmaceutical industry, which we have strived to overview in this chapter. In particular, we have seen significant advances in high-throughput analytics which in turn has alleviated the data analysis bottleneck, and enabled the development and routine application of medium- to high-throughput screening activities. The ability to run many experiments in parallel has also led to more routine use of multivariate data analysis and in turn this is now commonly applied to address important regulatory areas such as the delineation of design space. Some tools have not evolved greatly, but nonetheless are still as valuable as ever for the precise parameter control they afford when collecting multivariate data for regulatory purposes. Intensification tools (continuous flow, microwave heating, *in situ* use of PAT tools) have clearly become more routinely adopted; moving forward, the use of micro-fluidic screening and optimisation tools, whilst still in its infancy today, is likely to play an increasingly important role in the industry. In conclusion, the enabling technologies available today have been at least partly successful in "closing the gap", that is to allow increased productivity and understanding in the wake of an environment with ever decreasing resource. We have little doubt that this gap will continue to close in line with the continual evolution of process enabling technology in the coming years.

References

1. (a) B. Hirschler and K. Kelland, Reuters, Special Report, 2010; (b) T. A. Roberson, N. V. Smith and J. Scherer, *Outsourcing R&D*, Oct 2009, 90.
2. F. Bravo, Z. Cimarosti, F. Tinazzi, D. Castoldi, P. Stonestreet, A. Galgano and P. Westerduin, *Org. Process Res. Dev.*, 2010, **14**, 832.
3. *Pharmaceutical cGMPs for the 21st Century – A Risk-Based Approach*, Final Report; US Department of Health and Human Services, Food and Drug Administration, Center for Drug Evaluation and Research (CDER), Silver Spring, MD, 2004.
4. T. Laird, *Org. Process Res. Dev.*, 2002, **6**, 337.
5. F. Stazi, G. Palmisano, M. Turconi and M. Santagostino, *Tetrahedron Lett.*, 2005, **46**, 1815.
6. E. W. Kirchhoff, D. R. Anderson, S. Zhang, C. S. Cassidy and M. T. Flavin, *Org. Process Res. Dev.*, 2001, **5**, 50.
7. (a) R. W. Wagner, F. Lei, H. Du and J. S. Lindsey, *Org. Process Res. Dev.*, 1999, **3**, 28; (b) A. R. Frisbee, M. H. Nantz, G. W. Kramer and P. L. Fuchs, *J. Am. Chem. Soc.*, 1984, **106**, 7143.
8. U. C. Dyer, D. A. Henderson and M. B. Mitchell, *Org. Process Res. Dev.*, 1999, **3**, 161.
9. M. Owen, presented at the symposium on The Evolution of a Revolution – Laboratory Automation in Chemical Process R&D, Chester, UK, 1998.
10. P. D. Higginson and N. W. Sach, *Org. Process Res. Dev.*, 2001, **5**, 331.
11. D. F. Emiabata-Smith, D. L. Crookes and M. R. Owen, *Org. Process Res. Dev.*, 1999, **3**, 281.
12. M. A. Armitage, G. E. Smith and K. T. Veal, *Org. Process Res. Dev.*, 1999, **3**, 189.
13. M. R. Owen, C. Luscombe, L. Lai, S. Godbert, D. L. Crookes and D. Emiabata-Smith, *Org. Process Res. Dev.*, 2001, **5**, 308.
14. D. Lendrum, M. Owen and S. Godbert, *Org. Process Res. Dev.*, 2001, **5**, 324.
15. C. Simms and J. Singh, *Org. Process Res. Dev.*, 2000, **4**, 554.
16. M. Harre, U. Tilstam and H. Weinmann, *Org. Process Res. Dev.*, 1999, **3**, 304.
17. M. E. Van Loo and P. E. Lengowski, *Org. Process Res. Dev.*, 2002, **6**, 833.
18. M. Pollard, *Org. Process Res. Dev.*, 2001, **5**, 273.

19. E. J. Delaney, M. L. Davies, B. D. Karcher, V. W. Rosso, A. E. Rubin and J. J. Venit, in *Fundamentals of Early Clinical Drug Development: From Synthesis Design to Formulation*, ed. A. F. Abdel-Magid and S. Caron, Wiley, Hoboken, NJ, 2006, ch. 9.

20. S. L. Morissete, O. Almarsson, M. L. Peterson, J. F. Remenar, M. J. Read, A. V. Lemmo, S. Ellis, M. J. Cima and C. R. Gardner, *Adv. Drug Delivery Rev.*, 2004, **56**, 275.

21. A. E. Rubin, S. Tummala, D. A. Both, C. Wang and E. J. Delaney, *Chem. Rev.*, 2006, **106**, 2794.

22. http://www.ichemexplorer.com (last accessed 26 December 2010).

23. http://www.us.mt.com, http://www.systag.ch (last accessed 26 December 2010).

24. P. D. Higginson and N. W. Sach, *Org. Process Res. Dev.*, 2004, **8**, 1009.

25. G. ten Brink, I. W. C. E. Arends, M. Hoogenraad, G. Verspui and R. A. Sheldon, *Adv. Synth. Catal.*, 2003, **345**, 497.

26. M. Hoogenraad, J. B. van der Linden and A. A. Smith, *Org. Process Res. Dev.*, 2004, **8**, 469.

27. C. Cai, J. Y. L. Chung, C. McWilliams, Y. Sun, C. S. Schultz and M. Palucki, *Org. Process Res. Dev.*, 2007, **11**, 328.

28. Y. Hsiao, N. R. Rivera, T. Rosner, S. W. Krska, E. Njolito, F. Wang, Y. Sun, J. D. Armstrong III, E. J. J. Grabowski, R. D. Tillyer, F. Spindler and C. Malan, *J. Am. Chem. Soc.*, 2004, **126**, 9918.

29. D. M. Tellers, J. C. McWilliams, G. Humphrey, M. Journet, L. DiMichele, J. Hinksmon, A. E. McKeown, T. Rosner, Y. Sun and R. D. Tillyer, *J. Am. Chem. Soc.*, 2006, **128**, 17063.

30. K. R. Campos, A. Klapars, J. C. McWilliams, C. S. Shultz, D. J. Wallace, A. M. Chen, L. F. Frey, A. V. Peresypkin, Y. Wang, R. M. Wenslow and C. Chen, PCT Int. Appl. (2006); CODEN: PIXXD2 WO 2006017045.

31. B. T. Grau, P. N. Devine, L. N. DiMichele and B. Kosjek, *Org. Lett.*, 2007, **9**, 4951.

32. K. M. Bullock, M. B. Mitchell and J. F. Toczko, *Org. Process Res. Dev.*, 2008, **12**, 896.

33. J. A. Corona, R. D. Davis, S. B. Kedia and M. B. Mitchell, *Org. Process Res. Dev.*, 2010, **14**, 712.

34. A. D. Broske, R. D. Ricker, B. J. Permar, W. Chen and M. Joseph, Agilent Technical Note 5988-9251EN, 2003.

35. A. Gratzfeld-Hüsgen and E. Naegele, Agilent Technical Note 5990-5602EN, 2010.

36. H. Pennemann, P. Watts, S. J. Haswell, V. Hessel and H. Löwe, *Org. Process Res. Dev.*, 2004, **8**, 422.

37. J. Wang, G. Sui, V. P. Mocharla, R. J. Lin, M. E. Phelps, H. C. Kolb and H. Tseng, *Angew. Chem. Int, Ed.*, 2006, **45**, 5276.

38. P. W. Miller, L. E. Jennings, A. J. de Mello, A. D. Gee, N. J. Long and R. Vilar, *Adv. Synth. Catal.*, 2009, **351**, 3260.

39. J. Le Bars, T. Häußner, J. Lang, A. Pfaltz and D. G. Blackmond, *Adv. Synth. Catal.*, 2001, **343**, 207.

40. D. G. Blackmond, T. Rosner and A. Pfaltz, *Org. Process Res. Dev.*, 1999, **3**, 275.
41. I. M. Olsson, J. Gottfries and S. Wold, *Chemom. Intell. Lab. Syst.*, 2004, **73**, 37.
42. I. M. Olsson, J. Gottfries and S. Wold, *J. Chemom.*, 2004, **18**, 548.
43. R. Carlson, G. Simonsen, A. Descomps and J. E. Carlson, *Org. Process Res. Dev.*, 2009, **13**, 798.
44. (a) M. Hoogenraad, G. M. Klaus, N. Elders, S. M. Hooijschuur, B. McKay, A. A. Smith and E. W. P. Damen, *Tetrahedron: Asymmetry*, 2004, **15**, 519; (b) J. B. van der Linden, E. J. Ras, S. M. Hooijschuur, G. M. Klaus, N. T. Luchters, P. Dani, G. Verspui, A. A. Smith, E. W. P. Damen, B. McKay and M. Hoogenraad, *QSAR Comb. Sci.*, 2005, **24**, 98.
45. N. Fey, J. N. Harvey, G. C. Lloyd-Jones, P. Murray, A. G. Orpen, R. Osborne and M. Purdie, *Organometallics*, 2008, **27**, 1372.
46. R. Osborne, presented at the Conference on Optimising Organic Reactions, Basel, Switzerland, 2007.
47. D. M. Roberge, *Org. Process Res. Dev.*, 2004, **8**, 1049.
48. H. Bjørsvik, *Org. Process Res. Dev.*, 2004, **8**, 495.
49. M. Bollyn, A. van den Bergh and A. Wright, presented at the 7th International RC1 User Forum, Interlaken, Switzerland, 1995.
50. K. B. Hansen, Y. Hsiao, F. Xu, N. Rivera, A. Clausen, M. Kubryk, S. Krska, T. Rosner, B. Simmons, J. Balsells, N. Ikemoto, Y. Sun, F. Spindler, C. Malan, E. J. J. Grabowski and J. D. Armstrong III, *J. Am. Chem. Soc.*, 2009, **131**, 8798.
51. L. M. Oh, H. Wang, S. C. Shilcrat, R. E. Hermann, D. B. Patience, P. G. Spoors and J. Sisko, *Org. Process Res. Dev.*, 2007, **11**, 1032.
52. E. R. Strieter, D. G. Blackmond and S. G. Buchwald, *J. Am. Chem. Soc.*, 2003, **125**, 13978.
53. P. H. Phua, S. P. Mathew, A. J. P. White, J. G. de Vries, D. G. Blackmond and K. K. Hii, *Chem. Eur. J.*, 2007, **13**, 4602.
54. S. Eyley, AstraZeneca, Laboratory Automation and New Technologies in Process Research and Development, London, 2003.
55. D. G. Blackmond, *Angew. Chem. Int. Ed.*, 2005, **44**, 4302.
56. S. B. Kedia and M. B. Mitchell, *Org. Process Res. Dev.*, 2009, **13**, 420.
57. C. Bernlind and C. Urbaniczky, *Org. Process Res. Dev.*, 2009, **13**, 1059.
58. http://www.ich.org, Pharmaceutical Development, Q8(R2) (last accessed 6 January 2011).
59. Z. Cimarosti, F. Bravo, D. Castoldi, F. Tinazzi, S. Provera, A. Perboni, D. Papini and P. Westerduin, *Org. Process Res. Dev.*, 2010, **14**, 805.
60. T. Y. Zhang, *Chem. Rev.*, 2006, **106**, 2583.
61. http://www.ich.org, Pharmaceutical Development, Q9 (last accessed 6 January 2011).
62. http://www.ich.org, Pharmaceutical Development, Q10 (last accessed 6 January 2011).
63. M. Collier, M. Creed and R. Owens, *P&G*, Laboratory Automation and New Technologies in Process Research and Development, London, 2003.

64. Z. Cimarosti, F. Bravo, P. Stonestreet, F. Tinazzi, O. Vecchi and G. Camurri, *Org. Process Res. Dev.*, 2010, **14**, 993.
65. J. Huang, G. Kaul, C. Cai, R. Chatlapalli, P. Hernandez-Abad, K. Ghosh and A. Nagi, *Int. J. Pharm.*, 2009, **382**, 23.
66. I. Molnar, H.-J. Rieger and K. E. Monks, *J. Chromatogr. A*, 2010, **1217**, 3193.
67. F. Visentin, G. Puxty, O. M. Kut and K. Hungerbühler, *Ind. Eng. Chem. Res.*, 2006, **45**, 4544.
68. M. Berry, *presented at ACS Prospectives*, Process Chemistry in the Pharmaceutical Industry with Special Emphasis on Continuous Manufacturing, Durham, USA, 2009.
69. C. F. Carter, H. Lange, S. V. Ley, I. R. Baxendale, B. Wittkamp, J. G. Goode and N. L. Gaunt, *Org. Process Res. Dev.*, 2010, **14**, 393.
70. Z. Qian, I. R. Baxendale and S. V. Ley, *Chem. Eur. J.*, 2010, **16**, 12342.
71. F. Stazi, D. Cancogni, L. Turco, P. Westerduin and S. Bacchi, *Tetrahedron Lett.*, 2010, **51**, 5385.
72. N. W. Sach, presented at Optimising Organic Reactions, 2008, Vancouver, Canada.
73. A. R. Bogdan and N. W. Sach, *Adv. Synth. Catal.*, 2009, **351**, 849.
74. R. V. Jones, L. Godorhazy, N. Varga, D. Szalay, L. Urge and F. Darvas, *J. Comb. Chem.*, 2006, **8**, 110.
75. B. Clapham, N. S. Wilson, M. J. Michmerhuizen, D. P. Blanchard, D. M. Dingle, T. A. Nemcek, J. Y. Pan and D. R. Sauer, *J. Comb. Chem.*, 2008, **10**, 88.
76. S. Caddick and R. Fitzmaurice, *Tetrahedron*, 2009, **65**, 3325.
77. J. D. Moseley, P. Lenden, M. Lockwood, K. Ruda, J. Sherlock, A. D. Thomson and J. P. Gilday, *Org. Process Res. Dev.*, 2008, **12**, 30.
78. J. R. Schmink, C. M. Kormos, W. G. Devine and N. E. Leadbeater, *Org. Process Res. Dev.*, 2010, **14**, 205.
79. M. Damm, T. N. Glasnov and C. O. Kappe, *Org. Process Res. Dev.*, 2010, **14**, 215.
80. J. D. Moseley, R. F. Sankey, O. N. Tang and J. P. Gilday, *Tetrahedron*, 2006, **62**, 4685.
81. J. A. Marafie and J. D. Moseley, *Org. Biomol. Chem.*, 2010, **8**, 2219.
82. M. D. Argentine, T. M. Braden, J. Czarnik, E. W. Conder, S. E. Dunlap, J. W. Fennell, M. A. LaPack, R. R. Rothaar, R. B. Scherer, C. R. Schmid, J. T. Vicenzi, J. G. Wei and J. A. Werner, *Org. Process Res. Dev.*, 2009, **13**, 131.
83. F. Sistare, L. St. Pierre Berry and C. A. Mojica, *Org. Process Res. Dev.*, 2005, **9**, 332.
84. J. Tewari, V. Dixit and K. Malik, *Sens. Actuators B: Chem.*, 2010, **144**, 104.
85. J. Burgbacher and J. Wiss, *Org. Process Res. Dev.*, 2008, **12**, 235.
86. P. Frake, D. Greenhalgh, S. M. Grierson, J. M. Hempenstall and D. R. Rudd, *Int. J. Pharm.*, 1997, **151**, 75.

CHAPTER 12

The Analytical Interface and the Impact on Pharmaceutical Process Development

SIMON HAMILTON AND ALEXIA BERTRAND

MSD Ltd (a subsidiary of Merck & Co., Inc), Hertford Road, Hoddesdon, Hertfordshire, EN11 9BU, UK

12.1 Introduction

The challenges associated with the development of organic syntheses are numerous. Determination and eradication of side reactions to increase yields and reduce impurities is critical for the development of a robust, cost-effective process. The analytical function is no longer just a confirmation of the desired product and assurance of quality, but a critical support function guiding all aspects of the synthetic endeavour. As a result, a deeper understanding of the reaction pathways and associated kinetics taking place within the reactor can significantly reduce the number of experiments required to successfully optimise a synthetic process. The valuable information generated in well-planned analytical experiments not only assists the process chemist in developing a more efficient process, which impacts the overall cost of goods, but also impacts operator and patient safety.

12.2 Evolution of Analytical Techniques

The field of analytical chemistry is rapidly changing, with new techniques and existing methodology being improved by both instrument vendors and

RSC Drug Discovery Series No. 9
Pharmaceutical Process Development: Current Chemical and Engineering Challenges
Edited by A. John Blacker and Mike T. Williams
© Royal Society of Chemistry 2011
Published by the Royal Society of Chemistry, www.rsc.org

academics alike. These improvements are translating directly to improvements in the efficiency of analytical techniques and also in the robustness and sensitivity of analytical instrumentation. The highly competitive nature of the pharmaceutical industry means that they are ever more focused on the reduction of development cycle times. Manufacturing drug substances for early stage clinical trials, using fit-for-purpose chemistry and large scale-up factors, can carry significant financial risks if reactions do not proceed as expected. Understanding and monitoring the process successfully is critical. While chromatographic separations are the mainstay of analytical testing, the majority of sampling and sample preparation is still performed manually. Recent years have seen a trend towards both automated sampling and preparation and the use of on-line spectroscopic techniques to monitor a process in real time. As high-throughput reaction screening has become commonplace in process chemistry laboratories, so the number of analytical samples has increased to the point that analytical testing is in danger of becoming a bottleneck, leaving the analytical chemist with no choice but to go faster or perform the less satisfactory operation of triage. Producing data in a shorter timeframe can be achieved in two ways: either through reducing the analysis cycle time or running multiple experiments in parallel. Historically, chromatographic separations have taken in excess of 60 minutes, with a rapid separation being around 30 minutes. High-throughput screening experiments may utilise 96, or 384 in some cases, well plates. To analyse this number of samples using a 30 minute chromatographic run would take many days. This is clearly unacceptable and has been partly responsible for the increases in the speed and separation efficiency of both gas and liquid chromatography. Generating suitable chromatographic methods to assess impurity profiles and conversion rates is a key part of this process. Fast chromatographic techniques have enabled more thorough screening of mobile and stationary phase conditions in very limited timeframes.

The theory behind improved efficiencies in chromatographic analysis is nothing new and had been published as far back as the 1950s.[1] van Deemter's equation demonstrates clearly the critical factors affecting the separation of a packed chromatographic column, which with improvements in hardware capabilities have provided numerous opportunities for the analytical chemist to improve separation efficiency and reduce cycle times. A basic understanding of this theory and its limitations is paramount to the generation of robust, rapid methodology.

12.3 HPLC Theory and the van Deemter Equation

The height of a theoretical plate (or number of plates in a column) has long been the measure of both chromatographic and distillation efficiency.[2] The greater the number of plates in a column, the more efficient will be the separation. The number of theoretical plates is typically derived experimentally and displayed as a van Deemter plot, although a number of other approaches to determine column efficiency are now employed (Knox, Poppe).

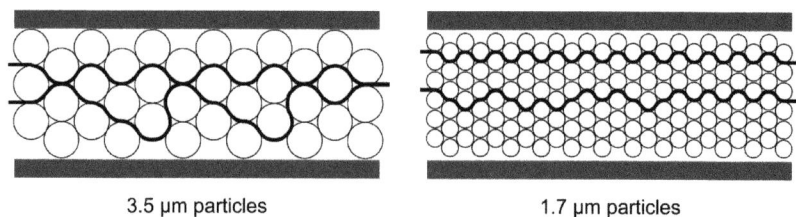

3.5 μm particles   1.7 μm particles

Figure 12.1 Schematic showing the cross section of two HPLC columns packed with different sized stationary phase particles and the impact on eddy diffusion. The difference in length of the flow path taken by the analyte is reduced with a reduction in particle size diameter.

Equation (12.1) shows the abbreviated van Deemter equation for plate height (*H*):

$$H = A + \frac{B}{u} + Cu \qquad (12.1)$$

where A = eddy diffusion, B = longitudinal diffusion, C = mass transfer and u = linear velocity.

12.3.1 Eddy Diffusion

Eddy diffusion occurs as a result of irregularities in the path lengths taken by the analyte molecules in packed columns. As the analyte molecules take different paths, dispersion occurs, causing some molecules to arrive at the detector later than others. When the particle size of the stationary phase is reduced, the closer packing arrangement results in a reduction in path length differences and therefore less dispersion and associated band broadening. Figure 12.1 shows the differences in flow path through columns of different particle sizes. These show an almost theoretically perfect packing configuration. In reality there are irregularities in the packing quality and the particle size distribution, which leads to an even greater increase in band broadening.

12.3.2 Longitudinal Diffusion

Longitudinal diffusion is a function of the time that the analyte is present in the mobile phase, which is a function of mobile phase flow rate (velocity). The analyte concentration also contributes to the rate of longitudinal diffusion, since a more concentrated band will diffuse faster than a less concentrated band.

12.3.3 Mass Transfer

Resistance to mass transfer refers to the rate at which the analyte molecule transitions from the stationary phase to the mobile phase during a chromatographic separation. A large fully porous particle (3.5 μm) will effectively display

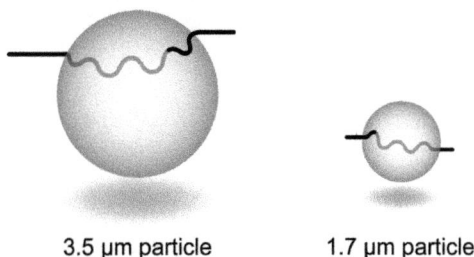

3.5 µm particle   1.7 µm particle

Figure 12.2   Rate of mass transfer is proportional to the diameter of the stationary phase particle. The reduction in diameter leads to a decreased resistance to mass transfer and a corresponding reduction in band broadening, at the expense of elevated back pressure.

a slower rate of mass transfer, as the analyte molecule can percolate further into the particle than it can with a smaller particle (1.7 µm), as illustrated in Figure 12.2.

12.4   Rapid HPLC Analysis

The simplest approach to reducing HPLC run times is to reduce the column length and/or increase the flow rate of the mobile phase. This approach has been employed in many laboratories and has enabled the reduction of run-times to less than 5 minutes. In HPLC, the efficiency of the separation is directly proportional to column length,[3] so any reduction in length results in a reduction in efficiency. In many separations there is often an excess of efficiency, *i.e.* there is excess resolution between the critical pair of peaks in a chromatographic separation. In this case a reduction in column length is acceptable provided all peaks are still resolved. Doubling the flow rate of a chromatographic system will decrease the run time by one half, in most cases with a reduction in efficiency. Over the past 10 years, many advances in HPLC column and hardware technology have enabled not only a reduction in the analysis time while maintaining the same level of efficiency, but in some cases there is a also a gain in efficiency. As generic HPLC methodology has been implemented in many pharmaceutical companies to support process controls, the ability to maximise efficiency and increase the probability of a successful separation in a short cycle time has become paramount.

12.4.1   The Effect of Particle Size

The particle size of the stationary phases will affect both the *A* (eddy diffusion) and the *C* (mass transfer) terms. The result of reducing these terms has a marked effect on the van Deemter plots and thus the efficiency of the separation. This is shown in Figure 12.3. The combination of the *A*, *B* and *C* terms give us the van Deemter curve; the linear velocity associated with the maximum

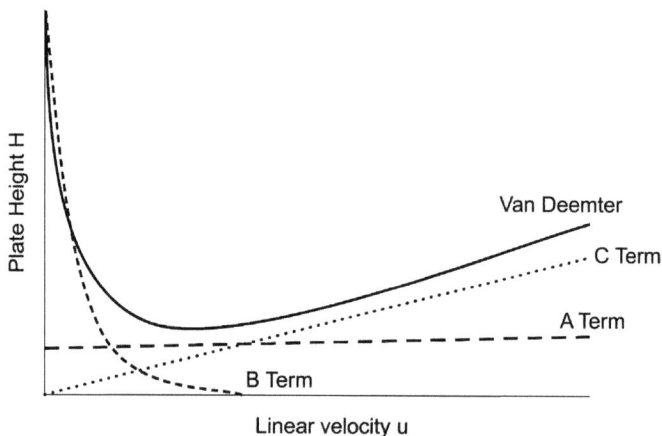

Figure 12.3 The three terms of the van Deemter equation are combined to generate a graphical representation of efficiency *vs.* linear velocity of the mobile phase. The lowest point of the curve represents the linear velocity that affords the most efficient separation.

efficiency is at the lowest point of the curve, or the point that gives the lowest plate height. As the velocity deviates from this lowest point a corresponding loss in efficiency is observed.

12.4.2   Sub 2 μm HPLC Packing Material

van Deemter clearly demonstrated the efficiency gains associated with reducing the particle size of chromatographic stationary phases as far back as the 1950s, yet the back pressures associated with a reduction in particle size have exceeded that of available HPLC equipment, which in almost all cases was limited to a maximum pressure limit of 400 bar ($\sim 300 \times 10^3$ Torr). The HPLC system parameters, such as maximum pumping pressure, extra column volume and detector collection rate, have long been the limiting factors preventing the chromatographer from exploiting the potential of short, narrow-bore columns packed with small diameter particles. The launch of the first ultra high performance liquid chromatography (UHPLC) instrument, by Waters in 2004, has opened the flood gates for instruments with extended pressure limits of up to 1200 bar ($\sim 900 \times 10^3$ Torr). HPLC columns packed with particles of less than 2 μm in diameter are now available from multiple vendors and are commonplace in analytical applications globally. The design of the stationary phase and the column hardware have been adapted to tolerate the increases in back pressure, and a number of stationary phases of different selectivities are now available in sub 2 μm dimensions. One of the main benefits of reducing the particle size is the effect of reducing the *C* term (resistance to mass transfer) of the van Deemter equation. Reducing the *C* term causes the curve to become flatter as the velocity increases, allowing higher flow rates without the associated loss in efficiency, as demonstrated in Figure 12.4.

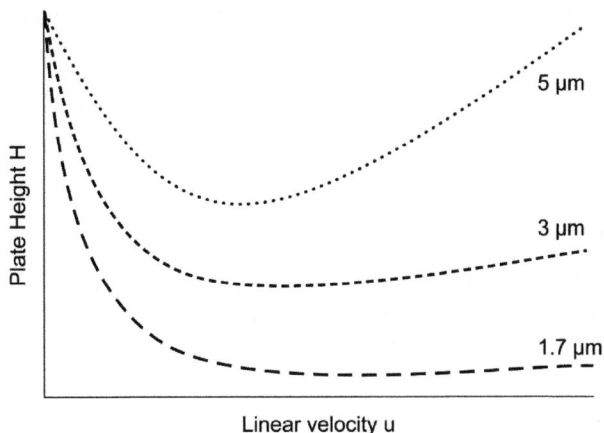

Figure 12.4   Combined van Deemter curves of HPLC columns packed with three different particle size stationary phases. The plate height is reduced (increased efficiency) as the particle size reduces. It should also be noted that the reduction in efficiency associated with increasing linear velocity is almost negated when using sub 2 μm particles.

12.4.3   Semi-porous HPLC Stationary Phases

Sub 2 μm particles have clearly marked the path forward for improved efficiency in HPLC analysis. However, there are some disadvantages, the most serious being the increase in back pressure. Reduction in particle size by a factor of two results in a four-fold increase in back pressure. This leaves the analyst with the choice of either using shorter columns, which often do not provide the efficiency gains required, or to purchase an HPLC system with high-pressure capability [>600 bar ($>450 \times 10^3$ Torr)]. Semi-porous stationary phases have revolutionised the speed and efficiency of HPLC separations at pressures that can be utilised by almost all HPLC instruments purchased in the past 10–15 years.

Semi-porous stationary phases with particle sizes less than 5 μm became widely available in 2007. Until this time the benefits of using semi-porous silica particles for small molecule separation were not commonly realised. Semi-porous phases consist of a solid spherical silica core surrounded by a thin porous layer of silica. This technology is not new and has been in use since the late 1960s.[4] However, the benefits of this type of stationary phase were not exploited for small-molecule applications until 2007, when the first semi-porous phase came to market in 2.7 μm particle size.[5] The semi-porous particles benefit in the same way that a sub 2 μm particle does, with respect to the reduction in band broadening, due to mass transfer effects, as the path through the particle that the analyte can take is restricted by the solid core. The process by which the silica particle is "grown" is also very well controlled as a result of starting from a solid silica core of known particle size. The resulting particles demonstrate a very tight particle size distribution, which enables incredibly high efficiency of packing. This has the effect of reducing eddy diffusion, since the path variation is reduced. The key benefit of using a semi-porous particle is that highly efficient separations can be achieved using particles of almost conventional size, resulting in significantly lower back pressures than observed with sub 2 μm particles.[6] The lower back pressures associated with fused core phases allows conventional HPLC equipment to be utilised at conventional flow rates.

12.5 Gas Chromatography

Owing to the non-volatile, often polar, nature of pharmaceutical compounds, HPLC is often the preferred separation technique for monitoring impurity profiles and determination of assay/yield. HPLC may have stolen some of the limelight from gas chromatography, but this has not prevented the evolution of GC equipment and applications.

GC is the workhorse for the analysis of volatile organic impurities present in drug substances and has remained relatively unchanged since the transition from packed columns to open tubular capillaries in the 1980s.[7] Although the column efficiencies, in plates/metre, cannot compare to modern HPLC applications, the sheer length of the columns available still make GC one of the most efficient separation techniques available, in terms of peak capacity. GC benefits from a range of available detectors, the most common being flame ionisation detection (FID). FID is almost universal and detects virtually all organic compounds, with high sensitivity. For applications where the analyte of interest does not contain a suitable chromophore, and is invisible to HPLC-UV detection, GC becomes a very attractive option. Other available detectors, such as electron capture (ECD), can be powerful tools for the detection and quantification of halogen containing species to ppm levels. This is increasingly more pertinent as the control on potential genotoxic alkylating agents, such as alkyl halides, becomes more heavily scrutinised.

12.5.1 Low Thermal Mass Gas Chromatographic Instruments

Low thermal mass GC instruments offer the analytical chemist the facility to perform very rapid thermal gradients, which results in reduced cycle times. Traditional GC equipment has always employed an oven for control of the column temperature. Although there have been many advances in the type of ovens used, and therefore improvements in heating rate, the technology is relatively antiquated, taking significant energy to heat and subsequent time to cool. Recent trends have seen multiple vendors supplying upgrades to their existing equipment that enables the use of low thermal mass heating techniques. In most cases an electrically operated heating element is located within the column, enabling resistive heating with rates of up to ~2000 °C min^{-1}. This technology provides the ability to perform extremely rapid gradients, thus reducing the run-time significantly. Other distinct advantages are the increase in cooling rate, which further reduces the method cycle time. Early iterations of this technology only permitted the use of short GC columns, such as 5 or 10 m, but the latest offerings enable 30 m columns to be used, so there is virtually no loss in efficiency.

12.6　Chromatographic Method Development

In their quest to increase productivity and efficiency through rapid analysis, process chemistry laboratories often use a standardised approach such as generic chromatographic methods to support routine chemical synthesis. Although this is sufficient in many cases, there will always be situations where these generic analytical procedures do not provide the appropriate separation. In these cases a specific chromatographic method is required. This is particularly true in the last stages of API synthesis, where the chemical reactions and the quality of the manufactured material need to be fully controlled. As most chemical reactions are monitored using chromatographic techniques, and since compound purity (including chiral purity) is primarily assessed using HPLC or GC, chromatographic method development is a key activity in analytical laboratories.

12.6.1　HPLC Column Screening

One of the tools available to the analyst who needs to rapidly develop a compound-specific HPLC method, or is seeking to optimise the separation of the components in a reaction mixture, is HPLC column screening. Screening is widely used within the pharmaceutical industry to rapidly identify the most selective method for given chemical substances, and provides an excellent starting point to any HPLC method development. HPLC systems dedicated to screening capabilities typically consist of a system fitted with a column switching valve, as illustrated in Figure 12.5. This enables the successive

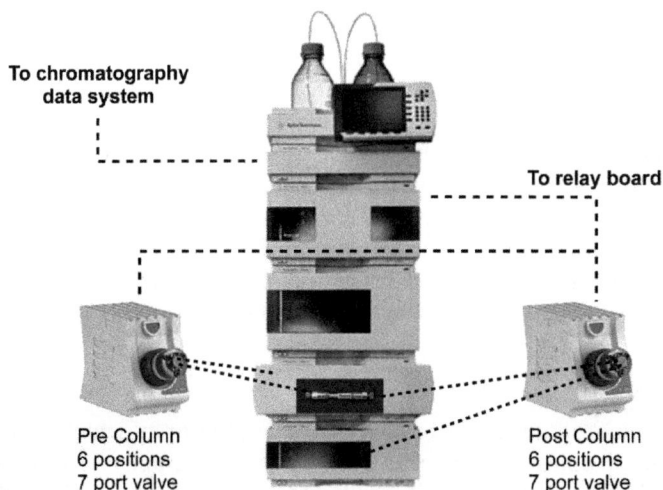

Figure 12.5　Automated HPLC column selection system configured with six stationary phases of orthogonal selectivity. Different mobile phase pH and organic modifier can also be selected to afford even greater selectivity.

injection of a sample onto several columns exhibiting different selectivities. In order to maximise the chances of successfully identifying the most suitable stationary phase, it is recommended to fit the screen with an "orthogonal" set of columns, *i.e.* to use a matrix of columns providing the largest choice of selectivities. In combination with a column switching valve, it is usual to screen different mobile phases. For instance, a sample could be run using successively acidic and neutral to basic mobile phases, or organic solvents of different strengths could be used.

Screening is generally used for the development of normal and reverse-phase HPLC. However, the principle can be applied to other chromatography techniques, such as supercritical fluid chromatography (SFC).

With the arrival on the market of UHPLC instrumentation, as well as small particle columns, it is now possible to screen a number of stationary phases and mobile phase compositions in a relatively short period of time, typically within a working day.

12.6.2 Achiral HPLC Screening

Currently, most chromatography methods for achiral compounds use reverse-phase liquid chromatography (RPLC) mechanisms. That is why the RPLC screen is usually the method development tool of choice. Achiral HPLC screening is generally performed if a mixture of compounds is proving difficult to separate, or when method optimisation is required. For instance, it is common practice to perform achiral HPLC screening if the HPLC method is used to assess the purity of the final API. It is critical in this case that the analytical method is able to separate all impurities from the manufactured product and also that it is stability indicating.

A typical screen is usually fitted with silica-based HPLC columns bearing bonded ligands such as C_{18}, C_8, phenyl, cyano, or pentafluorophenyl (PFP), to provide a suitable variety of selectivities. These modifying functional groups are available from most HPLC column manufacturers and are available in a variety of particle sizes and column geometries. As mentioned in Section 12.5.1, running RPLC screens at various pH values or using different organic modifiers (acetonitrile, methanol, tetrahydrofuran) can also be useful, as these parameters will more likely have an impact on the separation. However, as silica-based columns are pH sensitive, their use should be limited to the pH range 2.0–8.0 unless otherwise stated by the manufacturer.

12.6.3 Chiral HPLC Screening

Chiral chromatography is a branch of chromatography which focuses on the separation of chiral substances. Chiral chromatography is essential to the analysis of drug substances as it is critical to assess enantiomeric purity early in the lifecycle of a compound.

Typically, enantiomeric separations can be achieved through an appropriate choice of stationary and mobile phases. However, unlike achiral separations for which appropriate stationary and mobile phases can sometimes be predetermined by the nature of the sample, chiral separations are the result of complex and not always fully understood mechanisms. It is therefore extremely difficult to predict suitable chromatographic conditions and method development is usually conducted through a trial-and-error approach. Although literature suggests chiral modifiers added to the mobile phase have been successfully applied in chiral separations, it is far more common to achieve enantiomeric separation through appropriate choice of chiral stationary phases (CSPs).

At present, CSPs are available to use with normal-phase liquid chromatography (NPLC), RPLC and SFC and are available in a wide range of selectivities. The first step in the screening process is to identify a reduced set of CSPs.[8] Polysaccharide CSPs have been a favourite of the pharmaceutical industry for some time owing to their high success rate; most commonly used CSPs include the immobilised cellulose tris(3,5-dimethylphenylcarbamate), amylose tris(3,5-dimethylphenylcarbamate) and cellulose tris(4-methylbenzoate). These three CSPs are the subject of numerous publications and have demonstrated complementary separation of acidic, basic and neutral compounds.[9]

12.6.4   Multi-channel Column Screening

Screening HPLC stationary phases has traditionally been a linear process, the sample being injected in turn onto a set of columns. Although automation and the introduction of UHPLC has revolutionised this process and enabled cycle times to be measured in hours rather than days, there is still scope for further reduction in analytical cycle time. An eight-channel microfluidic HPLC has been demonstrated, which permits the screening of eight stationary phases in parallel.[10] As two miniature pneumatic pumps serve each channel, the instrument is capable of generating different gradients with different mobile phases on each chromatographic column, further enabling the optimisation of chromatography conditions. In theory this approach has the potential to reduce the screening time from hours to minutes. Figure 12.6 shows a typical column screen for flavanone using Chiralpak AD-H, Chiralpak AS-H, Chiralcel OD-H and Chiralcel OJ-H stationary phases (Chiral Technologies, West Chester, PA, USA). The screen was conducted using propan-2-ol (IPA) and ethanol as organic modifiers in heptane. The system was able to identify a suitable separation in just 30 minutes, with conditions that were scalable to conventional HPLC flow rates and column dimensions.

12.6.5   Screening Using Supercritical Fluid Chromatography

As mentioned earlier, chiral separations can be performed using NPLC or RPLC, in which case they are comparable to traditional chromatography. However, in recent years, the use of SFC for chiral separations seems to have become the preference throughout the industry. It is easy to understand why: SFC systems are now commercially and readily available. The fact that they use

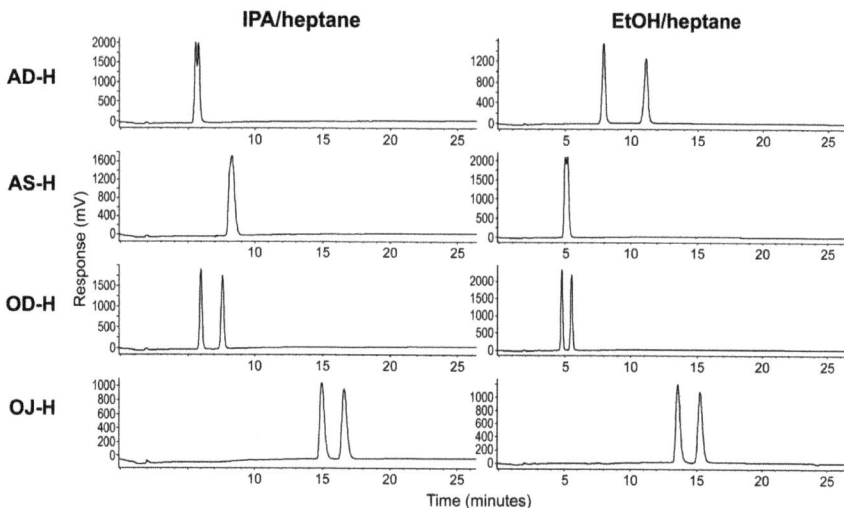

Figure 12.6   Chiral method development for flavanone enantiomers using standard gradient elution approach on an eight-channel microfluidic HPLC, using four different columns and two mobile phase configurations.

carbon dioxide as the main mobile phase is a definite advantage as it results in a significant reduction in environmental impact, compared to traditional solvents. The low density of the supercritical carbon dioxide also enables higher flow rates through the system than when using more traditional approaches, which accounts for faster equilibration and analysis times.

Method screening using SFC follows the same principles as other HPLC screenings: separation is attempted on a set of columns, using carbon dioxide as the non-polar solvent, and a polar organic modifier. In general, SFC screening systems have two switching valves: a column switch and a modifier switch. This allows screening of a set of organic modifiers onto the set of columns. As carbon dioxide is miscible with most common organic solvents (unlike normal-phase non-polar solvents such as hexane or heptane), it is possible to use solvents such as methanol or acetonitrile as organic modifiers in addition to ethanol or IPA, which are commonly used in NPLC.[11]

12.6.6   GC Column Screening

Predicting the selectivity of GC stationary phases, especially chiral ones, can be very difficult. Multiple stationary phases must typically be evaluated before a suitable separation is obtained. Manually changing GC columns is laborious and time consuming and to dedicate a single instrument to a single stationary phase would not be cost effective. Like HPLC it is therefore beneficial to automate the column screening process. Screening GC stationary phases is not a practice typically automated. The use of high-temperature valves in GC ovens has been demonstrated, but there are significant downsides to the use of these

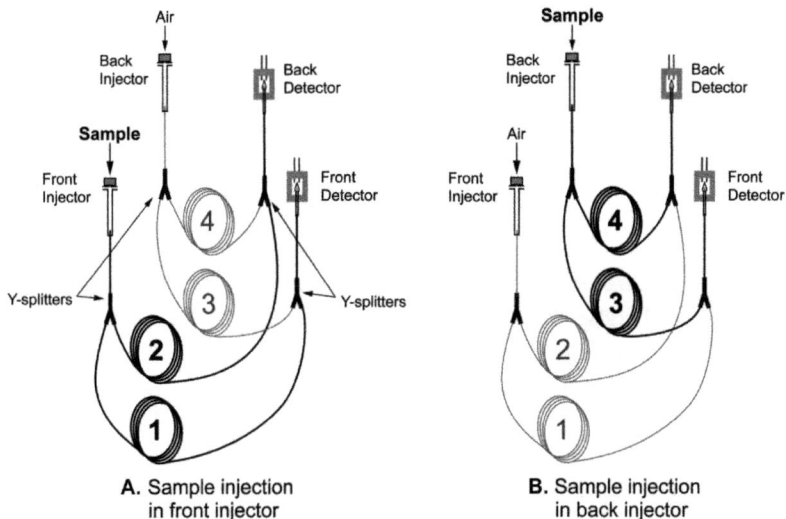

Figure 12.7 (A) Initial injection is made on front injector, where the analyte is split into columns 1 and 2 for separation and detection on the front and back detectors. (B) A second injection is made on the back injector, where the analyte is split into column 3 and 4 for separation and detection on the front and back detectors.

valves. The materials must be inert and demonstrate stability and sealing capability at a range of temperatures. There is also a thermal lag in the temperature of the valve when compared to the oven temperature. Merck demonstrated a low-cost chiral screening system, utilising four orthogonal chiral columns with simple Y splitters. The instrument features two split/splitless injectors and two FID detectors. The use of static splitting negates the need for expensive switching valves and enables the screening of two columns simultaneously. A total of four columns can be screened in just two chromatographic runs, as illustrated in Figure 12.7.

Figure 12.8 shows the typical output from a chiral GC column screen experiment. The racemic diol is analysed on all four columns in approximately 30 minutes, generating four chromatograms. The analytical chemist can quickly review the data and select the stationary phase which provides the best separation. In cases where incomplete separation is obtained, the data can still be a useful starting point, highlighting conditions worthy of further optimisation.

12.7   Preparative Chromatography

The vast majority of chromatographic separations today are for the purpose of analytical determinations. The analytical chemist has embraced the technique and put it at the forefront of the arsenal of techniques for reaction/process monitoring; however, the first reported applications of chromatography are

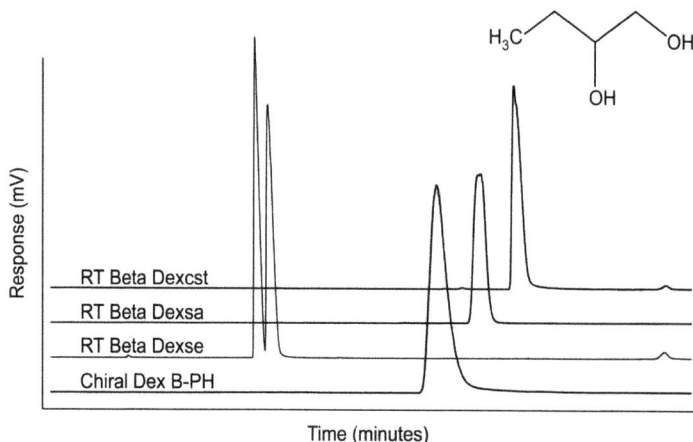

Figure 12.8 Overlaid chromatograms demonstrate the selectivity differences in chiral GC stationary phases. The separation of butanediol enantiomers was established in less than 1 h using a multicolumn screening approach.

accredited to the botanist Mikail Tswett. Tswett used low-pressure (mobile phase driven by gravity) chromatography for the separation, and subsequent collection, of complex mixtures of plant extract. Preparative chromatography remains mostly unchanged to this day and is a powerful tool for the rapid purification of microgram to kilogram quantities of an API or intermediates.[12] The technique is typically employed to upgrade the purity of materials or as a means of delivering material of suitable enantiopurity where a classical resolution cannot be achieved or is not practical. The use of preparative chromatography over more traditional purification techniques has become increasingly more widespread as the industry has realised both the reduction in cycle times and the cost saving benefits. While the typical chromatographic parameters (described earlier) apply to preparative chromatography in the same way in which they do to analytical chromatography, there is the additional critical parameter of productivity. Productivity is typically expressed as the quantity of material that can be purified for a given quantity of stationary phase in a given time period. The most common unit of measurement is kg kg^{-1} d^{-1}. An estimation of productivity can be obtained through loading experiments on an analytical chromatographic system. Increasingly larger quantities of material are injected onto the column of interest until the separation is lost. From these data the productivity of columns of different diameters can be readily calculated with a high level of accuracy. The resulting information can then be used to determine whether the time and labour costs are viable for a given purification. Another critical parameter that must also be considered is solvent usage. Is the quantity of solvent required acceptable to achieve the desired quantity of purified material? Both the cost and ability to remove and dispose of the solvent should be considered. In addition, the environmental impact may also be of concern.

12.7.1 Supercritical Fluid Preparative Chromatography

SFC is becoming more widespread across the pharmaceutical industry for purification up to low kg quantities. SFC utilises carbon dioxide, in a super-critical fluid state, as the mobile phase in place of more common solvents, such as heptane or hexane. Organic modifiers can be added in the same way as normal phase separations to enhance selectivity. There is not only a speed increase, in chromatographic terms, associated with SFC, but also the desol-vation of the material is significantly more efficient as carbon dioxide is a gas at room temperature and pressure. The fractions are typically collected and the major component of the mobile phase, carbon dioxide, is allowed to evaporate, leaving the compound dissolved in the organic modifier of the mobile phase. The environmental impact of using carbon dioxide as a mobile phase is sig-nificantly less than using organic solvents such as heptane and hexane. The organic modifiers are typically relatively innocuous alcohols such as ethanol or IPA, which further encourages the use of SFC as the "green" option when chromatographic purification is desired on scale.

12.8 On-line/In-line Analytical Techniques

Reaction monitoring is commonly used as part of process development, as it provides an insight into the kinetics of the formation of product, intermediates or by-products. Typically, reaction monitoring and in-process testing in the pharmaceutical industry is carried out using off-line techniques such as HPLC, GC or NMR, where aliquots from the reaction mixtures are manually sampled, processed by, for example, filtration, quench and dilution prior to analysis. However, these conventional methods can sometimes be time consuming, labour intensive or can involve potential exposure to toxic substances; this is why the use of automated on-line/in-line techniques is being encouraged for the development of efficient processes in some cases. At present, the majority of on-line/in-line techniques used for reaction monitoring or in-process testing are spectroscopic techniques, with the exception of on-line HPLC.

12.8.1 On-line HPLC

The most common technique for reaction monitoring is HPLC, due to its ability to resolve complex mixtures of compounds in a variety of matrices in a highly sensitive fashion. Reaction monitoring through HPLC allows, for instance, the chemist to monitor the formation of low-level impurities or to monitor the enantiopurity of materials in a reaction to ensure that there is no erosion of the enantiopurity under the reaction conditions. However, HPLC analysis can be labour intensive as the sample needs to be taken from the reaction mixture, quenched and diluted, then analysed. This can also be dis-ruptive for the reaction if a vacuum is involved for example, or if the reaction is performed on a small scale using low volumes.

Although many applications of on-line HPLC analysis have been demonstrated, the technology never met with significant uptake due to sampling/sample preparation issues and system size/complexity. To some extent these issues have been mitigated with the development of microfluidic HPLC, which has been applied to on-line analysis and has resulted in the development of new, easy to use, portable systems which can be located in the laboratory or directly in pilot/production plant where the reaction of interest is conducted.[13] Additionally, as the flow rates used for microfluidic systems are quite low (typically $4 \, \mu L \, min^{-1}$ for a 0.3 mm diameter column), only small aliquots of sample are needed (typically 10–40 μL) and solvent requirements are minimal. This is a definite advantage in the pilot/manufacturing plant as these low-volume systems present less safety hazards. Commercial microfluidic on-line HPLC systems enable fully automated sampling, filtration and dilution of an aliquot from a reaction mixture, followed by injection on a microfluidic HPLC system. Data collection and analysis are also automated.

The small size of the aliquot required for analysis enables the collection of multiple successive samples from reactions without affecting the overall volume. It is possible to not only monitor the reaction completion during the chemical processing, but also to establish the reaction time course by plotting reactant concentration *vs.* time when the process is in development stages, as shown in Figure 12.9. As a wide selection of stationary phases is commercially available, including both achiral and chiral phases, a large variety of reactions can be studied.

12.8.2   In-line Near/Mid-infrared Spectroscopy

One of the disadvantages of on-line HPLC is that, despite the very small sample aliquots, it still requires the removal of a sample from the reaction mixture, followed by sample preparation and analysis. This can result in a potential delay of

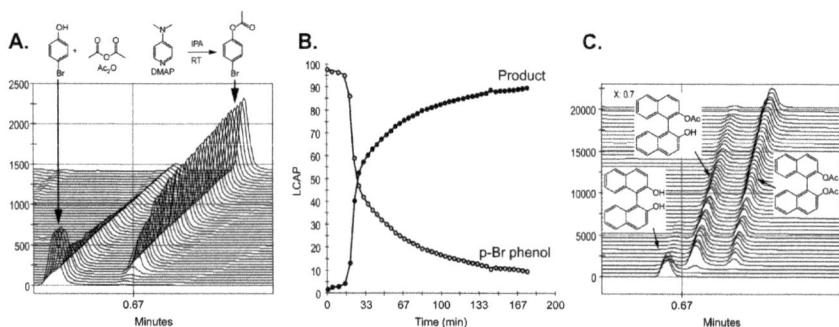

Figure 12.9   Monitoring the time course of DMAP-catalysed acylation of phenols with acetic anhydride. (A) Waterfall plot of chromatograms *vs.* time showing acylation of *p*-bromophenol. (B) Plot of area% *vs.* time, showing a slowing of the reaction rate over time. (C) Waterfall plot showing acylation of β-binaphthol (1,1'-binaphthalene-2,2'-diol) (reproduced from reference 13 with permission from American Chemical Society).

up to 1 hour in data generation associated with the HPLC analysis cycle time. This also poses an issue when handling potent, unstable compounds or reaction intermediates. Therefore there has been a growing interest in techniques which allow *in situ* analysis, such as near-infrared spectroscopy (NIRS) and Fourier transformed infrared (FTIR) in-line techniques. These spectroscopic techniques are suitable for the qualitative and quantitative analysis of reaction mixtures without any sample preparation, allowing the measurement to be taken in real time. This is achieved through the implementation of a measuring device (typically a remote fibre optic probe) in the reaction vessel or *via* a pumped loop.[14] Moreover, unlike HPLC-UV techniques, NIRS and FTIR techniques can be used to analyse compounds which do not display suitable UV absorbance and are not amenable to HPLC-UV analysis. The different components of the mixture can simply be differentiated by monitoring the appropriate characteristic wavelengths. Furthermore, as the measuring device is placed at the site of the reaction, analysis can be performed without interrupting the ongoing reaction (by remaining under vacuum for instance), or placing the operator under chemical hazard.

However, there are some limitations to spectroscopic in-line analyses: there is generally a need for external calibrations (often performed using HPLC or GC), which can be labour intensive, and the low sensitivity of the technique means it is best suited to reaction monitoring, that is product formation rather than impurity monitoring. Applications of in-line spectroscopy are nonetheless numerous and their use in support of pharmaceutical process research is gaining in popularity.[15–18]

NIRS also provides a powerful analytical advantage in that it can differentiate compounds in the solid state and even those in different polymorphic or salt forms, providing real time physical property data. NIRS has also been utilised in the monitoring of drug substance drying, for which it has shown an ability to differentiate between surface and bound water, and allowed drying to the desired trihydrate of an API, as shown in Figure 12.10.[19] Another example is the salt formation of a monohydrate salt, which was also monitored by in-line FTIR spectroscopy, overcoming the limitations of HPLC or titration by allowing differentiation between the mono- and the bis-hydrochloride salts of the API and allowing determination of the end-point very accurately.[20] This is shown in Figure 12.11

In-line FTIR was also found to be applicable to the analysis of air-sensitive compounds and highly corrosive materials, where the actual act of taking a sample from the vessel may invalidate the data or be potentially hazardous. The titration of Grignard reagents and the conversion of a highly toxic critical intermediate using in-line FTIR spectroscopy has been demonstrated, and displays good agreement with results generated by titration and HPLC.[21]

12.9 Validation of Analytical Procedures

While pharmaceutical process laboratories are constantly driving for increased productivity and efficiency through rapid analysis, there is also a continuous

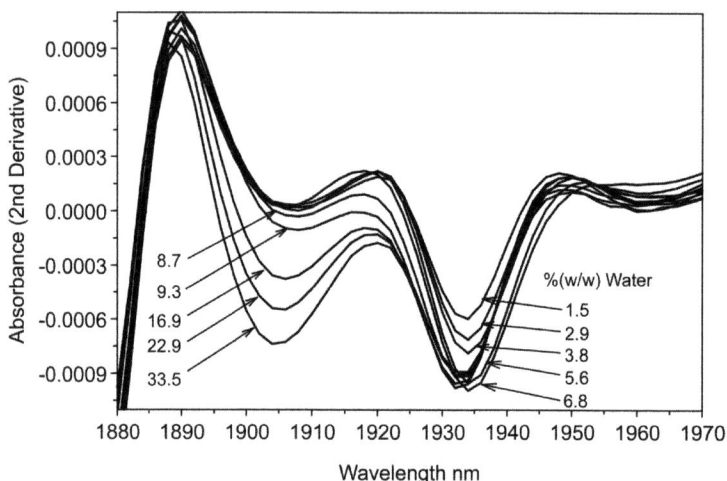

Figure 12.10 Second-derivative NIR spectra of API samples showing differentiation between levels of unbound (surface water) and bound water (incorporated in the crystal lattice) upon drying (reproduced from reference 19 with permission from John Wiley and Sons).

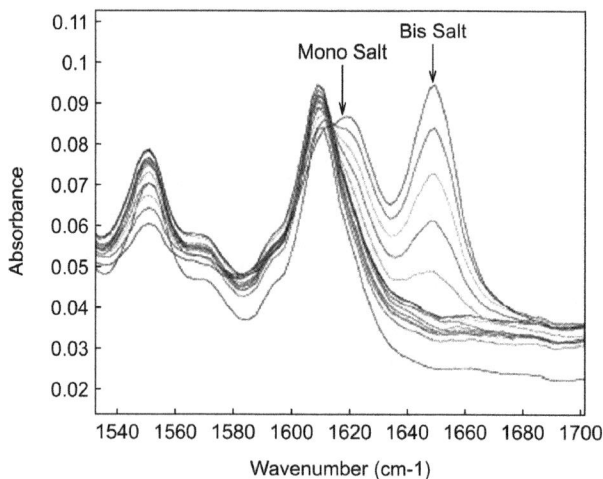

Figure 12.11 FT-IR reaction monitoring of mono-HCl salt formation. The hydrochloric acid charge is optimised to minimise the formation of the bis-HCl salt (reproduced from reference 20 with permission from Elsevier).

concern that the rapidity with which analytical data are generated must be matched by the quality with which they are produced. There cannot be any compromise to the safety of the individuals receiving the API; this is why it is essential that not only the analytical testing be thorough, but also that the analytical procedures utilised be fully validated according to regulatory guidelines.

The validation of analytical procedures is a regulatory requirement for all investigational or commercialised drug substance. Its purpose is to demonstrate the suitability of the analytical procedure for its intended purpose. Guidelines for the validation of analytical procedures are available from the International Conference on Harmonisation (ICH) of technical requirements for registration of pharmaceuticals for human use and should be referred to when performing methods validation.[22] The validation of an analytical procedure generally requires the evaluation of the following method characteristics: accuracy, precision, specificity, limit of detection (LOD), limit of quantitation (LOQ), linearity, range and robustness. It is generally incumbent on the analyst to determine the appropriate validation studies required to evaluate these characteristics (provided they are in agreement with ICH guidelines for clinical study materials); these studies can vary depending on the stage of development of the drug substance. Generally, full validation testing is performed for regulatory analytical procedures at late development stages when the manufacturing process is in place, that is when no changes in impurity profiles or analytical data are expected from batch to batch.

Validation is required for all analytical procedures used to assess the quality of a chemical entity, such as identification methods, purity methods and assays. The validation characteristics requiring assessment also depend on the nature of the analytical method.

12.9.1   Accuracy

As defined in the ICH guidelines, "The accuracy of an analytical procedure expresses the closeness of agreement between the value which is accepted either as a conventional true value or an accepted reference value and the value found". The measurement of accuracy is instrumental in assessing parameters such as the solubility of the chemical entity in the assay solvent, or the effects caused by the matrix on the analytical technique.

In practice, accuracy can be assessed by submitting a well-characterised material (reference standard) to the analytical method and verifying the agreement between the reference value and the value experimentally determined, or by comparing the experimental result to a result obtained using an alternative procedure. Accuracy of purity methods can also be assessed using spiked samples to determine the percentage recovery, for instance. The acceptance criteria for accuracy will be determined prior to the validation being performed and should be set at an appropriate level, commensurate with the measurement being assessed. For example, it cannot be expected to have the same level of accuracy when determining impurities at, or near, the LOQ as when determining the assay of a drug substance.

12.9.2   Linearity/Range

A linear relationship should be demonstrated across the range of the analytical method, confirming that the test results are directly proportional to the amount of analyte in the sample. Typically, linearity should be demonstrated using a minimum of five concentrations over the range. In general, linearity should be demonstrated at all stages of development for all purity methods. The range is generally assessed depending on the nature of the analytical procedure.

12.9.3   Precision

The precision of an analytical method expresses the scatter of results between a series of measurements under unchanged conditions. Precision is generally evaluated at three levels: repeatability (intra-assay precision,), intermediate precision (intra-laboratory precision) and reproducibility (inter-laboratory precision) for most analytical procedures (identity, purity and assay tests).

Often, in the early development of a pharmaceutical entity, only the repeatability of the analytical method will be tested. Intermediate precision and reproducibility will, in general, only be evaluated at later stages of development when the pharmaceutical process has been designed and the API has reached clinical development phases II or III.

Precision is generally evaluated using the standard deviation or relative standard deviation of the series of measurements. Acceptable specifications for precision are set by the analyst and should take into account the incidence of the reported analytical result.

12.9.4   LOD/LOQ

The LOD and LOQ are essential variables for an analytical method and should always be determined, no matter at what stage of development the analytical procedure is applied. These limits define the lowest amount of analyte which can be detected or quantified with precision and accuracy by the method and are particularly relevant when determining purity or stability.

The approaches for determining LOD/LOQ are numerous and depend on the nature of the analytical procedure. The most common approaches include but are not restricted to:

- Visual evaluation of the lowest level at which the analyte can be determined.
- Determination of the signal-to-noise ratio. This is only applicable to procedures such as chromatography, or spectral methods which exhibit baseline noise. Generally, a signal-to-noise ratio of 3 is required for LOD and 5–10 for LOQ. The signal-to-noise ratio is generally determined using the root mean square (RMS) or peak-to-peak (as per Figure 12.12) approaches. Often, these parameters can be calculated (eqns 12.2 and 12.3) through the data acquisition software.

$$\text{LOD} = \frac{3.3\sigma}{S} \qquad (12.2)$$

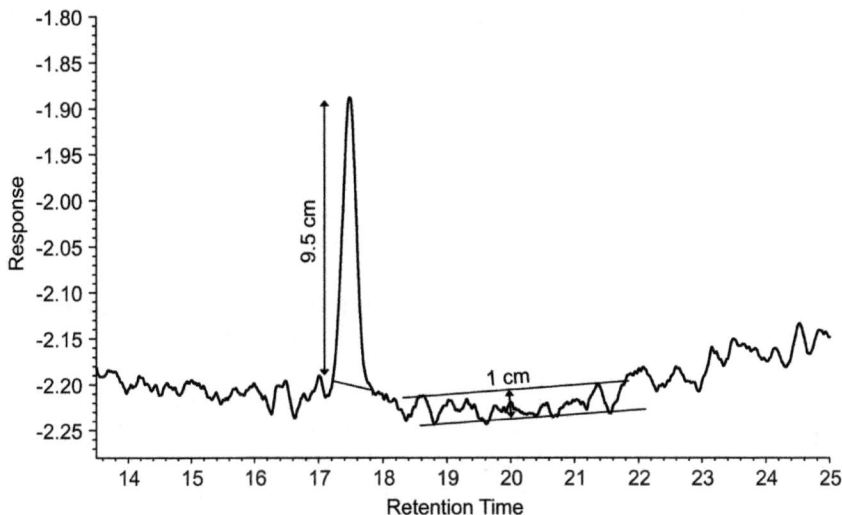

Figure 12.12 Peak-to-peak signal-to-noise determination (s/n = 9.5/1 = 9.5); the background noise (1 cm) and the intensity of the signal (9.5 cm) are determined manually or through data acquisition software by measuring the amplitude of the detector output over a range of baseline noise and at peak maxima.

$$LOQ = \frac{10\sigma}{S} \qquad (12.3)$$

where σ = standard deviation of the response (based on the blank or calibration curve) and S = slope of the calibration curve.

- LOD/LOQ determination based on the standard deviation of the response and the slope from the linearity/calibration curve.
- Percent recoveries for a pre-determined spike level. Appropriate level of recovery is subjective and should be defined carefully, as very low level determinations may result in recoveries that are not comparable to those generally observed at the optimum range of the analytical method.

The means of calculating the signal-to-noise ratio should be carefully documented in the analytical reports, as different approaches can give significantly different results.

12.9.5 Specificity

As defined in ICH guidelines, demonstrating the specificity of an analytical procedure will demonstrate "the ability to assess unequivocally the analyte in the presence of components which may be expected to be present". Approaches to specificity assessment can vary greatly, depending on the stage of development of the analyte, though specificity is a validation parameter which is assessed for most analytical procedures. Specificity is typically assessed by comparing the analytical results obtained for the chosen analyte to the response of the analytical method towards impurities, starting materials or known degradation products.

The specificity of an analytical procedure should be assessed, because not only is it important to ensure that reported analytical results are specific to the analyte and are not affected by the potential impurities or degradates, but also it can be a means to verify that the method is stability indicating and is able to quantify any degradation occurring.

In practice, the specificity of the analytical procedure towards degradates is assessed by deliberate degradation of the compound under chosen stressing conditions. The stressing conditions can be various, but typically cover a wide range that includes conditions representative of those that the drug substance may be exposed to upon subsequent formulation. This includes submitting samples to acidic, basic, oxidative, light and thermal stresses for instance, and verifying that the products of degradation can be detected by the analytical procedure. Advances in laboratory instrumentation mean stress degradation studies can now be performed using automated systems. Not only does this allow the analyst to save time and effort, but it also enables a standardised approach to forced degradation of chemical entities.

12.9.6 Robustness

The robustness of an analytical method expresses its ability to remain unchanged by small variations in parameters and provides information on the reliability of the procedure. Robustness is typically assessed for APIs in late phases of development, when the manufacturing process of the chemical entity is well defined and when the typical final impurity profile and analytical testing plan have been determined. Robustness is typically assessed by measuring the effects of deliberate variations in the method parameters; the nature of these variations should be adapted to the analytical procedure undergoing validation and can be determined using a design of experiments (DOE) approach.

12.10   Genotoxic Assessment

In recent years, pharmaceutical regulators have turned their focus on potential genotoxic compounds used, or generated in synthetic processes. Since the first white paper was published,[23] there has followed a number of guidance documents from regulators on the subject of potential genotoxic impurities (PGIs) in APIs, listing the structural alerts. The list includes moieties such as nitro containing species, aldehydes and alkyl halides. PGIs are required to be controlled to ppm levels in drug substances, depending on the dose and dosing duration.

PGIs are often identified by screening through knowledge-based computer systems, such as DEREK (deductive estimation of risk from existing

Table 12.1 Staged TTC taken from the EMEA guidelines.[25]

	Duration of exposure				
	Single dose	≥1 month	≥3 months	≥6 months	≥12 months
Allowable daily intake	120 μg	60 μg	20 μg	10 μg	5 μg

knowledge) or MCASE. The computer programs assess the molecular structure and compare it to a database of "structural alerts" that flags the compound, should it contain functional groups that may be of concern. This type of screening is common practice across the entire pharmaceutical industry. Screening compounds present in the process is critical, but it is also the responsibility of the chemist to assess the reaction and its conditions in case there is the potential to generate PGIs. For example, preparing a sulfonate salt in the presence of an alcohol has the potential to form sulfonate esters, which are known to be genotoxic. In some conditions, hydrochloric acid present in alcohols can generate alkyl chlorides, which also flag as concerns in many of the software screening tools.

A hit from the computer database is not a guarantee that the compound will be genotoxic. Where possible, compounds flagged as PGIs should be submitted to *in vitro* testing such as the AMES and chromosomal aberration to ensure the DEREK assessment is correct. Often sufficient quantities of the compound are not available, and the only option is to apply analytical testing to ensure that the levels are below a safe limit. The common *in vitro* tests do not provide sufficient information to set specific acceptance limits. In these cases the staged threshold of toxicological concern (TTC) approach is applied.[24] The threshold is based on a statistically valid number of carcinogens. Using the probability distribution of carcinogenic potencies, an estimation of the daily exposure level is calculated that would amount to a one in a million lifetime risk of cancer. The EMEA guideline proposes a TTC of $1.5 \mu g \, d^{-1}$. However, higher concentrations may be acceptable based on the duration of the exposure or the dose regimen, for example once weekly *versus* daily. For compounds in early development, higher concentrations may be acceptable since most clinical studies are of relatively short duration. This approach is often referred to as the "staged TTC". This is illustrated in Table 12.1.

12.11 Mass Spectrometric Detection

Early phase approaches to PGI analysis often utilise mass spectrometry (MS) as a chromatographic detector, coupled to either gas or liquid chromatography systems. The highly selective quadrupole MS detectors enable the detection of very low concentrations of analyte, in very complex matrices. The most common types of MS scan used in quantitative analysis are full scan, selected ion monitoring (SIM) and selected or multiple reaction monitoring (SRM or MRM). SRM and MRM are essentially the same experiment. In full scan mode the quadrupole is set to scan over a wide mass range and enables the operator to collect data for all ions observed within the given mass range. This is very useful in qualitative analysis, but for quantitative analysis the selectivity is poor, which can lead to decreased sensitivity. The quadrupole mass filter can only detect a single m/z value at a given time. The wider the mass range, the less time is spent monitoring a specific m/z value, which leads to a reduction in sensitivity. An example of a full scan chromatogram is shown in Figure 12.13.

Single quadrupole instruments are capable of running SIM scans. In this mode the quadrupole is set to monitor at a specific m/z value. This has two advantages over a full scan: one is the increase in sensitivity as a result of the increase in the duty cycle, since now only the transmission of ions of a single m/z value is achieved; the other is the increase in selectivity. SIM chromatograms can be very simple, and in cases where the sample matrix is uncomplicated can feature just a single peak. An example of a SIM chromatogram is shown in Figure 12.14.

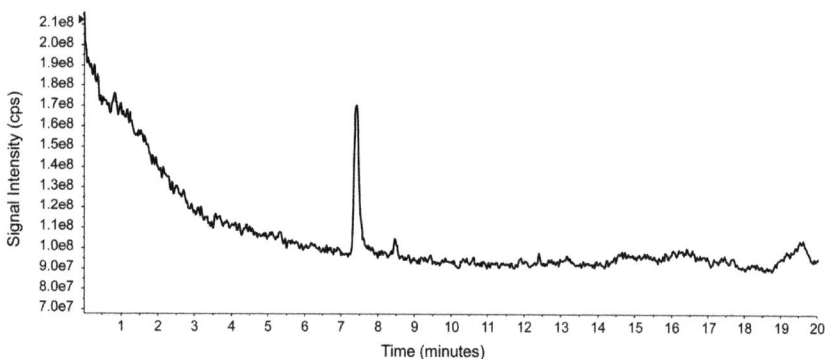

Figure 12.13   Full scan chromatogram of $1\,mg\,mL^{-1}$ solution of *p*-toluenesulfonic acid methyl ester. Electrospray MS scan from m/z 100 to 300.

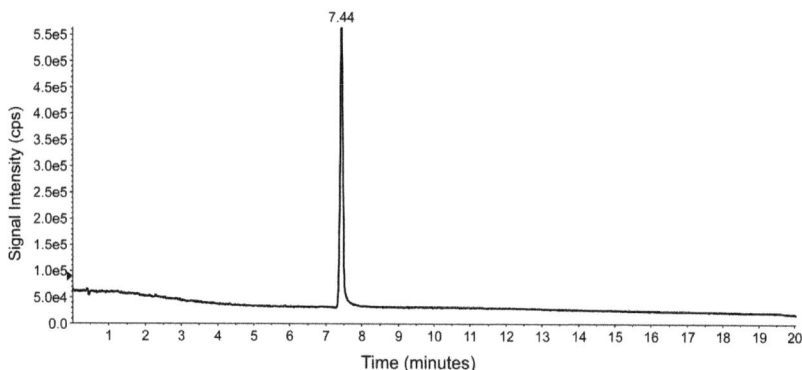

Figure 12.14   SIM chromatogram of $0.1\,mg\,mL^{-1}$ solution of *p*-toluenesulfonic acid methyl ester. Electrospray MS SIM of m/z 187.

For very low level determinations (ppb), triple quadrupole MS instruments can be utilised. While a single quadrupole instrument in SIM mode offers a high level of selectivity, there is still the potential for interference from isobaric (same nominal molecular weight) compounds present in the sample matrix. By running an MRM experiment, much of this interference can be eradicated. The essence of MRM is to set the first quadrupole (Q1) to monitor a specific m/z value, known as the precursor ion. The ions transferred into the second quadrupole (Q2), or collision cell, are then subjected to collision induced dissociation (CID). The resulting fragment ions are then transferred to the third quadrupole (Q3), which is again set to monitor a specific m/z value or product ion. The transition from precursor to product ion is specific for a given molecule and adds a further level of selectivity. A schematic for an MRM mechanism and an example of an MRM chromatogram are given in Figure 12.15 and Figure 12.16, respectively.

Figure 12.17 shows the MRM transition of m/z 187 > 91 of p-toluenesulfonic acid methyl ester, a common PGI, which can potentially form when using p-toluenesulfonic acid in the presence of methanol.

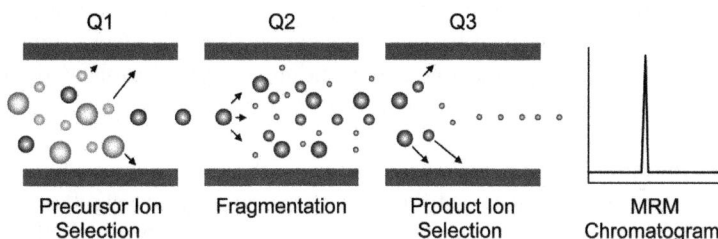

Figure 12.15 Multiple reaction monitoring (MRM). Precursor ion selection occurs in quadrupole 1 (Q1), allowing only ions of a specified mass to pass. The ions enter quadrupole 2 (Q2, collision cell) and are subjected to collisionally induced dissociation (fragmentation). The desired product ion is selected in quadrupole 3 (Q3) for detection

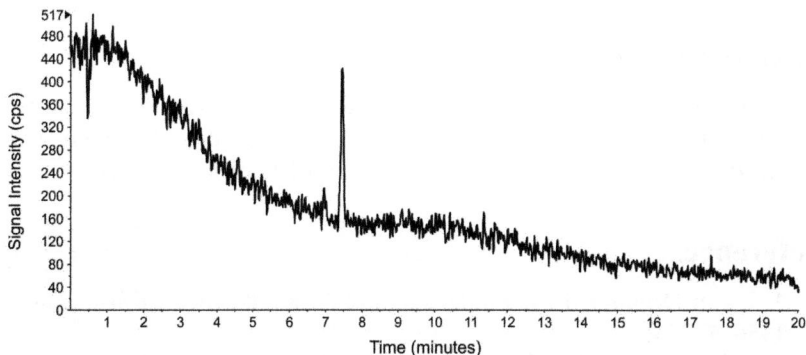

Figure 12.16 MRM chromatogram of 0.0001 mg mL^{-1} solution of p-toluenesulfonic acid methyl ester. Electrospray MS MRM of m/z 187 > m/z 91.

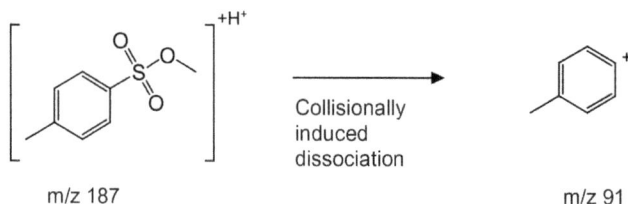

m/z 187   m/z 91

Figure 12.17   MRM collisionally induced dissociation. The precursor ion is selected
and undergoes fragmentation to form the product ion. This transition
adds an additional level of selectivity to the analysis.

12.12   Conclusion

The field of pharmaceutical process development is an ever changing landscape. As the drive to reduce cycle time increases, the need to gain as much information about a process, and fully exploit the available data generated, becomes more critical. Analytical chemistry is constantly evolving to address the change in paradigm and provide more detailed assessment of reaction kinetics and impurity profiling. The ability to develop analytical methodology in hours rather than days has enabled the synthetic chemist to move from fume hood to pilot plant more rapidly, and the use of on-line analytical tools at the location of the reaction, collecting data in real time, has provided greater confidence in the scaling of reactions.

While the analytical chemist has embraced the challenges of reduced cycle times, the need for validated methodology to support agency regulations has not relaxed. The data generated is still required to meet the strict guidelines of the numerous global regulators and the increased focus on genotoxic impurities in pharmaceutical drug substance has propagated the application of complex mass spectrometric detection in analytical areas, where previously these were the pursuits of "mass spectrometrists". The synergy between process and analytical chemistry grows ever stronger, and the importance of a flexible interface between analytical and process even more critical in addressing the associated challenges.

Acknowledgements

The authors would like to thank Sophie Strickfuss for her spectroscopy contribution, Michele McColgan for graphic design and also Tony Davies, Gareth Pearce and Chris Welch for review and technical guidance.

References

1. J. J. van Deemter, F. J. Zuiderweg and A. Klinkenberg, *Chem. Eng. Sci.*, 1956, **5**, 271.
2. A. J. Martin and R. L. Synge, *Biochem. J.*, 1941, **35**, 1358.
3. L. R. Snyder, J. J. Kirkland and J. L. Glajch, in *Practical HPLC Method Development*, Wiley-Interscience, New York, 2nd edn, 1997.

4. J. J. Kirkland, *Anal. Chem.*, 1969, **41**, 218.
5. J. J. Kirkland, F. A. Truszkowski, C. H. Dilks Jr. and G. S. Engel, *J. Chromatogr., A*, 2000, **890**, 3.
6. A. Abrahim, M. Al-Sayah, P. Skrdla, Y. Bereznitski, Y. Chen and N. Wu, *J. Pharm. Biomed. Anal.*, 2010, **51**, 131.
7. W. G. Jennings, K. Yabumoto and R. H. Wohleb, *J. Chromatogr. Sci.*, 1974, **12**, 344.
8. Z. Pirzada, M. Personick, M. Biba, X. Gong, L. Zhou, W. Schafer, C. Roussel and C. J. Welch, *J. Chromatogr., A*, 2010, **1217**, 1134.
9. C. Perrin, V. A. Vu, N. Matthijs, M. Maftouh, D. L. Massart and Y. Vander Heyden, *J. Chromatogr., A*, 2002, **947**, 69.
10. P. Sajonz, W. Schafer, X. Gong, S. Shultz, T. Rosner and C. J. Welch, *J. Chromatogr., A*, 2007, **1145**, 149.
11. M. Maftouh, C. Granier-Loyaux, E. Chavana, J. Marini, A. Pradines, Y. Vander Heyden and C. Picard, *J. Chromatogr., A*, 2005, **1088**, 67.
12. T. D. Nelson, C. J. Welch, J. D. Rosen, J. H. Smitrovich, M. A. Huffman, J. M. McNamara and D. J. Mathre, *Chirality*, 2004, **16**, 609.
13. W. A. Schafer, S. Hobbs, J. Rehm, D. A. Rakestraw, C. Orella, M. McLaughlin, Z. Ge and C. J. Welch, *Org. Process Res. Dev.*, 2007, **11**, 870.
14. G. X. Zhou, Z. Ge, J. Dorwart, B. Izzo, J. Kukura, G. Bicker and J. Wyvratt, *J. Pharm. Sci.*, 2003, **92**, 1058.
15. Z. Ge, B. Buchanan, J. Timmermans, D. de Tora, D. Ellison and J. Wyvratt, *Process Contr. Qual.*, 1999, **11**, 277.
16. Y. Chen, G. X. Zhou, N. Brown, T. Wang and Z. Ge, *Anal. Chim. Acta*, 2003, **497**, 155.
17. G. X. Zhou, L. Crocker, J. Xu, J. Tabora and Z. Ge, *J. Pharm. Sci.*, 2006, **95**, 2337.
18. M. Cameron, G. X. Zhou, M. B. Hicks, V. Antonucci, Z. Ge, D. R. Lieberman, J. E. Lynch and Y J. Shi, *J. Pharm. Biomed. Anal.*, 2002, **28**, 137.
19. G. X. Zhou, Z. Ge, J. Dorwart, B. Izzo, J. Kukura, G. Bicker and J. Wyvratt, *J. Pharm. Sci.*, 2003, **92**, 1058.
20. Z. Lin, L. Zhou, A. Mahajan, S. Song, T. Wang, Z. Ge and D. Ellison, *J. Pharm. Biomed. Anal.*, 2006, **41**, 99.
21. Y. Chen, T. Wang, R. Helmy, G. X. Zhou and R. LoBrutto, *J. Pharm. Biomed. Anal.*, 2002, **29**, 393.
22. ICH Harmonised Tripartite Guideline, Validation of analytical procedures: Text and methodology, Q2 (R1), ICH, Geneva, Switzerland.
23. L. Muller, R. J. Mauthe, C. M. Riley, M. M. Andino, D. De Antonis, C. Beels, J. DeGeorge, A. G. M. De Knaep, D. Ellison, J. A. Fagerland, R. Frank, B. Fritschel, S. Galloway, E. Harpur, C. D. N. Humfrey, A. S. Jacks, N. Jagota, J. Mackinnon, G. Mohan, D. K. Ness, M. R. O'Donovan, M. D. Smith, G. Vudathala and L. Yotti, *Regul. Toxicol. Pharmacol.*, 2006, **44**, 198.
24. European guideline on the limits of genotoxic impurities: CHMP/SWP/5199/02, European Medicines Agency, London, 2006.
25. EMEA toxicological guidelines: EMEA/CHMP/SWP/431994/2007, European Medicines Agency, London, 2007.

CHAPTER 13

Materials Science: Solid Form Design and Crystallisation Process Development

KEVIN ROBERTS,[a] ROBERT DOCHERTY[b] AND STEFAN TAYLOR[b]

[a] Institute of Particle Science and Engineering and Institute of Process Research and Development, School of Process, Environmental and Materials Engineering, University of Leeds, LS2 9JT, UK; [b] Pharmaceutical Sciences, Pfizer Global R&D, Ramsgate Road, Sandwich, Kent, CT13 9NJ, UK

13.1 Introduction and Context

The selection of the commercial solid form and associated crystallisation process is one of the key milestones in the development of any new chemical entity (NCE). It is critical not only from an active pharmaceutical ingredient (API) manufacturing standpoint, but also from a drug product (DP) processing, performance and stability perspective. The regulatory landscape associated with the solid form of the API and dosage form development has already been described.[1,2] The issues associated with the development of an unexpected solid form[3] and the importance of intellectual property (IP) has also been well documented.[4] The progress of automation and structural informatics that allows development scientists to search and identify the solid form with optimal properties has also been reported.[5,6]

RSC Drug Discovery Series No. 9
Pharmaceutical Process Development: Current Chemical and Engineering Challenges
Edited by A. John Blacker and Mike T. Williams
© Royal Society of Chemistry 2011
Published by the Royal Society of Chemistry, www.rsc.org

In 1987 the Nobel Prize for chemistry was awarded to Cram, Lehn and Pedersen for their work on supramolecular chemistry. Since then, publications[7–9] have charted the evolution of pharmaceutical materials science. Pharmaceutical materials science has emerged as a foundation of quality by design (QbD),[10] with the solid form, crystallisation and particle engineering being core elements linking the product attributes to the final steps of the synthetic pathway of the API. Whilst increasing interest in the crystallisation of pharmaceutical entities within academia has resulted in substantial progress over the last decade, the challenge for the process chemist and pharmaceutical scientist in tackling the crystallisation of highly complex chemical entities remains a significant one because:

- Increasing molecular complexity results in a complicated solid form space (salts, cocrystals, polymorphs, hydrates and solvates).
- Multiple molecular conformational degrees of freedom can result in complex solid form structures and consequently significant barriers to crystallisation.
- Highly anisotropic external particle morphology with different crystal faces exhibiting different surface chemistry, interactions with solvents and process impurities.
- Different solid forms may have different chemical and physical stabilities, biopharmaceutical properties and product processing behaviour.

In this chapter we will attempt to bridge cutting-edge academic progress to the best current industrial practices that crystallisation scientists can apply to advance the development of NCEs to medicines consistent with the emerging QbD landscape.

13.2 The Crystal

13.2.1 Crystallography

Crystals may be considered as three-dimensional (3-D) repeating patterns of atoms or molecules. As with any other pattern, they can be described by defining the item to be repeated (the motif) and the way in which it is repeated (symmetry operations). Extending this general concept to crystal structures, the motif is an atom, a molecule or a collection of molecules or ions. The lattice describing the scheme of repetition is now a 3-D array of points and the unit cell is the smallest repeating unit within this 3-D structure. The unit cell is fully described by six lattice parameters, comprising three lengths of the unit cell (a, b and c) with the three inter-axial angles (α, β and γ). Consideration of the relative magnitude of these parameters gives rise to the definition of the seven crystal systems. Having extra lattice points in the face- or body-centred sites produces the 14 unique Bravais lattice types spread over these seven systems. Examination of these seven crystal systems reveals that the unit cells become progressively less symmetrical upon moving from cubic through orthorhombic to triclinic.[11]

It is also possible to refine the seven crystal systems in terms of the symmetry elements which they possess. These elements represent various combinations of rotation, mirror, translation and inversion and together with the Bravais lattice type define the full 3-D arrangement of atoms, molecules or ions within a given structure as expressed by its space group. The symmetry exhibited by a unit cell is also reflected by the physical and chemical properties of the resulting macroscopic crystal. Symmetry is evident in properties such as crystal growth rates and crystal shape and surface chemistry (see Sections 13.2.4 and 13.5.4). Visualisation of the crystal structure of a drug molecule can be challenging, particularly when their molecular weight is high. Figure 13.1 (middle) shows two unit cells of paracetamol along the crystallographic b-direction. Each unit cell contains four paracetamol molecules.

The crystallographic planes that define the external growth morphology of the "as-grown" crystal can be described by the Law of Rational Indices as expressed through their Miller indices. The former enables the 3-D nature of the crystal lattice to be expressed through integer variables where the indices normally have only small values (usually 0 or 1 and sometimes higher) of either parity. Crystal planes are thus defined through their Miller indices (hkl) as the reciprocals of the fractional intercepts which the plane makes with the crystallographic axes. Miller indices are important as they allow the process chemist to link the internal molecular structure to the chemical functionality on the external surface structure. Figure 13.1 (bottom) shows the observed morphology for paracetamol[12] with the Miller indices labelled.

13.2.2 Crystal Chemistry and Crystal Packing of Drug Molecules

Molecules can essentially be regarded as impenetrable systems whose shape and volume characteristics are governed by the molecular conformation and the radii of the constituent atoms. The atomic radii are essentially exclusion zones in which no other atom may enter except under special circumstances, such as bonding. Figure 13.1 (top) shows a comparison between a ball-and-stick and van der Waals (space-fill) representation of paracetamol. The structures and crystal chemistry of molecular materials are often classified into different categories according to the type of intermolecular forces present. A number of factors are of particular importance in assessing the influence of intermolecular bonding on the physicochemical properties of organic solids. These include the strength of the interaction, the distance over which the interaction exerts an influence and the extent to which the interaction is directional or not.[13,14]

Organic molecules, in general, and drug molecules, in particular, are mostly found in only a limited number of low-symmetry crystal systems. The generally uneven shape of their molecular structures tends to result in unequal unit cell parameters. A further consequence of their unusual shape is that organic molecules prefer to adopt space groups which have translational symmetry elements, as this allows the most efficient spatial packing of the protrusions of

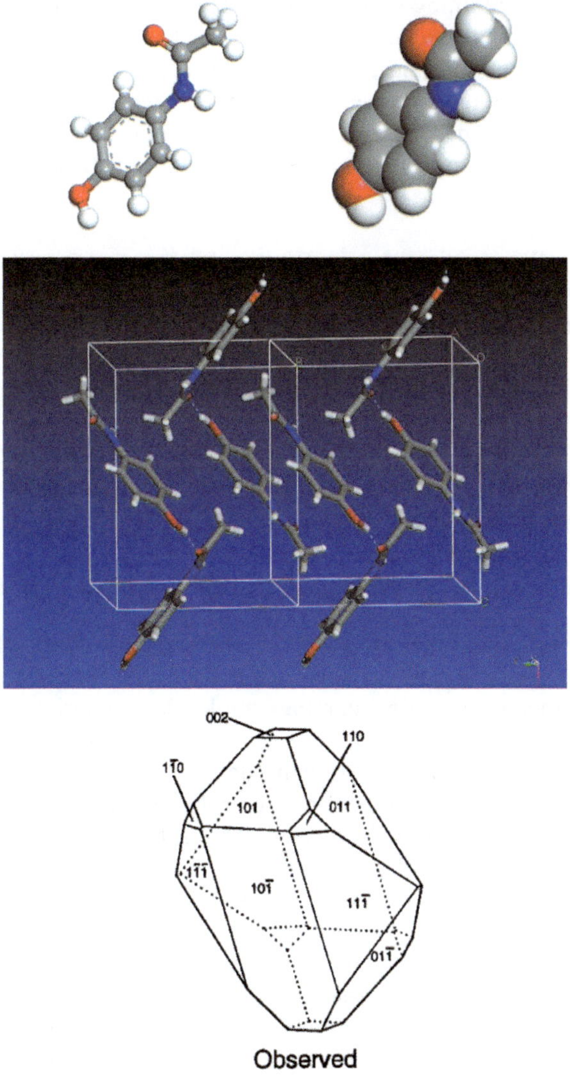

Figure 13.1 The molecule (*top*), the crystal chemistry (*middle*) and the crystal morphology (*bottom*) of paracetamol.

one molecule into the gaps left by the packing arrangements of its neighbours. The vast majority of the organic structures reported prefer the triclinic, monoclinic and orthorhombic crystal systems.[11,15] In order to understand the principles which govern the wide variety of solid state properties and structures of drug molecules, it is important to describe both the energy and direction of interactions of molecules. As a result of pioneering work in the development of atom–atom intermolecular potentials,[16–18] it is now possible to interpret intermolecular packing effects in organic crystals in terms of their interaction energies. Through summing up the intermolecular interactions it is possible to calculate the crystal lattice energy[19] and so relate the molecular structure to the solid state packing and physical stability.[20] Powder X-ray diffraction (PXRD), which examines the angular dependence of X-rays when scattered from crystal lattice planes, is the main characterisation technique for identifying crystal structures and for probing the crystallinity and structural integrity of the packing arrangement:

- The unit cell dimensions govern the angular occurrence of the peaks in the PXRD trace.
- The position of the molecular species within the unit cell govern the relative intensity of these peaks.
- The angular width of the peaks is roughly proportional to the quality and perfection of the crystals.

The use of PXRD in the characterisation of the solid state has been described elsewhere.[21]

13.2.3 Polymorphism, Thermodynamic Stability and Solubility

In general, for molecular materials it is the desire to pack effectively in the solid state that is the single biggest driving force towards the formation of a selected structural arrangement.[15] For complex molecular materials such as drug molecules there will be notable exceptions, where the need to form complex H-bonding arrangements will override this desire.[13] Weaker interactions such as special H-bonds and polar interactions are probably not primary drivers in the packing arrangements adopted by drug molecules, but will tend to be optimised within potential arrangements. Crystallisation and the properties of the solid state are a result of molecular recognition processes on a grand scale and polymorphism is due to balancing these subtle intermolecular interactions.[7]

Despite considerable debate in the scientific literature, a comprehensive definition of the term polymorphism is by no means straightforward. It can be defined as the existence of a compound in at least two different crystal structures. While this definition is not comprehensive, it is sufficient for the discussion here. The use of the term pseudopolymorphism to describe the relationship between such solids and their associated solvated forms is misleading. Solvates and non-solvated solids differ in composition and, therefore, should not be considered to be polymorphs. The formation and behaviour of polymorphs is determined by the relative free energy of the different structures, and the free energies of the barriers associated with their formation and interconversion. The polymorph with the lowest free energy (lowest lattice energy) under a given set of conditions will be the most stable. Particular properties such as density, melting point, solubility and mechanical properties can all be impacted by different solid state structures. The influence of polymorph stability on solubility can be understood by considering two polymorphs, A and B, where B is the more stable form. The solubility of a given drug molecule is ultimately a balance between the energy of solvation (how much the molecule likes to be in a solvent environment) and the lattice energy (how much the molecule likes to be in the solid state).[22] The energy of solvation is a unique molecular property and constant for a given solvent. Given B is thermodynamically more stable at room temperature, it will have a lower lattice energy and stronger packing, resulting in a lower relative solubility. Differential scanning calorimetry (DSC)[21] is a thermal method to determine the melting point and enthalpy of melting (sometimes referred to as fusion enthalpy). The solid state structure that has the larger enthalpy of fusion for a given compound tends to be the most stable polymorphic form. Data for two polymorphs of chloramphenicol palmitate[23] show that form A is the more stable: it has the higher melting point (by 6 °C), the greater heat of fusion (which is a surrogate measurement of the lattice energy) and less than half the solubility of form B. It should be noted that the lattice energy difference between the polymorphs of 3.8 kcal mol^{-1} is not unusual and neither is the resultant two-fold change in solubility (see Section 13.4.2).

13.2.4 Particle Morphology and Surface Structure

Early crystallographers were fascinated by the flat and symmetry related external faces observed in both natural and synthetic crystallised solids. This led them to postulate that the ordered external arrangement was a result of an ordered internal arrangement. The external shape of a crystal, referred to as the crystal habit, is determined by the relative growth rates of the various faces, but is bounded by the slowest growing ones. Crystal habit is traditionally described using a variety of qualitative terms such as plate-like, prismatic and needle-like. In the current discipline we should be moving to more quantitative descriptions of the shape using Miller indices to allow a greater understanding of the different surface chemistry being exposed (see Section 13.5.4). Morphological simulations based on crystal lattice geometry were initially proposed.[24] Subsequent work focused on quantifying the crystal morphology in terms of the interaction energies between crystallising units.[25] Attachment and slice energies can be calculated directly from the crystal structure by partitioning the lattice energy in certain crystallographic directions.[26] The calculated attachment energies for crystal faces can be used as a measure of their relative growth rates and so the theoretical morphology may be computed by determining the smallest polyhedron that can be enclosed by these faces and their relative growth rates.[27]

The shape of the growing crystals is influenced by a wide variety of factors such as the supersaturation, the nature of the solvent, the presence of impurities and the nature, number and distribution of crystallographic defects. The shape of crystals not only has a considerable influence on the properties of the solid, but also has an impact on the handling of particulate materials and their interactions with the excipients used in product design.

Given the complex nature of crystal growth, it is not surprising that a variety of mechanisms exist through which heteroatomic species present within the crystallisation environment, such as solvent and impurities, can influence the growth process. The dominant effect of such species is to interact with a crystal surface, retarding the growth of that face. Symmetrically non-equivalent crystal faces have distinct structure and surface chemistry and such species will interact to varying extents. This retards growth to different degrees and results in a change in morphology. Figure 13.2 shows an example of the action of a tailor-made additive, benzoic acid, on the crystal growth of a structurally similar compound, benzamide. The mechanism proposed is that a benzoic acid molecule is readily incorporated into the growing crystal face by virtue of its similarity to the host molecule and its capacity for H-bonding (shown by the dashed lines). Once incorporated, the addition of further benzamide molecules is, however, prevented by the termination of the infinite chain of H-bonds along the *b*-axis due to repulsion between the lone pairs of electrons on the oxygen of the benzoic acid hydroxy group and the carbonyl of benzamide. Overall, this leads to a retardation of growth and a concomitant increase in the relative surface area of the (010) face.[27]

The modern process chemist has tools at his disposal that describe the link between the bulk crystal and the surface chemistry of faces, as well as the

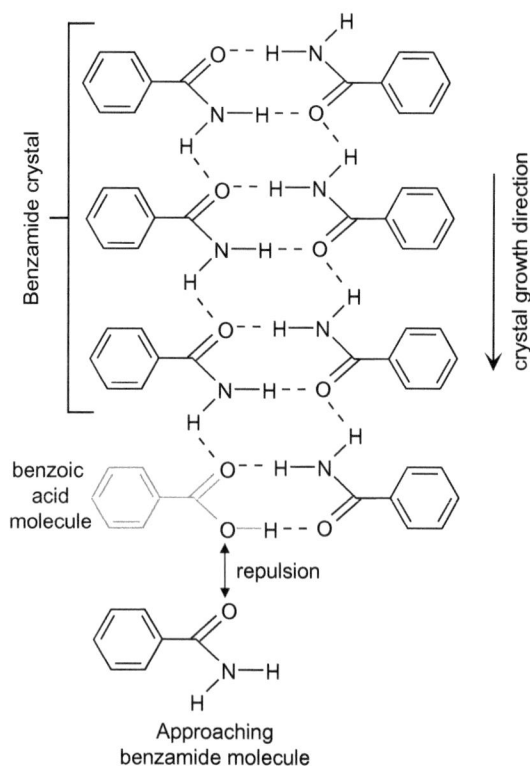

Figure 13.2   The blocking of the crystal growth of benzamide along the *b*-axis due to the incorporation of a benzoic acid molecule (reproduced by permission of Elsevier from *J. Crystal Growth*, 1994, **135**, 331).

potential interactions with solvents and impurities.[28,29] Dynamic vapour sorption (DVS) is a powerful technique that monitors the mass of a sample as a function of the partial pressure of vapour in contact with it.[21] This provides an understanding of water or solvent sorption on different crystal surfaces, hygroscopicity and non-solvated/solvate interconversions.

13.2.5   Particle Size

Particle size influences a wide range of physical and chemical properties. The most relevant to pharmaceutical material behaviour includes solubility, dissolution rate, bioavailability, flow characteristics, bulk density and the segregation behaviour of powder mixtures. The dimensions of a particle can be approximated to the diameter of an equivalent sphere which has the same particle volume. This is a crude approximation, ignoring the shape and surface chemistry aspects that contribute to the properties of the material. A collection of particles of the same equivalent diameter is referred to as mono-sized. More typically, a pharmaceutical powder consists of a range of particle sizes which can be described by a particle size distribution curve. A size distribution which is symmetrical about the maximum frequency value (the mode) is a normal distribution. A distribution may exhibit asymmetry due to either an increased proportion of smaller (fines) or larger particles resulting in a negatively or positively skewed distribution, respectively. A distribution showing two maxima is referred to as a bimodal distribution, which is not atypical since the process development scientist has to balance control of nucleation, growth and agglomeration. In addition, fines may be generated during the isolation processes.

A wide range of particle size measurement techniques exist,[30] including sieving, microscopy (optical or electron microscope), electric stream sensing (Coulter counter), laser diffraction, image analysis, sedimentation (gravity or centrifugal) and ultrasonic attenuation. The most common techniques utilised are laser diffraction and image analysis. The regulatory perspective on particle size has recently been reviewed.[31]

13.3   Crystal Formation

13.3.1   Solubility, Supersaturation and the Metastable Zone

The crystallisation process can be viewed as a two-step process involving the dissolution of the API, and then changing some attribute of the system. This might involve adjusting temperature, pH, solubility or solvent content, to induce crystallisation. Alternatively, chemical reaction might form a precursor to crystallisation. At a given temperature and pressure there is a maximum amount of solute that can dissolve in a given amount of solvent. When this maximum is reached, the solution is said to be saturated. The amount of solute required to make a saturated solution at a given condition is the solubility.[32] During crystallisation the solute molecules must de-segregate from their solvated state and self-assemble, aligning structural elements such as molecular conformation and packing in order to produce a stable 3-D ordered crystallographic array of molecules. A solution in which the solute concentration exceeds the equilibrium saturation at a given temperature is known as a supersaturated solution. Supersaturated solutions are metastable, implying that crystallisation will ultimately occur, albeit after time has elapsed, but that the process is inhibited by a kinetic barrier. Every solution has a maximum limit that it can be supersaturated to before it becomes unstable and crystallisation spontaneously occurs.[32] The region between the saturation curve and this unstable boundary is called the metastable zone (MSZ), and it is within this that all crystallisation operations normally occur. If we plot concentration *versus* temperature behaviour we find three regions (Figure 13.3):

- A stable or undersaturated region where crystal growth is not favoured.
- A metastable region where the solution is supersaturated to a degree and where crystallisation will take place after a time.
- An unstable region where the solution is more supersaturated and where spontaneous crystallisation with no time delay is expected.

Within the MSZ the nucleation stage is quite controlled and crystals are able to grow with a steady supply of solute molecules without the formation of other nuclei. The MSZ width (MSZW) should be large enough to provide a stable region for crystal growth, but not so large that it leads to a barrier for growth. In the unstable region, controlled crystal growth to macroscopic dimensions is not possible. Thus, in this region, depending on the degree of supersaturation, very

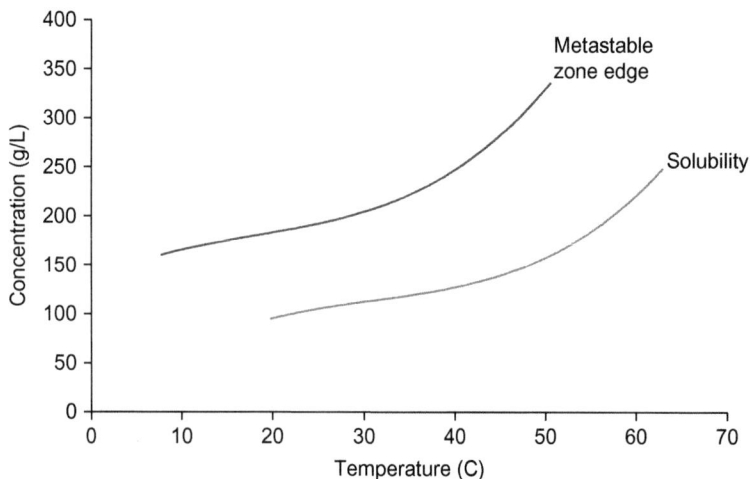

Figure 13.3   An overview of the metastable zone width. The region below the solubility curve is stable and undersaturated. The region above the metastable zone curve is unstable and spontaneous crystallisation will occur. The region in the middle is the metastable zone width.

Table 13.1 Common approaches to supersaturation generation.

Generation of supersaturation (through a change of solution concentration)			
Change the solution temperature (increase or decrease depending on sign of solubility coefficient)	Change the solution pH in cases where the solution is an electrolyte	Evaporate the solvent	Add a low solubility but highly miscible second solvent component to a saturated solution (drowning-out)
Most commonly used in industrial crystallisation for both batch and continuous processes	Used for many pharmaceutical products in cases where the material is formulated as a salt	Simple technique, ideal for the production of non-speciality bulk chemical (salt lakes)	Generates very high supersaturations but commonly used in many precipitation reactions

small crystal particles will be produced. From a practical perspective, supersaturation may be realised by a number of methods, as described in Table 13.1.

Figure 13.4 shows an optical turbidometric plot describing the change in transmittance associated with the crystallisation of nortriptyline hydrochloride, a tricyclic antidepressant.[33] It reveals the typical hysteresis behaviour of such measurements, showing both the onset of nucleation (decrease in transmittance) and the onset of dissolution (increase in transmittance) as the temperature is cycled above and below the saturation temperature of the solution. The MSZW is the difference between the temperature of dissolution and that of crystallisation. This needs to be carefully characterised and understood in order to produce optimal crystals. Very high cooling rates may result in an unwanted outcome, such as formation of a metastable polymorph, precipitation of an amorphous phase or the formation of a colloidal dispersion (oiling out, see 13.4.4.2).

Driven by supersaturation, crystallisation proceeds through two distinct and closely inter-related steps. Nucleation, associated with the formation of stable 3-D solute clusters, plays a key role in defining particle size, polymorphic form and crystallinity. Growth, associated with the formation and subsequent development of a set of 2-D surfaces, plays a key role in defining particle shape, product purity and inter-particle agglomeration.

13.3.2 Nucleation Processes

In the nucleation stage, small clusters of solute molecules are formed; some of these clusters may grow sufficiently to form stable nuclei and subsequently form crystals. Others fail to reach adequate dimensions before they dissolve again. The kinetics of the nucleation process are governed by the balance between the surface and bulk volume free energies associated with cluster formation. Within the MSZ the induction time to the onset of crystallisation is related to the supersaturation. Although very much the idealised case, homogeneous nucleation is useful in that it provides a full derivation of the parameters

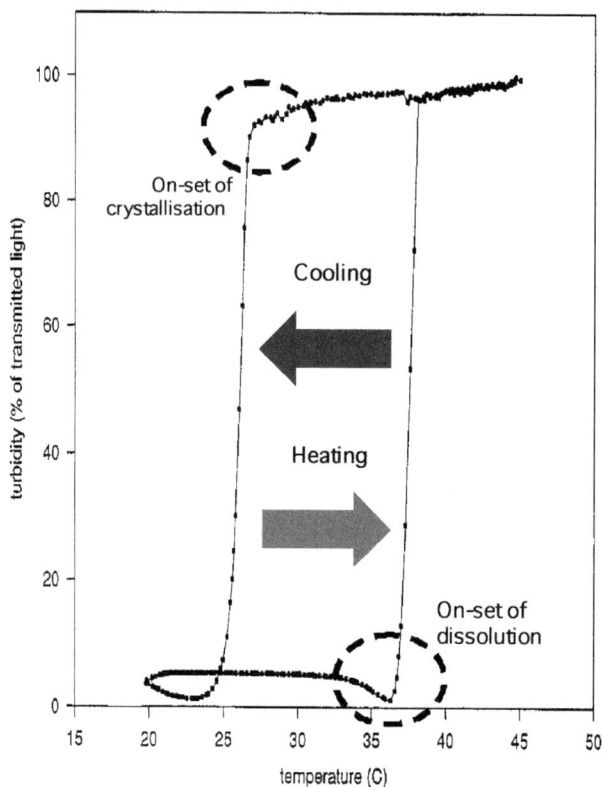

Figure 13.4   Change in solution transmittance as a function of temperature recycling associated with the crystallisation and dissolution of nortriptyline hydrochloride (reproduced by permission of Elsevier from *J. Crystal Growth*, 1996, **166**, 189).

important in nucleation theory and hence provides a useful benchmark to the process. Using this as a basis, the more representative heterogeneous case can be considered as a modification to the homogeneous case.[34] Heterogeneous nucleation is the case when nucleation is induced by other particles which are able, in turn, to act as structural templates by lowering the interfacial tension to encourage nucleation within the MSZ. In practical crystallisation systems, particulate materials are often present in the solution and act to lower the interfacial tension, so reducing the induction time. Secondary nucleation is also a significant problem in industrial crystallisation, where an aggressive environment for soft particles is provided by their interaction with mechanical elements such as reactor surfaces, pumps, baffles, stirrers, *etc*. This results in the production of attrition fragments which are deleterious due to the fact that their growth rate is often less than that produced by primary nucleation. Such a dispersion of growth rates can result in a product with a variable particle size distribution, leading to problems in downstream processes. These include issues with isolation, drying, particle flow and variability in product performance attributes such as content uniformity.

13.3.3 The Crystal Growth Process

Following nucleation, crystals grown from solution typically exhibit regular, planar facets characterised *via* their Miller indices. Although appearing flat to the naked eye, these crystalline surfaces are rarely so at the molecular level. The various features which make up the nanoscale surface topography of crystal faces are intimately involved in the mechanisms by which they grow. Figure 13.5 provides a schematic representation of a crystal surface in which the molecules, or growth units, are represented by cubes. Examining the various

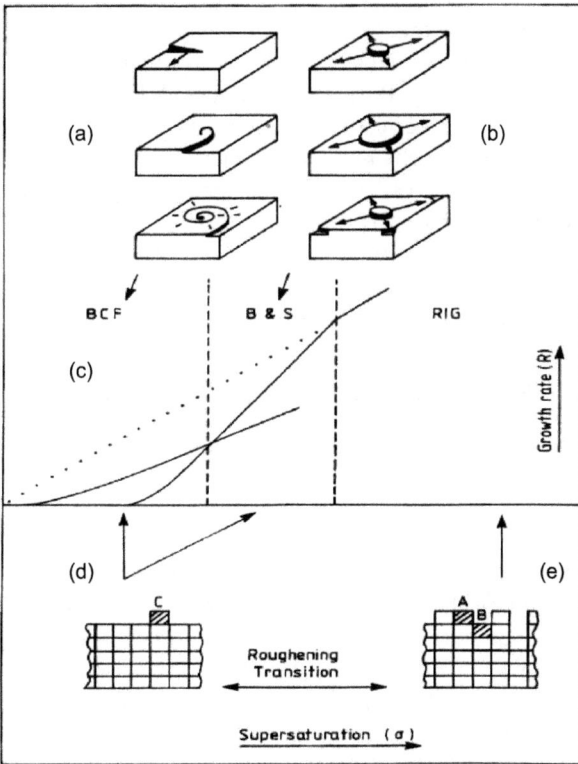

Figure 13.5 Schematic representation of the crystal growth mechanism taking place at the interface between the crystal surface and the growth environment associated with the molecular recognition needed to effect high species selectivity, notably involving screw dislocation (a) and 2-D surface nucleation (b) together with the associated growth rates as a function of supersaturation (c) and the transition between stable growth on atomically smooth surfaces (d) and unstable growth at the roughened growth interface (e) (reproduced by permission of the Institute of Physics from *J. Phys. D: Appl. Phys.*, 1993, **26B**, 7).

sites on the schematic crystal surface A, B and C, we can see that each cube is capable of forming a different number of bonds. A cube from solution which adsorbs onto a flat part of a crystal face (C) is relatively weakly bound to the surface as the cube is bound on only one side. There is, therefore, a high probability that the cube will be desorbed again in a relatively short period of time. To remain on the flat surface, molecules must form stable clusters or nuclei. This consecutive birth and spread (B&S) of new nuclei on the crystal face is shown in Figure 13.5(b). This process is the 2-D analogue of the 3-D nucleation process discussed in the previous section. This 2-D nucleation is controlled by the competition between the free energy reduction produced by molecules moving from solution to the crystal and the increase in free energy due to the formation of a new surface.[35] At low supersaturation, growth by 2-D nucleation is difficult since a large number of molecules are required to form a stable nucleus. Detailed analysis reveals that at low supersaturation the rate of growth by 2-D nucleation is negligibly small. Experimental studies show that crystals do in fact grow at appreciable rates even at very low supersaturation. This apparent contradiction between theory and experiment was resolved by Burton, Cabrera and Frank (BCF),[36] with the proposition that crystal imperfections, known as screw dislocations, provide a permanent surface step, thereby removing the need for 2-D nucleation as part of the growth process. Figure 13.5 shows a schematic diagram with a screw dislocation intersecting a surface. The step extends only part way across the surface. As the step grows by addition of molecules, it adopts the shape of a spiral and consequently the screw dislocation acts as a continuous source of surface steps and 2-D nucleation is not required for the crystal to grow.

13.3.4   Growth Stability and Interface Roughening

The structure of the crystal growth interface depends on the supersaturation, which in turn gives rise to three types of growth mechanism. In the roughened growth (RIG) model[37] shown in Figure 13.5, surface roughness provides ample sites for surface integration. It provides more kink binding sites (B) and concomitantly a much higher growth rate. Such an inherently rough surface provides only a limited mechanism for molecular recognition (see Section 13.2.4), resulting in irregular surface growth and poor rejection of solvent and/or impurities. As shown in Figure 13.5, growth at low supersaturation is dominated by the BCF mechanism and as supersaturation increases there is greater importance of the B&S mechanism. There is no specific transformation point between these two regions. In contrast, the transformation from the BCF/B&S regions to the RIG regime at high supersaturation is abrupt. The transformation point to RIG dominated growth is referred to as the roughening transition. Crystallisation and thus crystallisation processes, where purification is a prime function, are usually optimised away from the RIG region.

The tendency of the growth interface to roughen depends on the surface structure of the crystal face. Hence when optimising a crystallisation process, conditions need to be established such that the fastest growing, and hence least stable, surface grows below the roughening transition. Generally speaking, faster growing surfaces with smaller relative areas are most likely to be prone to surface roughening and, for example, fast growing needle-shaped crystals may tend to incorporate impurities selectively at their facet ends if the growth process on these interfaces is not carefully enough controlled.

13.3.5   Nucleation and Growth Control

The crystalline form in which a material is obtained can potentially be controlled through manipulation of three main aspects of polymorphic behaviour: nucleation, crystal growth and phase transformations. The nucleation process involves the formation of aggregates of molecules. In a polymorphic system, the situation can be more complex. It is assumed that a number of different types of aggregate may exist in solution. Each type of aggregate is connected by competing equilibria and may subsequently develop into one or more different polymorphic forms. Nucleation is seen by many leading academics as the key to crystal engineering.[38] If there are substantial differences in the rate of nucleation of the possible polymorphs, kinetics rather than thermodynamic stability will determine the polymorph obtained. An alternative perspective is that the kinetic process is a second-order phenomenon with respect to polymorph selection and that size-dependant polymorphic stability is more important, reflecting the fact that cluster size is controlled by nucleation kinetics and that different polymorphs might be stable at different cluster sizes. If the nucleation kinetics of the various forms are similar, the relative thermodynamic stability may then determine which form crystallises. Finally, if all the above parameters are similar for all potential polymorphs, a number of structural forms may be obtained together, so-called concomitant polymorphs.[39] In this case, careful selection of temperature and supersaturation may, therefore, provide control over which polymorphic form would be obtained. Eliminating, rather than controlling, nucleation can also influence the polymorphic form obtained. The addition of seed crystals of a given polymorphic form to a saturated solution can promote the preferential formation of that polymorph.

Table 13.2   Qualitative summary of the balance between nucleation *versus* growth factors in crystallisation from solution phases.

Nucleation versus growth		
Solutions with small MSZWs	*Solutions with moderate MSZWs*	*Solutions with large MSZWs*
Material easily nucleates, growth phase likely to be limited by interface kinetics, with nucleation at quite low supersaturation	Material not too easy to nucleate and growth preferred at expense of nucleation	Poor nucleator and the solution likely to supersaturate substantially, resulting in eventual nucleation at high supersaturation
Crystals expected to be small to medium in size	Should form large crystals	Very small, poorly formed, crystals crystallised under non-equilibrium conditions

Qualitatively, the balance between nucleation and growth is summarised in Table 13.2. From this we can see that, provided we work within the MSZ, it should be energetically more favourable to nucleate on an existing crystal surface than nucleate in the bulk mother phase. However, for a material with a wide MSZW, supersaturation will build up in the solution, resulting in growth conditions beyond the MSZ boundary, which causes spontaneous nucleation and yields poorly formed crystals in many instances.

13.4 Industry Practices

13.4.1 Salt Screening and Selection

The predominant reason for screening for salts of APIs is to overcome any undesirable chemical or physical properties of the free acid/base. Such properties may include poor solubility in bio-relevant media, exposure limiting dissolution rate, chemical instability and poor mechanical properties. In addition, salts typically exhibit good crystallinity due to the additional electrostatic intermolecular interactions compared to the free acid/base, which assists in achieving good isolation, purification and stability. In the case of enantiomeric compounds, optical resolution can be achieved by diastereoisomeric salt formation in order to obtain high enantiomeric purity, although more often chiral resolution is achieved earlier in the synthetic route with key intermediates. An overview of acceptable salts and development factors that influence this selection process has been covered elsewhere.[40]

Most pharmaceutical compounds contain acidic or basic functional groups; the majority will undergo salt screening and selection, and it is estimated that around half of all APIs are utilised in the form of salts. A wide variety of organic and inorganic anions and cations are used in forming salts of drug compounds. While final selection of the most suitable salt former involves assessment of an extensive range of physical and chemical properties, the primary concern is that potential counterions must not exhibit any adverse physiological effects, so the safety of the counterion selected in the proposed dosage form regime is of critical importance. The most frequently used salts and their precedence in different delivery routes have already been described.[41] When designing a salt screen there are some important factors to consider. Each drug molecule is different, so screening all counterions could prove wasteful. In the literature, guidance is provided for salt formation with APIs such that there should be a pK_a difference (ΔpK_a) of 2–3 units between the compound and its potential salt former. Constraining the selection of counterions by using this approximation ignores the impact that solvent and temperature can have on potential salt-forming reactions. Figure 13.6 shows the relative usage of the most commonly utilised acidic and basic counterions, revealing that hydrochloride (43%) and sodium (62%) are the most common salt formers for acid and basic drug compounds, respectively.

When designing a salt screen for multi-basic or multi-acidic compounds and counterions, the potential stoichiometries should be considered. Even for

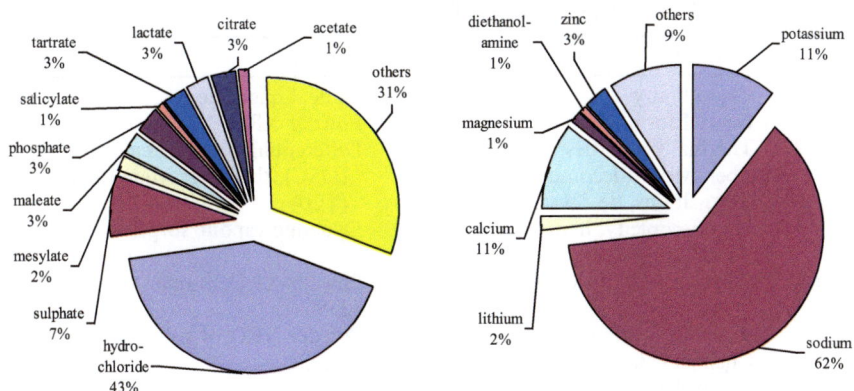

Figure 13.6   The relative usage of the most commonly utilised acidic and basic counterions (data from Table 8.4 in *Pharmaceutics: The Science of Dosage Form Design*, ed. M. E. Aulton, Churchill Livingstone, Barcelona, 2nd edn, 2001, p. 117).

monobasic or monoacidic compounds, multiple salt stoichiometries may be achieved. Similarly, when screening for salts of a chiral API, unique solid forms may be achieved by the use of single enantiomer salt formers.

During selection the physical properties of the candidate salts must be assessed, a task which requires comprehensive characterisation of selected salts and their polymorphs or hydrates. The characterisation and assessment of the physical forms of salts is demanding in terms of experimental effort.[41] While it is desirable to screen a range of salts, it is impractical to attempt characterisation of all aspects of every potential salt form. For this reason, salt selection is typically carried out using a tiered approach. After characterisation at each level, a decision is made as to whether the salt proceeds to the next level of assessment. The range of potential salts is, therefore, progressively narrowed down. The depth of characterisation increases from one level to the next. The number of levels required typically depends upon the number of potential salts. In this way, a substantial number of salt forms can be investigated while minimising experimental effort. Table 13.3 shows a schematic representation of a possible multi-level salt selection process, drawing down upon the desirable properties of the candidate salts.

The importance of having a robust salt selection process from a development standpoint was recently highlighted through the concerns of the partial breakdown of the salt of prasugrel back to its freebase and the potential impact that had on the bioavailability and efficacy of the product.[42]

13.4.2   Polymorph Screening

The importance of polymorphism to the pharmaceutical industry can be traced back to reviews by McCrone[43] and Byrn.[44] The temporary withdrawal of the

Table 13.3   Multi-tier evaluation of salt candidates.

	Properties	*Techniques*
Tier 1	Crystallinity	Optical microscopy
	Crystal form	Powder XRD
Tier 2	Thermal properties and thermal behaviour (decomposition, phase transitions, desolvation)	Differential scanning calorimetry (DSC), thermogravimetric analysis (TGA)
	Hygroscopicity, hydrate/solvate formation	Dynamic vapour sorption (DVS)
Tier 3	Polymorph and hydrate screening	Powder XRD, Raman microscopy, DSC
	Aqueous & pH solubility	Powder XRD of solubility residues
	Chemical stability testing (*e.g.* hydrolysis, oxidation, photolysis)	HPLC
	Accelerated physical stability testing	Powder XRD, DSC
Tier 4	Humidity/temperature induced changes in crystal form	Environmental powder XRD
	Influence of processing conditions (*e.g.* milling, micronisation, compaction) on solid form	Powder XRD, DSC, DVS
	Compatibility with excipients	HPLC

protease inhibitor ritonavir just over a decade ago[3] and the recent appearance of a new polymorph (observed as "snowflakes") in the rotigotine transdermal patch[45] highlighted the potential impact of the appearance of a more stable polymorph on solubility, dissolution and ultimately bioavailability. The potential impact on product manufacturing robustness and development timelines together with the potential need for repeated clinical and stability studies means that most pharmaceutical companies have incorporated selection and screening programmes and practices within their product development strategies. In this, the key elements of screening practices and the appropriate timing of these activities during drug development need clear definition. For the development scientist the key questions are:

- How often do these new polymorphs or solvates appear?
- What will be the impact on biopharmaceutics and stability?
- How many experiments are needed in order to ensure robustness?

These questions are difficult to give definitive answers to, but Figure 13.7 consolidates some data based around recent reviews in this area.[46,47] The summary suggests that the vast majority of APIs exhibit multiple polymorphs. It also shows that hydrates are more prevalent for salts, and polymorphs more likely for non-salts. Polymorph pair solubility ratios generally involve less than three-fold solubility changes, but there are instances of greater than five-fold changes, especially for highly complex molecular structures.

In preclinical development, only a limited amount of material is available. Early screening is focused on the definition of the form being used and how to

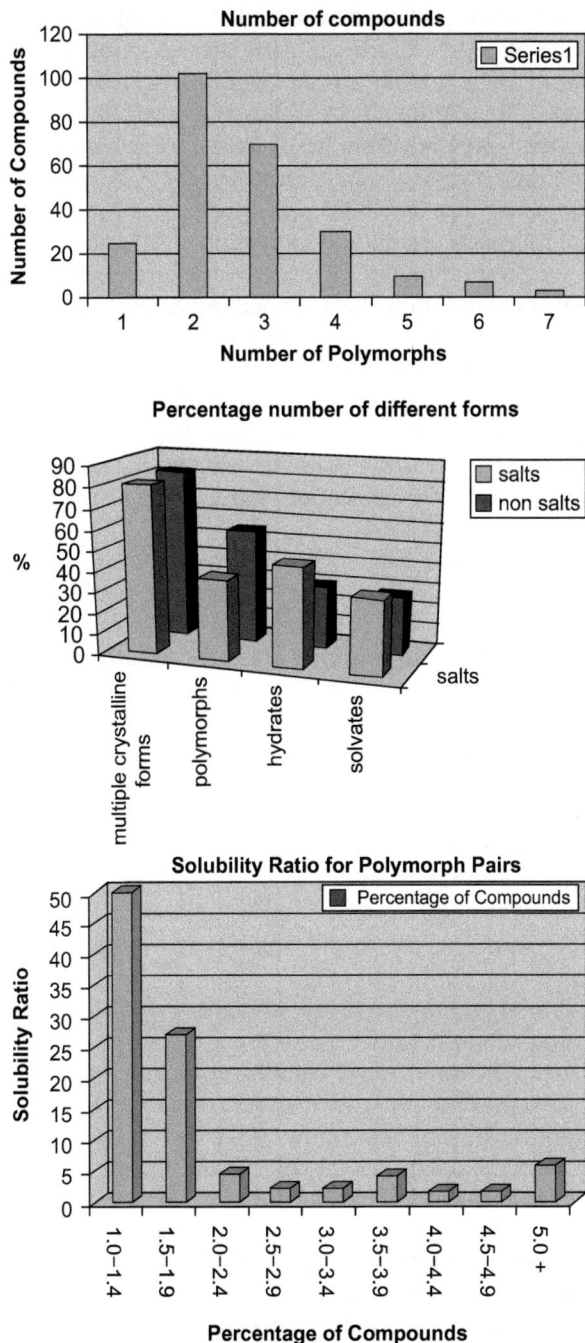

Figure 13.7 The number of compounds with a given number of forms (*top*), the frequency of forms found for salts and non-ionisable materials (*middle*) and the cumulative plot of solubility ratios for 180 polymorph pairs of salt candidates (*bottom*) (data from references 46, 47 and 68).

ensure consistent delivery of that form with the enabling chemistry. Limited screening will be targeted towards an early awareness of the potential forms accessible from the salts that have been identified. A selection of techniques is used to provide a fingerprint of the solid state chemistry.[21] As a candidate progresses into the early clinical studies there is a commensurate increase in the effort to find all the polymorphs and solvates of the API and to link this to the product development strategy. This will involve a variety of crystallisation experiments and physical stability investigations of the bulk drug in different humidity and temperature ranges. In recent years the concepts of "stable form" screens have emerged and these have gained increasing importance, especially at this point in development. These screens are essentially slurries in solvents with suitable solubility to facilitate the transformation from a given metastable solid state structure to a potentially more stable structure.[48] This solvent-mediated phase transformation involves two steps: the dissolution of the metastable polymorph to form a solution supersaturated with respect to the stable form followed by the nucleation and growth of the new more stable phase from this solution.

The rate of such solvent-mediated transformations is driven by solubility in the selected systems. The number of solvents used, their solubilities, temperature and the duration of the slurry are often company practice dependent. Whilst there is no definitive standard industry practice, the slurry experiments have become commonplace as they can rapidly help identify the most stable form and therefore baseline the clinical exposure from a bioavailability perspective. A number of publications have articulated these principles, but have also highlighted that these slurries are not a panacea for polymorph screening as there are caveats that can effectively reduce the value of the screen. Small solubility differences between polymorphs may limit the transformation and a lack of solubility/stability in preferred solvents may limit the screen design. Also, impurities that may be present in the early clinical batches could inhibit the more stable form from appearing.

In the later clinical stages, screens are designed to find all known polymorphs to ensure a comprehensive understanding of the solid form space and the potential impact on the commercial manufacturing process. Screens underwrite final process solvent variations, temperature excursions and the impact of impurities. Given the pivotal importance of the correct polymorph on the final crystallisation step, the design of the DP and the stability of the product, the industry continues to enhance screening practices through both better informed screen design[49] and the application of high-throughput technologies. A review of high-throughput publications suggests that 52 500 crystallisations on 51 APIs identified 155 new solid forms.[5]

Increasingly, the fate of the solid form in the DP is emerging as a focus area and many groups are actively embedding milling and compaction tests into their polymorph screens/solid form selection practices. Given the potential regulatory and IP impact, research on product processing induced transformations is only likely to increase given the issues recently highlighted.[42,45]

13.4.3 Hydrate Screening

As with polymorphs, it is known that a hydrated solid form of a drug can have a potentially significant negative impact upon bioavailability.[50] It is also possible for the presence of water in the crystal structure to lead to chemical and physical instability and may result in undesirable chemical reactivity with the excipients within a solid dosage form. In a previous section we identified DVS as a key technique for assessing hygroscopicity. This approach may also be used to screen for potential hydrate formation, identifying the key temperatures and relative humidity for anhydrous–hydrate transitions. The method may not be relied upon to prepare hydrates for all new drug candidates, since conversion to a hydrated form may be a kinetically slow process. Also, the anhydrous form may need dissolution and recrystallisation steps in order to incorporate water molecules into a new hydrate crystal structure. Even comparatively high levels of relative humidity (RH), such as 90% RH, may be insufficient to rapidly ensure such structural rearrangements.

One way to try to overcome these kinetic barriers to hydrate formation is to recrystallise or slurry in water. Whilst, in theory, this would seem the most applicable approach, the drug molecule may have insufficient aqueous solubility to make this practicable. An alternative approach is to utilise organic solvent/water mixtures, of known water activity (α_w), to enhance the solubility of the drug candidate. Recrystallisation or slurry equilibration in solvent systems of high water activity such as $\alpha_w = 0.90$ (equivalent to 90% RH) would be a typical approach. The formation of a thermodynamically stable hydrate by this technique is reported to have a high success rate.[51] In addition, equilibration at multiple temperatures for these slurry conversions allows a broad picture of the anhydrous–hydrate phase diagram to be produced,[52] which is a useful tool when assessing the risk of isolating the incorrect phase or of having a solid form change during downstream processes.

For the case where an anhydrous form is developed, if hydrated forms are identified, it is important to consider their potential impact upon bioavailability, stability or other physicochemical properties that may be affected by an anhydrous–hydrate conversion. In such cases it is important to take a broad product overview to ensure the secondary processing and stability aspects have been considered.

13.4.4 Crystallisation Process Design

Crystallisation processes are complex phenomena, which are ultimately the result of an interplay between several process factors.[32,34,53,54] Throughout the development of a new drug, it is crucial to have a strategy in place to progressively increase the understanding of the factors involved, ultimately resulting in greater control and robustness of the process. The impact of well-designed crystallisation should not be underestimated, in terms of reduction of API processing cost (efficient filtration, removal of milling operations) and on the product in terms of safety, efficacy and processing.

Crystallisation proceeds by a combination of two mechanisms: nucleation and growth. Nucleation is the creation of the new crystalline phase, while growth occurs on existing nuclei. The ratio of nucleation and crystal growth determines the particle size distribution. Because of the interplay between these competing mechanisms, the design and development of crystallisation processes are best achieved through a systematic approach using a good knowledge of fundamentals.[34,53] The actual approach will depend on the stage of development, the acquired knowledge and the amount of the material available.

Particular consideration is given to solvent selection for the commercial process. The solvent or solvent combination should have a number of key properties. It should have the ability to dissolve the reactants used, have the right solvation properties enabling easy crystallisation with appropriate purge of impurities, and be chemically inert under the reaction conditions. In addition, it should have the right surface adsorption properties for optimal API morphology, be non-toxic, environmentally friendly and readily available at low cost.[55]

Solubility measurements, as a function of potential process conditions, such as temperature, solvent and pH, are the crucial first steps in the development of crystallisation processes. Solubility data should be acquired as soon as sufficient material is available. The first crystallisation is usually performed in preclinical development when little information on the properties of the API is known. The crystalline material obtained in the first crystallisation is used in subsequent investigations of polymorph, hydrate and solvate screening. In selecting potential crystallisation routes the following guidelines are helpful:

- If a solubility change over a limited temperature range is high, then cooling can be employed solely as the supersaturation generation method.
- If solubility change is lower over the selected temperature change, then cooling can be followed by anti-solvent addition.
- If the solubility change is poor over the selected temperature range, the anti-solvent could be added before cooling is initiated.
- Crystallisation can also be induced by solvent evaporation.

The measurement of the MSZW for a solute/solvent system is also required to better understand the appropriate crystallisation approach. Control of the metastability of a mixture is crucial, as this will have implications on purity, particle size, shape and crystal form.

13.4.4.1   Cooling Crystallisation

A cooling crystallisation is usually preferred because of its relative ease of control and scale-up. Perhaps the most important factor in a cooling crystallisation is the cooling rate. If the rate is high, the mixture will precipitate very quickly in order to avoid excessive levels of supersaturation, and the control over the properties of the crystallising solid will be very limited. Furthermore, the rapid formation of solid can cause solvent or other impurity molecules to be occluded in the solid. Three types of cooling profiles are normally used: natural, linear or cubic. The natural cooling rate simulates the temperature profile of a mixture, which is allowed to cool down without any control. Natural cooling profiles, which are characterised by an initial steep cooling rate, followed by a much slower cooling rate in the latter part, can result in poor quality crystals being formed and are not generally used. Linear cooling profiles can be used very efficiently if the rate is adjusted to suit the purposes of the crystallisation. A steep linear cooling can be used to generate small particles (primary nucleation will predominate as a means to rapidly decrease the supersaturation), but will not be as effective as a purification approach. This type of cooling profile is useful as it can be easily transferred to a different plant or vessel. During a cubic cooling profile, an initial slow period of cooling is followed by a steep cooling period. This type of profile is optimal for crystallisation because the initial slow cool prevents the mixture from reaching excessive supersaturation levels and allows existing super-saturation to deplete *via* crystallisation, resulting in a more controlled crystallisation.

13.4.4.2   Anti-solvent Crystallisation

During anti-solvent (drown-out) crystallisation an anti-solvent miscible with the process solvent, but in which the API has a significant lower solubility, is added to a solution of the API in order to create supersaturation and drive crystallisation. Careful attention to addition rates and mixing are required for this type of process, as the addition of small amounts of anti-solvent can result in a dramatic increase in local supersaturation. A rapid increase in supersaturation may result in the preferential crystallisation of metastable forms as well as purity and particle size challenges. Another aspect of anti-solvent crystallisation that must be considered is the potential for these mixtures of solvents and solute to "oil out", where a phase separation between two mixtures of similar thermodynamic stability, but different che-mical composition, is observed. If oiling out, or the formation of amorphous materials, is observed on anti-solvent addition, then a solvent with different polarity or a lower volume of anti-solvent should be used. A preliminary study of the dependence of solubility on the composition of the solvent mixture at both the temperature of anti-solvent addition and isolation can help identify the best parameters for this process. Care should be taken when anti-solvent crystallisations are scaled-up, as inefficient mixing in larger scale vessels could result in local supersaturation hotspots where rapid crystal-lisation occurs, often of highly agglomerated multi-crystal particles (which may be of lower purity).

13.4.4.3   Reactive Crystallisation

In reactive crystallisation, a compound is prepared by reaction in a solvent/ temperature environment above its thermodynamic solubility. When this process is particularly fast, it is sometimes referred to as precipitation. The reaction and precipitation typically follow the addition of one of the reactive species (usually the lower molecular mass component) to a homogeneous solution of the compound. In reactive crystallisation, the rate of nucleation and growth depends on the supersaturation levels of the precipitating species. Factors which influence the rate of formation of the precipitate include the addition rate of the reagent, the mixing/mass transfer and the rate of the reaction. Clearly the identification of the rate-limiting step is key for process optimisation. In order to determine the relative impact of these factors, experiments involving different addition rates and mixing rate/impeller types should be performed. Generally speaking, reactions that involve the formation of salts are rapid if all species are in solution. At the laboratory scale, where efficient mixing is easily attained, the rate of crystallisation can be controlled through the rate of reagent addition. However, care should be taken when scaling-up to ensure that inefficient mixing does not result in rapid nucleation, leading to agglomeration and encrustation. These factors may lead to an addition rate profile that mimics a controlled cooling in terms of super-saturation control.

13.4.4.4   Other Crystallisation Processes

Other crystallisation approaches include evaporative, pH swing and salting out methods. During an evaporative crystallisation, the concentration of the product in solution is increased by distilling off the solvent, with the supersaturation consequently increasing until the metastable limit at that temperature is reached and crystallisation starts. In a pH swing crystallisation, the pH of a solution is adjusted, since the solubility of a compound will change if the product has acidic or basic centres. Salting out crystallisations rely on the common ion effect where the solubility of a compound can be decreased if a different salt with the same counterion is added to the mixture.

13.4.4.5   Seeding

Seeding is a very powerful tool in crystallisation process design.[56] By providing a seed of the appropriate type, shape or size, a template for crystallisation is provided. Homogeneous seeding can be used to control crystallinity, particle size distribution, purity and polymorph. Additionally, seeding can be used to initiate crystallisation away from the edge of the MSZ during a cooling process. By using a well-defined seeding protocol, the batch-to-batch variability intrinsic to spontaneous crystallisation is reduced. Seed concentrations typically used lie in the 2–5 wt% range and seed particle size is usually optimised to ensure sufficient surface area to promote nucleation and growth. Caution is needed when using dry milled powder as seeds, since these may have suffered mechanical damage which may either impede their regrowth or encourage impurity ingress. To remediate this, seeds are often added as a slurry in solvent after annealing out any damage induced by the milling process.

13.4.5   Particle Reduction Techniques

If the particle size and its distribution produced by crystallisation are unsuitable, additional processing in the form of grinding or crushing may be required to modify these characteristics. A material subjected to a load, as in the case of grinding or crushing, will respond in one of three ways, depending upon the physical properties of the material and the magnitude of the load applied. Firstly, the material could deform elastically, which means it returns to its original shape when the load is removed. At higher loads, the material may deform plastically (permanently) or, alternatively, it may fracture, like glass. Particle size reduction during grinding or crushing occurs through fracture. Cracks form and propagate through the particles, breaking them into smaller pieces. Elastic and plastic deformation do not assist this process; rather, they hinder it. This behaviour can be characterised by two properties of the material: toughness and surface hardness. Toughness relates to the ability of a material to resist cracking and fracture. Hardness, in contrast, is a measure of the ease with which a material deforms, either elastically or plastically. Soft materials tend to deform rather than fracture and so particle size reduction is typically problematic.[57] The mechanisms involved in particle reduction for solid dosage forms may be divided into three categories: cutting, impact and attrition. The technique selected for size reduction will depend upon the initial and required final particle sizes, mechanical properties such as hardness and toughness and whether the material is sticky or abrasive. Since materials generally become more brittle at low temperatures, soft materials may be milled successfully under liquid nitrogen. The cost of milling increases as the particle size decreases. Consequently, appropriate selection of both methods and conditions are required such that particle size is reduced only to the level required.

13.5   Future Outlook

13.5.1   Changing the Drug Product Design Paradigm

Over the last decade, through embracing both academic and technological advances, significant progress has been made in defining relationships between the API properties and the formulation design aspects of new products. Examples of established progress include API particle size distributions and content uniformity,[58] flow,[59] mechanical properties,[60] dissolution[61] and milling behaviour.[57]

Models have been built that allow pharmaceutical scientists to model the impact of particle size variation on dissolution rate and bioavailability. The biopharmaceutics classification system (BCS)[62] is used to define classes of compounds based on their solubility and permeability. Permeability is a molecular property, but solubility and dissolution rate are related to the internal structure (salt and polymorph) and particle size distribution/surface area. These relationships, combined with institutionalised corporate knowledge of formulation design practices,[63] have opened up the potential of a fully integrated holistic product design process consistent with the emerging QbD philosophy.[12,64]

13.5.2   Cocrystals

Cocrystals are defined as neutral multi-component systems having extended molecular networks formed through strong H-bonding patterns. Components of a pharmaceutical cocrystal include at least one API and one or more ligand (coformer), all of which are neutral and solid at room temperature and atmospheric pressure.[65] The appeal of cocrystals is that they offer a pathway for altering the solid state properties of non-ionisable APIs. Cocrystal formation between an API and ligand(s) relies on complementary, non-covalent interactions such as H-bond, van der Waals, π–π stacking and electrostatic interactions. Research on H-bond motifs led to guidelines for the design of molecular assemblies.[13] The H-bond rules can be applied to the targeted design of cocrystals, while taking broader aspects into consideration such as crystallisation kinetics and thermodynamic properties. The ability of cocrystals to demonstrate different solubilities, dissolution rates and stabilities has also been described for various cocrystals. Cocrystals of itraconazole with diprotic carboxylic acids achieve and sustain 4- to 20-fold higher drug concentrations than crystalline itraconazole during aqueous dissolution.[66]

13.5.3   Solid Form Design

Risk is a combination of the probability of occurrence and the severity of the impact. The ICH Q6a guidelines[67] consolidate this risk framework into a decision tree on polymorphs. The first two decision points on this framework remain the key questions that need to be addressed by the development scientist:

- Decision point 1: PROBABILITY– "can different polymorphs be formed?"
- Decision point 2: IMPACT – "do the forms have different properties (*e.g.* solubility)?"

In this section we consider new computational/structural approaches to describe the probability of new forms and consolidated institutional knowledge to quantify the impact of a potentially different structure. By understanding the structural chemistry and the biopharmaceutics the risk can be quantified with greater rigour, and experimental plans shaped accordingly. In attempting to obtain a greater definition of the probability of a new form appearing, tools ranging from quantum chemistry analysis,[68] H-bonding statistics[69] and full polymorph prediction[70] can now be applied either individually or in combination. Recent developments in theoretical chemistry mean that from a molecular structure a 3-D structural optimisation can occur and charges can be visualised on the van der Waals surface. These developments can be used to quantitatively describe the relative strength of the H-bond donors and acceptors.

The Cambridge Crystallographic Data Centre (CCDC) has, for the last 40 years, consolidated organic crystal structure information and distilled this knowledge into tools and software that is routinely applied to drug design. Recently there has been an enhanced effort in the application of these sorts of tools in crystal engineering. The Logit model[69] carries out a statistical analysis of H-bonding patterns for a given structure to identify potential H-bonding patterns that are compared to those in the known crystal structures in order to rationalise the physical form stability.

The *ab initio* generation through computational methods of reliable solid state structural details, based only on molecular descriptors, remains a major scientific goal. The methods being developed for structure prediction usually involve the stages of generating, clustering and refining trial structures. Final refinement of the potential structures is carried out by minimising the lattice energy (see Section 13.2.2) with respect to the unit cell dimensions (a, b, c, α, β and γ).[70] Despite the inherent difficulties, predictions from first principles have been the subject of much elegant investigation through the last decade, with increasing application to pharmaceutical compounds.[71] Proponents of these methods have now become so confident in their approaches that they are prepared to engage in blind tests to assess the predictability of their methods.[72]

13.5.4 *In Silico* Particle Design

Traditionally, the solid form selection process has focused on two main factors, that is achieving an appropriate degree of product stability and bioavailability. However, increasing emphasis is now also being focused on selection of solid forms at the pre-formulation stage that have optimal physical properties such as mechanical behaviour, surface properties and particle shape. A number of emerging computational technologies are thus forming foundation elements of the modern paradigm of QbD strategy for the development and manufacture of advanced particulate products, notably:

- Predicting crystal surface–solvent interactions and the solvent mediation of the crystal habit of aspirin.[28]
- Understanding interparticle interactions associated with polymorphic transformation in L-glutamic acid.[73]
- Estimation of the enhancement of the solubility of aspirin as a function of particle size reduction and morphological change.[74]

- Estimation of the size dependence of polymorphic stability in relation to crystallisation.[75]

13.5.5 Process Analytical Technologies

Crystallisation provides a significant opportunity with respect to the application of process analytical technologies (PAT), owing to the number of material and process parameters which must be assessed in order to define and control crystallisation behaviour. The key information required for process monitoring and control in the case of crystallisation includes solubility, supersaturation, MSZW, particle size distribution, particle shape and polymorphic form. A simple but very effective means of determining the MSZW, as well as other aspects of nucleation behaviour, is by measuring the turbidity of the crystallising solution. As nuclei form and grow in an originally clear solution, the optical transmittance of the medium decreases. A simple turbidity probe comprises a light source, a solution gap, a mirror reflector and a detector. Fibre optics are typically the most convenient way of achieving such an arrangement in the minimum volume. Figure 13.4 shows the change in transmittance, as a function of temperature, associated with the cooling crystallisation and dissolution of nortriptyline hydrochloride.[33] Similarly, a number of techniques are available for the determination of particle size and size distribution, including using the focused beam reflectance method (FBRM),[76] which is most suitable for on-line characterisation. FBRM generates information on changes in chord length (indicative of particle size) and the number of particles generated at different times during standard processing. In combination with FTIR spectroscopy, for closed-loop reactor control, this has the potential to allow for the process understanding and control of the relative balance associated with the nucleation of new particles and growth of existing particles during all aspects of the crystallisation process.[77] A recent academic/industry consortium project spanned the evaluation, development and definition of a wide range of PAT techniques for examining crystallisation processes. Key outcomes in this area can be highlighted from this and related work, including the impact of on-line FTIR spectroscopy,[77] flow through ultrasound spectroscopy,[79] digital video-microscopy[78,80] and on-line PXRD.[81]

13.6 Concluding Remarks

Crystallisation is the final step of API manufacture and so from a regulatory perspective must be both controlled and reproducible. In particular, it must provide APIs of a suitable quality in terms of both purity and the appropriate physical properties for robust dosage form design and processing. In recent years a greater interest in the latter aspect has resulted in an emphasis of the link between the solid form, particle formation and formulation aspects to be considered in a more integrated fashion. This has the associated benefits of streamlined API solid form selection, rapid commercial product design and IP creation and product protection. In this chapter we have attempted to bridge the new cutting-edge academic progress to the best industrial practices that crystallisation scientists can apply to advance NCEs consistent with the emerging QbD landscape. The physical characteristics of the API ultimately have the potential to affect the safety, efficacy and manufacturability of the product being designed and manufactured. The emergence of a range of particle engineering technologies, coupled to state-of-the-art characterisation technologies, has allowed access to a greater range of desirable particle attributes. Harnessing this capability to new sophisticated small-scale materials testing and institutionalised product design rules has led to the creation of a "design by first intent" strategy for API particles with tailored physicochemical attributes and functionality.

References

1. W. H. DeCamp, *Am. Pharm. Rev.*, 2001, **4**(3), 70.
2. S. R. Byrn, R. Pfeiffer and J. G. Stowell, *Am. Pharm. Rev.*, 2002, **5**(3), 92.
3. S. R. Chemburkar, J. Bauer, K. Deming, H. Spiwek, K. Patel, J. Morris, R. Henry, S. Spanton, W. Dziki, W. Porter, J. Quick, P. Bauer, J. Donaubauer, B. A. Narayanan, M. Soldani, D. Riley and K. McFarland, *Org. Process Res. Dev.*, 2000, **4**, 413.
4. J. Bernstein, in *Polymorphism in the Pharmaceutical Industry*, ed. R. Hilfiker, Wiley-VCH, Weinheim, 2006, ch. 14.
5. R. Storey, R. Docherty and P. D. Higginson, *Am. Pharm. Rev.*, 2003, **6**(1), 100.
6. M. Ticehurst and R. Docherty, *Am. Pharm. Rev.*, 2006, **9**(7), 32.
7. *The Crystal as a Supramolecular Entity*, ed. G. R. Desiraju, Wiley, Chichester, 1997.
8. B. H. Hancock and J. Elliot, *MRS Bull.*, 2006, **31**, 869.
9. K. Chow, H. Y. H. Tong, S. Lum and A. H. L. Chow, *J. Pharm. Sci.*, 2008, **97**, 2855.
10. *Fed. Reg.* 2005, **70**, 134.
11. A. S. Myerson, *Molecular Modelling Applications in Crystallization*, ed. A. S. Myerson, Cambridge University Press, New York, 1999, ch. 2.
12. G. Nichols and C. J. Frampton, *J. Pharm. Sci.*, 1998, **87**, 684.
13. M. C. Etter, *Acc. Chem. Res.*, 1990, **23**, 120.
14. R. Taylor and O. Kennard, *Acc. Chem. Res.*, 1984, **17**, 320.
15. R. Docherty and W. Jones, in *Organic Molecular Solids: Properties and Applications*, ed. W. Jones, CRC Press, London, 1997, ch. 3.
16. A. I. Kitaigorodsky, *Molecular Crystals and Molecules*, Academic Press, New York, 1973.
17. D. E. Williams, *J. Chem. Phys.*, 1966, **45**, 3770.
18. S. Lifson, A. T. Hagler and P. Dauber, *J. Am. Chem. Soc.*, 1979, **101**, 5111.
19. F. A. Momany, L. M. Carruthers, R. F. McGuire and H. A. Scherega, *J. Phys. Chem.*, 1974, **78**, 1595.

20. A. Gavezzotti and G. Filippini, *J. Phys. Chem.*, 1994, **98**, 4831.
21. *Physical Characterization of Pharmaceutical Solids*, ed. H. G. Brittain, Dekker, New York, 1995.
22. *Solubility Behavior of Organic Compounds*, ed. D. J. W. Grant and T. Higuchi, Wiley-Interscience, New York, 1990.
23. A. Koda, S. Ito, S. Itai and K. Yamamoto, *J. Pharm. Sci. Jpn.*, 2000, **60**, 43.
24. J. D. Donnay and D. Harker, *Am. Miner.*, 1937, **22**, 446.
25. P. Hartman and W. G. Perdok, *Acta. Crystallogr.*, 1955, **8**, 49.
26. Z. Berkovitch-Yellin, *J. Am. Chem. Soc.*, 1985, **107**, 8239.
27. G. P. Clydesdale, K. J. Roberts and R. Docherty, in *Controlled Particle and Bubble Formation*, ed. D. J. Wedlock, Butterworth-Heinemann, Oxford, 1994, ch. 4.
28. R. B. Hammond, K. Pencheva, V. Ramachandran and K. J. Roberts, *Cryst. Growth Des.*, 2007, **7**, 1571.
29. E. Tedesco, D. Giron and S. Pfeffer, *CrystEngComm*, 2002, **4**, 1.
30. B. Y. Shekunov, P. Chattopadhyay, H. Y. Tong and A. H. L. Chow, *Pharm. Res.*, 2007, **24**, 203.
31. Z. Sun, N. Ya, R. C. Adams and F. S. Fang, *Am. Pharm. Rev.*, 2010, May/June, 68.
32. R. Davey and J. Garside, *From Molecules to Crystals*, Oxford University Press, Oxford, 1998.
33. K. J. Roberts and B. A. Hendriksen, *J. Cryst. Growth*, 1993, **128**, 1218.
34. *Handbook of Industrial Crystallisation*, ed. A. S. Myerson, Butterworth-Heinemann, Boston, 2002.
35. G. H. Gilmer and K. A. Jackson, in *Crystal Growth and Materials*, ed. E. Kaldis and H. J. Scheel, North-Holland, Amsterdam, 1977, p. 80.
36. W. K. Burton, N. Cabrera and F. C. Frank, *Phil. Trans. R. Soc.*, 1951, **243**, 299.
37. P. Bennema and J. P. van der Eerden, in *Morphology of Crystals*, Terra Scientific, Tokyo, 1987, ch. 1.
38. D. Erdemir, A. Y. Lee and A. S. Myerson, *Curr. Opin. Drug Discovery Dev.*, 2007, **10**, 746.
39. J. Bernstein, *Polymorphism in Molecular Crystals*, International Union of Crystallography/Oxford University Press, Oxford, 2007.
40. *Handbook of Pharmaceutical Salts*, ed. P. H. Stahl and C. G. Wermuth, Wiley-VCH, Weinheim, 2002.
41. S. M. Berge, L. D. Bighley and D. C. Monkhouse, *J. Pharm. Sci.*, 1977, **66**, 1.
42. (a) http://www.fda.gov/ohrms/dockets/ac/09/slides/2009-4412s1-01-FDA.pdf; (b) http://www.fda.gov/ohrms/dockets/ac/09/briefing/2009-4412b1-01-FDA.pdf (last accessed 27 January 2011).
43. J. Haleblian and W. C. McCrone, *J. Pharm. Sci.*, 1969, **58**, 911.
44. S. R. Byrn, R. R. Pfeiffer and J. G. Stowell, *Solid State Chemistry of Drugs*, SSCI Press, West Lafayette, IN, 2nd edn, 1999.
45. European Medicines Agency, London, 18 June 2008, doc. ref.: EMEA/265069/2008 rev. 1.

46. G. P. Stahly, *Cryst. Growth Des.*, 2007, **7**, 1007.
47. M. Pudipeddi and A. T. M. Surajaddin, *J. Pharm. Sci.*, 2005, **5**, 94.
48. J. M. Miller, B. M. Collman, L. R. Greene, D. W. Grant and A. C. Blackburn, *Pharm. Dev. Tech.*, 2005, **10**, 291.
49. N. Blagden and R. J. Davey, *Cryst. Growth Des.*, 2003, **3**, 873.
50. R. K. Khankari and D. J. W. Grant, *Thermochim. Acta.*, 1995, **248**, 61.
51. Y. Cui and E. Yao, *J. Pharm. Sci*, 2008, **97**, 2730.
52. M. D. Ticehurst, R. A. Storey and C. Watt, *Int. J. Pharm.*, 2002, **247**, 1.
53. J. W. Mullin, *Crystallisation*, Butterworth-Heinemann, Oxford, 2001.
54. *Theory of Particulate Processes*, ed. A. D. Randolph and M. A. Larson, Academic Press, San Diego, 1988.
55. D. Hsieh, A. J. Marchut, C. Wei, B. Zheng, S. S. Y. Wang and S. Kiang, *Org. Process Res. Dev.*, 2009, **13**, 690.
56. W. Beckmann, K. Nickisch and V. Budde, *Org. Process Res. Dev.*, 1998, **2**, 298.
57. L. J. Taylor, D. G. Papadopoulos, P. J. Dunn, A. C. Bentham, N. J. Dawson, J. C. Mitchell and M. J . Snowden, *Org. Process Res. Dev.*, 2004, **8**, 674.
58. S. H. Yalkowsky and S. Bolton, *Pharm. Res.*, 1990, **7**, 962.
59. M. P. Mullarney and N. Leyva, *Pharm. Technol.*, 2009, **33**, 126.
60. C. C. Sun, H. Hou, P. Gao, C. Ma, C. Medina and J. Alvarez, *Pharm. Sci.*, 2009, **98**, 239.
61. K. Johnson and A. C. Swindell, *Pharm. Res.*, 1996, **13**, 1795.
62. G. L. Amidon, H. Lennernas, V. P. Shah and J. R. A. Crison, *Pharm. Res.*, 1995, **12**, 413.
63. R. Roberts and R. Rowe, *Intelligent Software for Product Formulation*, Taylor and Francis, London, 1998.
64. R. Docherty, T. Kougoulos and K. Horspool, *Am. Pharm. Rev.*, 2009, **12**(6), 34.
65. C. B. Aakeroy, M. E. Fasulo and J. Desper, *Mol. Pharm.*, 2007, **4**, 317.
66. J. F. Remenar, S. L. Morissette, M. L. Peterson, B. Moulton, M. J. MacPhee, H. R. Guzman and O. Almarsson, *J. Am. Chem. Soc.*, 2003, **125**, 8456.
67. ICH Q6a, *Specifications: Test Procedures And Acceptance Criteria For New Drug Substances And New Drug Products: Chemical Substances*, http://www.ich.org/ (last accessed 27th January 2011).
68. Y. A. Abramov and K. Pencheva, in *Chemical Engineering in the Pharmaceutical Industry: R&D to Manufacturing*, ed. D.J. am Ende, Wiley, Hoboken, NJ, 2010, ch. 25.
69. P. T. A. Galek, F. H. Allen, L. Fábián and N. Feeder, *CrystEngComm*, 2009, **11**, 2634.
70. H. R. Karfunkel and R. J. Gdanitz, *J. Comput. Chem.*, 1992, **13**, 1771.
71. R. S. Payne, R. C. Rowe, R. J. Roberts, M. H. Charlton and R. Docherty, *J. Comput. Chem.*, 1999, **20**, 262.
72. G. M. Day, T. G. Cooper, A. J. Cruz-Cabeza, K. E. Hejczyk, H. L. Ammon, S. X. M. Boerrigter, J. S. Tan, R. G. Della Valle, E. Venuti,

J. Jose, S. R. Gadre, G. R. Desiraju, T. S. Thakur, B. P. van Eijck, J. C. Facelli, V. E. Bazterra, M. B. Ferraro, D. W. M. Hofmann, M. A. Neumann, F. J. J. Leusen, J. Kendrick, S. L. Price, A. J. Misquitta, P. G. Karamertzanis, G. W. A. Welch, H. A. Scheraga, Y. A. Arnautova, M. U. Schmidt, J. van de Streek, A. K. Wolf and B. Schweizer, *Acta Crystallogr., Sect. B: Struct. Sci.*, 2009, **65**, 107.

73. R. B. Hammond, K. Pencheva and K. J. Roberts, *Cryst. Growth Des.*, 2007, **7**, 875.

74. R. B. Hammond, K. Pencheva, K. J. Roberts and T. Auffret, *J. Pharm. Sci.*, 2009, **98**, 4589.

75. R. B. Hammond, K. Pencheva and K. J. Roberts, *J. Phys. Chem. B*, 2005, **109**, 19550.

76. A. R. Heath, P. D. Fawell, P. A. Bahri and J. D. Swift, *Part. Part. Syst. Char.*, 2002, **19**(2), 85.

77. H. Groen, A. Borissova and K. J. Roberts, *Ind. Eng. Chem. Res.*, 2003, **42**, 198.

78. A. Borissova, S. Khan, T. M. Mahmud, K. J. Roberts, J. D. P. Andrews, Z.-P. Chen and J. Morris, *Cryst. Growth Des.*, 2009, **9**, 692.

79. M. Li, D. Wilkinson, K. Patchigolla, P. Mougin, K. J. Roberts and R. Tweedie, *Cryst. Growth Des.*, 2004, **4**, 955.

80. R. F. Li, G. B. Thomson, G. White, X. Z. Wang, J. Calderon de Anda and K. J. Roberts, *AIChE J.*, 2006, **52**, 2297.

81. S. Dharmayat, R. B. Hammond, X. Lai, C. Ma, E. Purba, K. J. Roberts, Z.-P. Chen, E. Martin, J. Morris and R. Bytheway, *Cryst. Growth Des.*, 2008, **8**, 2205.

CHAPTER 14

Technology Transfer of an Active Pharmaceutical Ingredient

STEPHEN MCGHIE[a] AND STUART YOUNG[b]

[a] Technical Shared Service, Global Manufacturing and Supply,
GlaxoSmithKline, Shewalton Road, Irvine, KA11 5AP, UK
[b] New Product Introduction Centre of Excellence, Global Manufacturing and
Supply, GlaxoSmithKline, Temple Hill, Dartford, DA1 5AP, UK

14.1 Introduction

Technology transfer[1] is a key business process supporting the introduction of
newly developed products and technologies from development into full-scale
manufacture. It is often, in its first iteration, the culmination of scale-up from
pilot plant facilities into a scale more likely to meet the commercial volume
needs of the product. The same generic process can also be used to transfer
manufacture of products between different manufacturing sites, and in the
transfer from an innovator to an alternate supply site or to external supplier of
an active pharmaceutical ingredient (API).

The success or otherwise of any technology transfer is dependent on the
approach taken, and this can often directly influence the success or otherwise of
the product itself. For example, for an API, an effective approach to the
transfer is required to ensure continuity of supply to patients, no adverse
changes in safety, quality or efficacy of the drug product and compliance with
legal and regulatory requirements.

RSC Drug Discovery Series No. 9
Pharmaceutical Process Development: Current Chemical and Engineering Challenges
Edited by A. John Blacker and Mike T. Williams
© Royal Society of Chemistry 2011
Published by the Royal Society of Chemistry, www.rsc.org

However, what is it that we mean by technology transfer? It is simply the transfer of data, knowledge and understanding from one area to another. Often this can be a difficult process to get right: data transfer systems are becoming more and more advanced with the constant progression of software and hardware systems and computer interfacing; however, deriving knowledge and understanding as a follow-on from data sharing and transfer can be a stumbling block for both small and major players in the pharmaceutical industry.

It is key that the transferred process operates consistently to ensure that the development team or donor site reliably passes all relevant quality, technical, regulatory compliance, health and safety, and logistics information to the receiving site. In addition, the transfer of knowledge must ensure that the receiving site understands the design rationale, the principles of operation of the process, the critical control parameters and the critical steps in processing; essentially the receiving site must understand why the process of manufacture, and associated testing, have been designed the way they are.

Planning technology transfer correctly is paramount, not only to ensure that the process is transferred successfully but also to guard against surprises. Many technology transfers stumble or are delayed because of poor planning and a failure to predict what surprises may come along the way. The consequences (and indeed cost) of not doing it right are not only in time delays but certainly financial from either loss of opportunity, loss of sales or in capital remediation.

As a lead up to a transfer process, improved scientific understanding must be used to develop a process control and monitoring strategy for the manufacturing process, such that critical process parameters are defined and the impact of variation in these on the API quality attributes understood. This control strategy can be used to demonstrate the successful manufacture at the receiving site, and the increased knowledge gained provides a baseline for continuous improvement.

This chapter describes an outline process and discrete steps that can be implemented for transferring a process to manufacturing.

14.2   Technology Transfer Process Flow

Of course, many different technology transfer models exist both in the literature[2] and within the standard operating procedures of pharmaceutical companies, but they tend to default along the same principles of planning, communication and risk assessment prior to implementation.

Principally, technology transfer should follow a common framework that ensures consistency and the best chance for success. This framework does not need to be complex and any technology transfer of an API, whether involving many stages of chemistry, complex biotechnology or sterile manufacture, can easily fit into a simple process such as that shown in Figure 14.1. In this chapter we will use these four process steps to exemplify technology transfer of an API from development into manufacturing.

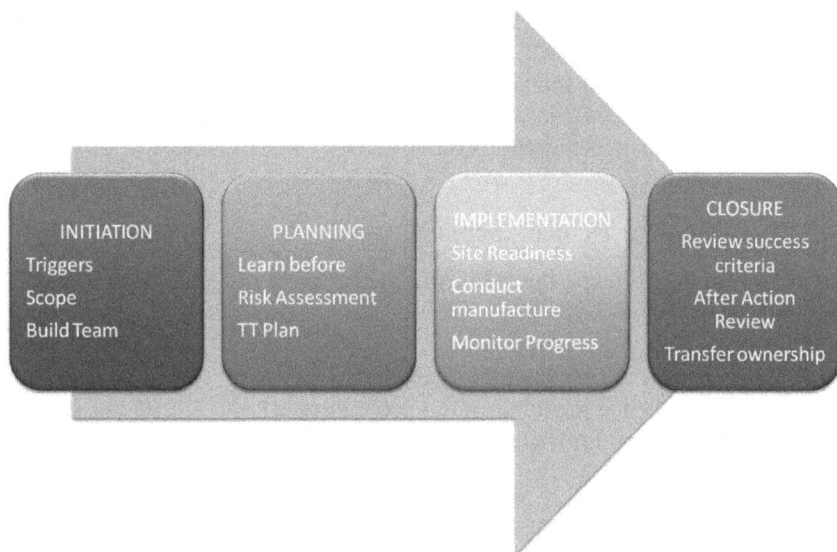

Figure 14.1 Simplistic technology transfer process flow.

14.2.1 Initiation: Triggers for Transfer

During the typical lifecycle of manufacture of an API there will be a trigger to transfer the manufacturing capability for the API or an intermediate stage from one development or manufacturing site to another. There are a number of triggers that could initiate the transfer of a process to manufacturing:

- New product introduction from development into commercial manufacturing scale plant.
- Increase in capacity required to satisfy commercial requirements through product lifecycle.
- Introduction of process improvements for quality, safety, sustainability or cost reasons.
- Lifecycle strategies for manufacturing sites, such as a drive to free up capacity for newer products and the concomitant move of existing products elsewhere.

The trigger initiates a technical evaluation of the API manufacturing process followed by the planned transfer of the technology from a donor site to a receiving site.

It is important to understand the reasons for technology transfer – these triggers – as it can put different requirements on the scope and planning of the project itself. For example, a complete transfer of an API process from one manufacturing site to another due to an increase in volume requirements may need to take place with best agility, whereas a technology transfer in late lifecycle to free capacity may require a period of dual manufacture to cover market requirements during the period of regulatory file changes or updates.

In this chapter we will focus on transfer of a process from development to manufacturing, where the success will depend on agility whilst ensuring that the quality of the product is not compromised.

14.2.2 Initiation: Project Scope

As with all major projects, the definition of the scope of technology transfer is fundamental:

- What is it that we are trying to achieve?
- What is and is not included in the project remit?
- What will success look like?

14.2.2.1 *What Is It that We Are Trying to Achieve?*

Typically in API process technology transfer the scope may be quite specific, but must include as a minimum what are the fundamental requirements of both the process and the manufacturing equipment. We will develop these themes further in the following planning and initiation steps, but some fundamentals in advance of technology transfer must be established:

- An evaluation of the understanding of the chemistry process.
- The intricacies of the equipment needs of the process *versus* the equipment availability in the receiving site.

For example, a process that utilises aggressive reagents, or where stage products or the API itself is corrosive, will typically require plant that can accommodate this.

These process fundamentals can often be the first step in choosing the receiving site location: what the site has *versus* the requirements of the process (and, of course, what may need to be purchased).

Take, as an example, a process that is known to suffer from poor filtration characteristics in the isolation of an API intermediate product. This may force a requirement for specific isolation and drying equipment or, alternatively, modifications to existing operational parameters to allow suitable processing in an alternative equipment type. All of this pre-work is essential to ensuring a defined scope of what is being transferred to where and what the objectives and measure of success will be. Ultimately this will help in understanding the risks created by the transfer proposed.

14.2.2.2 *What Is and Is Not Included in the Project Remit?*

It is important to include in the scope of technology transfer the boundaries of what is being transferred. Process technology transfer may sound fairly simplistic, but what supporting activities will be transferred with the process? Where will the supporting activity come from? From the donating site? From the receiving site? Indeed, from a third party?

So, for example, in support of the process, is analytical technology transfer accompanying the manufacturing process? Typically this should always be done well in advance of the process to ensure that the measurement systems will be in place to allow success criteria to be demonstrated.[3]

What logistical supporting criteria and functions will be required as part of the technology transfer? API manufacturing is a broad business; whilst the process chemistry may be at the heart of the technology transfer, the support systems are fundamental: whether it be packaging, shipping, storage arrangements, raw materials procurement, process safety considerations or quality testing (as a package for product release, for intermediate testing, for in-process check and control, stability protocols and so on), regulatory activity will be captured as part of the technology transfer. These broader considerations will help define the team required to carry out the technology transfer, wider than the technical team typically formed of chemists and process engineers.

14.2.2.3   What Will Success Look Like?

To facilitate any process technology transfer, a definition of the success criteria is essential. What is acceptable and can be used as a measure that technology transfer will be complete?

Often this is fairly self explanatory; for example, the material can be made to a suitable standard in the new location at the new scale, and at this stage of initiation the scope definition will be fairly high level. Only when process risk assessment and an understanding of the current state of process knowledge are carried out will defined and specific acceptance criteria be in place to measure success.

Quality target ranges for impurities, yield for acceptable financial targets, throughput for optimal plant usage and volume delivery are all typical simple targets included in acceptance criteria. However, it is important to include softer elements associated with knowledge transfer from one place to another (for example, the systems of data sharing, whether electronic or paper-based; the feeling of confidence in both donor and receiver in the technology transfer team; closure of risk items and actions associated with the transfer process; a review and agreement on the level of continued support post project closure) such that, if issues arise, swift resolution can be applied. Relationships and the ability to communicate across the team are paramount to ensuring that technical knowledge transfer is successful.

14.2.3   Initiation: Build the Team

As part of the transition from initiation of technology transfer to formal technology transfer planning, a technology transfer team should be set up.

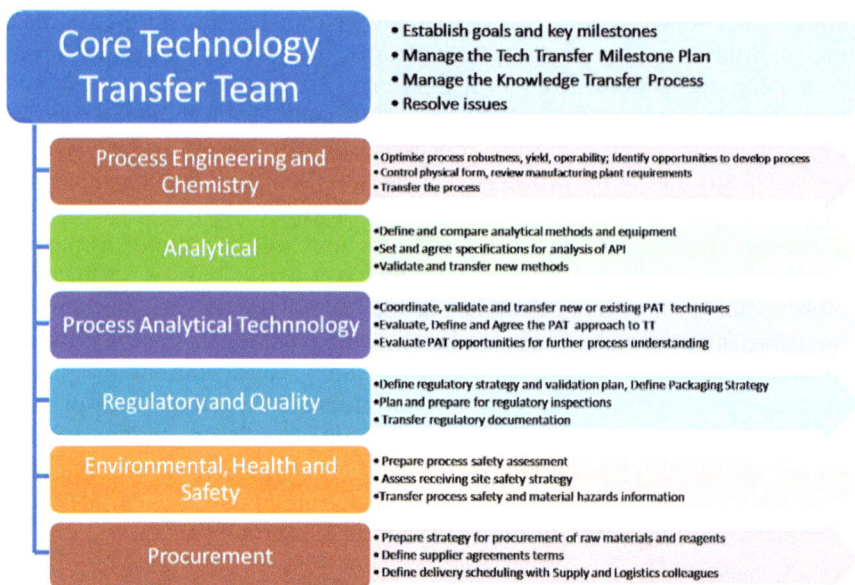

Figure 14.2   Example of a typical technology transfer core team and sub-team structure with responsibilities.

This core team will typically be made up of leads in each of the discipline areas required, as defined in the scope of the project. Within each of these discipline areas a sub-team will exist, made up of the functional groups that provide the deliverables identified for the technology transfer plan. An example of a typical core and sub-team organisation with responsibilities is outlined in Figure 14.2.

It is essential that detailed roles and responsibilities, objectives and expectations for both the core team and sub-team are defined as part of the team set-up. This should be defined based on the technology transfer scope and may be revised as the technology transfer process is followed. It is often prudent to ensure that these role definitions and responsibilities are recorded and agreed with the technology transfer sponsor and any subsequent modifications and amendment subject to necessary governance.

14.2.4   Planning: Learn before Doing[4]

Prior to the technology transfer team start-up, the core team and the leaders of the sub-teams should attend any necessary training on the technology transfer process and assess actions which may be required to ensure all sub-teams have the competency, skills and information required to perform their functions.

The core team and sub-team leaders should ensure that each member of the technology transfer team is fully familiar with the relevant documentation for their part of the technology transfer process.

As with all planning, a holistic review of previous similar activities should be carried out to ensure that the learnings are transferred and any necessary remediation or corrective actions highlighted are complete to ensure this current technology transfer is as successful as it can be. Often previous after action reviews provide a wealth of learnings and recommendations and these should always be available to the newly created team.

From a purely technical point the value of small-scale familiarisation batches in the laboratory of the receiving site should never be underestimated. It provides an opportunity for the receiving team to get their hands on the chemistry to see how it performs and how it looks and feels as a process, as well as ensuring that the appropriate knowledge has been transferred. This is, of course, not a simple exercise of proving that one person can copy what has been done, but is an opportunity to assess whether the tacit knowledge transfer is being done in tandem with the formal transfer of information and data.

Ultimately the ability of the receiving site to conduct the process at all scales, and to understand the intricacies of the process and to demonstrate their knowledge of the process, will ensure confidence that the process will ultimately be under good management. This demonstration that the receiving site can be self-sufficient in future operation of the process will enable the donating site to complete a timely handover.

14.2.5 Planning: Risk Assessment

The transfer of a process from laboratory or pilot plant into a manufacturing site at a new scale brings with it risk in terms of scale issues not previously predicted, or not previously expected. In preparation for the technology transfer the core technology transfer team should perform a risk or impact assessment, based on the available process knowledge and knowledge of the receiving site's manufacturing facility, infrastructure and resources. The assessment should take into account:

- An assessment of the risks to the process, the site and the product due to the complexity of the process under transfer. What could go wrong in the process of technology transfer and what mitigating actions will be put in place?
- The impact the technology transfer will have on the receiving site. This must include an assessment of the impact of any new or novel chemistry, technologies or manufacturing processes of which the receiving site has no experience. Of particular note is the requirement for new equipment, or changes to existing equipment, or ways of working at the receiving site.
- The suitability of the receiving site. This must include the receiving site requirements related to the technology transfer for facilities, work processes, environment and staff capabilities.

In all cases the risk assessment process must include core team members from the donating site and the receiving site, and it is key that the process operations staff at the receiving site be included in this review.

By carrying out these risk assessments, any potential surprises can be reduced and actions put in place to mitigate any major risks uncovered. This knowledge can then be used as the basis for technology transfer planning.

14.2.5.1 *Vive la Difference!*

The goal of technology transfer is to successfully demonstrate a similar product being made at a new site that is fit for its intended purpose (and poses no impact to the patient). However, aside from these essentials related to patient impact, it is essential that the process introduced does not present any new issues in manufacturing upon transfer (such as throughput differences because of scale effects, differences in trace levels of impurities due to new sources of raw materials and reagents or ways of working, or differences in the physical nature of the product due to differences in plant and equipment or control parameters). This can have untold consequences not only in the API manufacturing plant but more importantly in downstream processing to manufacture the drug product,[5] impacting a number of key metrics[6] such as throughput, shelf-life, handling of the API and increasing the incidence of waste in the value stream.

Often a customer of the API supply will request that the product is the "same" as previously provided and the term is often a key criteria used for technology transfer. However, experience tells us that processes are never the same. Difference is inevitable from batch to batch, from campaign to campaign and from site to site. More important is the understanding of the impact of that difference.

Risk assessment has increasingly become a prerequisite for any change programme in the pharmaceutical industry, and is a true test of whether we understand processes and whether we understand the science behind the changes we introduce. Technology transfer is often a significant change, but by investing time in risk assessment to understand the likely differences we will see in the manufacturing process, the API or in the downstream processing into the drug product, we will be better able to do this in an agile manner.

14.2.6 Planning: Technology Transfer Plan

The technology transfer leader should work with the sub-teams to produce an overall technology transfer plan. The plan should include the following:

- Technology transfer core team and sub-team organisation charts and agreed roles and responsibilities defined previously with the overall sponsors.
- Technology transfer team knowledge of the process and any actions from the before action review, including any necessary training.
- Actions identified during the risk and impact assessments.
- Incorporation of new technology or process into manufacturing and development of production and validation plans, if required. Often process technology transfer will involve suitable process validation to demonstrate process control and acceptance criteria have been delivered.

This chapter will not focus on process validation *per se*, except to note that it is typically part of the overall deliverables as part of technology transfer.

- Documentation and technology transfer deliverables, requirements and handover timelines across all team functions.
- Analytical and quality plan including analytical technology transfer timelines, API stability testing requirements, responsibility for batch release and the regulatory strategy for the receiving site for that API.
- Materials sourcing strategy, including the requirements of the raw materials, reagents and solvents and their associated impact assessments.
- Strategy for addressing any environmental, health or safety issues.
- Strategy for communication to ensure coordination between the sub-teams and with other areas within the company or companies involved in the technology transfer.
- Logistics between sites to support commercial demands or launch as required.
- Resources for technology transfer support, including where support will come into the team and leave the team.
- A plan to develop and agree the "acceptance criteria", and the point at which responsibility passes from the donating site to the receiving site.
- Strategy and timelines for completion of development work to meet the "acceptance criteria".
- Change control for the API process information.
- Reporting and communication.
- Developing technical strategy for manufacture post-transfer, for example a monitoring plan.

The technology transfer plan must be reviewed and approved by the technology transfer core team and the sponsors and re-issued following any significant changes.

14.2.7 Implementation: Site Readiness Actions

As a receiving site evaluates the process for technology transfer, it will understand the best available plant and technology within its current infrastructure and asset base. When comparing these against the requirements of the process, a number of fundamental decisions will need to be made:

- Can the process fit into the currently available plant and equipment?
- What additional plant and equipment is required?

Often these questions come down to finances, but as part of the risk assessment process these will also be important considerations: what are the risks of using the existing equipment; indeed, what are the risks in using new equipment?

Risk assessment should be seen as a living document throughout the lifetime of technology transfer and beyond, as changes are presented throughout, and further assessment will be required to adequately and successfully manage and control.

At a site level, decisions made on available plant and technology, *versus* procurement of new, will fundamentally affect the technology transfer timelines, and equally could impact the quality of the product to be made. Essential to timely completion is freezing the design to ensure completion can be done at an appropriate pace without the risk of scope creep!

Completion of site readiness actions will therefore always be on the critical path, whether as a result of engineering work, capital procurement and installation, safety risk assessment, raw material sourcing, analytical technology transfer in advance of process manufacture, training of key personnel or even as a result of more development work being required or further process understanding at the receiving site prior to manufacturing commencing.

14.2.8   Implementation: Conduct Manufacture

Often it seems like the most fundamental step in a technology transfer is to conduct manufacture, but in essence the planning that has gone ahead of this stage should ensure the best chance of success.

As defined during the planning stage, a number of batches are often included as part of initial manufacture within the scope of the technology transfer; these batches may be for evaluation in the drug product formulation, for clinical trials or for commercial purposes.

Typical manufacture will involve a plant trial in the new location prior to any formal validation; this can range from simulation batches with solvents only (to check plant integrity), to formal trials using real manufacturing conditions and materials to check all planning is successful and the plant, equipment and process are ready to be validated. This decision is often driven by complexity, cost and time available.

14.2.8.1   Process Validation

Process validation[7] is a fundamental framework that should be followed during technology transfer, with the simple purpose of proving that the plant, equipment and process and controls can provide a quality product that is fit for purpose. This chapter does not discuss the detail of process validation; however, the fundamentals provide an opportunity to formally record that predefined acceptance criteria have been met for the product. The framework also provides an opportunity to start the process of data management, where processing parameters, output quality parameters and quality critical attributes can be checked and compared with previous manufacture. This allows for different processes at different scales, sites or companies to be compared directly.

This formal comparison of output material can also allow for the reassessment of risk, such that an understanding is gained of how the material is likely to forward process, and any actions are progressed to enable any differences to be accommodated.

14.2.9 Implementation: Monitor Progress

Data systems in pharmaceutical manufacture range from the complex automated control and data storage platforms to simple data recording on batch manufacturing instructions.

Regardless of the stage of advancement in data systems, it is essential that close monitoring of the process that has been transferred continues postvalidation.

Lifecycle validation is commonplace in the industry, but it remains paramount that, even in companies not adopting a lifecycle approach, the control of quality can be demonstrated. Even in the simplest of terms, process trending of manufacturing data can provide confidence and assurance of control through continued manufacture, and highlight any deviation or trend where control is slipping.

From a statistical view, enough batches to be significant are an ideal position, where continued manufacture post-validation can be monitored closely; however, this is often dependent on other business factors including the demand volume of the product, and the manufacturing frequency.

A standardised approach for evaluating and reviewing data should apply statistical or quality tools to analyse the data, particularly where the resultant analysis is used to make observations and conclusions about the product or process:

- Step 1. Data identification and collation: to ensure that all information and data are available and in the correct format for required analyses.
- Step 2. Quantitative assessment: to describe and characterise quantitative review parameters and to assess the product performance with respect to process-driven control limits and process parameters, as well as the registered product specifications and quality attributes, and compare performance to previous review period results.
- Step 3. Qualitative assessment: to describe and characterise qualitative review parameters and to assess the product performance with respect to the registered product specifications and compare performance to previous review period results.
- Step 4. Change assessment: to assess the significance of identified process performance and/or process capability changes. Consult statistics expert if change assessment analyses are required.
- Step 5. Summary report generation: to create a summary document, appropriately store and document all analyses that support the conclusions of all statistical assessments.

14.2.10 Closure: Review Success Criteria

On closure of the product transfer, manufacture will continue and there will be a requirement to maintain the capability to make further changes or transfers to other manufacturing sites. To this end it is essential that the following are maintained:

- Clarity on roles and responsibilities.
- Continued updated process understanding.
- Continuous improvement plans.
- Knowledge management.

14.2.10.1   Roles and Responsibilities

Once the process transfer is complete, accountability for continuing manufacture will come to the appropriate technical functions on the receiving manufacturing site. Transfer of ownership of activities will fall to the site functions, who will have responsibility for any outstanding risks and ongoing risk management activities and the future technical plan, including continuous improvement.

14.2.10.2   Process Understanding

On completion of the transfer of the process, a review of current state of knowledge should be repeated and any new risks added to the risk assessment. The data trending requirements for quality attributes and process parameters can also be reviewed at this point and a strategy decided for continued manufacture.

A periodic review should then continue to maintain the process knowledge. These scheduled events might be triggered, for example, by:

- Number of batches manufactured or completion of a series of batches in a campaign.
- Commitments to a regulatory authority.
- A timed cycle (for example, quarterly, annually).
- Transfer of a process.

14.2.10.3   Continuous Improvement

The updated technical review will provide the current status of a process, listing the key risks together with appropriate mitigation plans. Additionally, opportunities for improvement that will satisfy the quality, safety and commercial requirements during the lifecycle of the product and process should be identified, and a prioritisation approach adopted to ensure that the activities with the highest benefits are progressed.

Examples of areas for improvement are:

- Process capability, such as CpK/PpK for quality attributes.
- Material usage.
- Yield.
- Capacity and operation expenditure.

14.2.10.4 Knowledge Management

The transfer should have a knowledge management plan to transfer the critical and supporting process knowledge, including a repository for storing the information. On completion of a process transfer, key technical functions should have access to all the relevant process knowledge. As manufacture continues the knowledge repository will be updated and roles and responsibilities for the ongoing management of the data and information defined and maintained.

The type of information to be maintained will come from the updated technical reviews and would typically include critical processing parameters during manufacture, critical quality attributes of the manufactured product (such as impurities, residual solvents and critical material properties that influence forward processing), throughput and cycle-times, and yield.

14.2.11 Closure: After Action Review

Essential to any future activity is to learn from the past (learn before!) and as a living process, technology transfer needs to evolve and improve.

The technology transfer team can share areas where the process was successful and allowed smooth handover, as well as identifying key areas where the process needs to be adapted or ways or working better defined.

As mentioned earlier, the key to success is planning and evaluating risk and ensuring actions are in place to mitigate these.

14.2.12 Closure: Transfer Ownership

As the project formally closes and confirmation has been received that the success criteria have been met and specific actions related to the process under transfer have been completed, the transfer should be formally closed. Typically this is a formal document recording that transfer of ownership from donating site to receiving site is recommended.

Typically the report will include:

- A review and confirmation of the achievement of the success criteria.
- Output from the project "after action review".
- Confirmation of the status of any actions relating to quality or technical review of the process following transfer to the new location.
- Arrangement for regulatory activities have been concluded.
- Arrangement of the transfer formally of the key contact for queries on the process from donor to receiver site lead.
- Agreement on the future support from the donor to the receiver.
- An assessment of the state of knowledge transferred to the receiver.
- An update to the risk assessment associated with the technology transfer and future accountability and ownership for outstanding risk mitigation action.

14.3   Conclusion

Technology transfer in the pharmaceutical industry can be a complex business process, where a failure to prepare, plan and implement properly can lead to delays and costs.

By using a standard framework, the transfer of API processes from R&D to manufacture can be better managed and success assured. This chapter has shown a typical process flow for transfer of technology and knowledge from one place to another that can be used generically between sites, functions and companies.

Acknowledgements

The authors wish to extend their thanks to Lindsay Lupton and Julie Lautens for the assistance in review of the GSK business processes for technology transfer. They also would like to recognise the contribution from the numerous technology transfer teams across various projects in GSK.

References

1. S. Green and P. Warren, *Technology Transfer in Practice*, Horwood, Storrington, West Sussex, UK, 2002.
2. *Technology Transfer: An International Good Practice Guide for Pharmaceutical and Allied Industries*, ed. M. Gibson, PDA, Bethesda, MD, USA/ DHI Publishing, River Grove, IL, USA, 2005.
3. S. Perry, *Pharm. Manuf.*, 2010, **9**(1), 16.
4. G. Pisano, *Res. Policy*, 1996, **25**, 1097.
5. (a) C. Vemavarapu, M. Surapaneni, M. Hussain and S. Badawy, *Int. J. Pharm.*, 2009, **374**, 96; (b) W. L. Hulse, I. M. Grimsey and M. De Matas, *Int. J. Pharm.*, 2008, **349**, 61.
6. K. J. Carpenter, *Chem. Eng. Sci.*, 2001, **56**, 305.
7. Many resources on process validation exist, based around key regulatory guidance documents, such as: *FDA Guideline on General Principles of Process Validation*, May 1987, prepared by Center for Drug Evaluation and Research, Center for Biologics Evaluation and Research, and Center for Devices and Radiological Health, Food and Drug Administration, http://www.fda.gov/Drugs/GuidanceComplianceRegulatoryInformation/ Guidances/ucm124720.htm (last accessed October 2010); P. Carson and N. Dent, *Good Clinical, Laboratory and Manufacturing Practices: Techniques for the QA Professional*, Royal Society of Chemistry, Cambridge, 2007.

CHAPTER 15

Future Trends and Challenges

A. JOHN BLACKER[a] AND MIKE T. WILLIAMS[b]

[a] University of Leeds, Institute of Process R&D, Leeds, LS2 9JT, UK
[b] 133 London Road., Deal, Kent, C14 9TY, UK

15.1 An Industry in Transition

For over a decade the pharmaceutical industry has been in flux as a result of a range of pressures and a growing realisation that the traditional drug discovery and development model was not working well. This chapter will examine some of the changes the pharmaceutical industry has been subject to, and how this changing environment has impacted chemists, chemical engineers and technologists working in the field of process research and development (R&D).

First amongst the inter-related pressures felt within the industry is the issue of drug development costs. The worldwide pharmaceutical market continues to grow at a compound rate of over 5% per annum, with projected global sales predicted to rise from $\sim$\$800 billion in 2009 to $\sim$\$1100 billion in 2014.[1] However, the average cost of developing each new chemical entity (NCE) through to market as a successful drug has also risen inexorably, and is now estimated at about \$1 billion.[2] So although the rewards of success are high, the discovery and development of NCEs is an extremely risky business.

These high costs, the continuing pressure to contain the rise of healthcare costs, and the expiry of patents on key "blockbuster" NCEs, have been key factors driving the merger and acquisition (M&A) activity that has led to consolidation within the industry. The 10 years to the end of December 2009 saw a total of 1345 mergers and acquisitions of pharmaceutical assets and companies, with total disclosed prices of \$694 billion (when deals within generic and consumer healthcare units are included),[3] with 2009 as the biggest single

RSC Drug Discovery Series No. 9
Pharmaceutical Process Development: Current Chemical and Engineering Challenges
Edited by A. John Blacker and Mike T. Williams
© Royal Society of Chemistry 2011
Published by the Royal Society of Chemistry, www.rsc.org

Table 15.1 Number of NCEs reaching the market in four-year periods.

	1985–1988	*1989–1992*	*1993–1996*	*1997–2000*	*2001–2004*
Synthetic agents	187	128	144	129	90
Total NCEs	204	141	160	136	113

year with deals totalling \$147 billion. The roll-call of once-prominent pharmaceutical names that have disappeared from view as a result of M&A activity includes American Home Products, Beechams, Ciba-Geigy, Hoechst, ICI Pharmaceuticals, Pharmacia, Rhone-Poulenc, Schering-Plough, Syntex, Upjohn, Warner-Lambert, Wellcome and Wyeth.

The second major pressure stems from the decline in productivity across the pharmaceutical industry, as measured by the number of NCEs reaching their first market each year. Table 15.1 shows the decline in both total launches and the number of small molecules (synthetic and semi-synthetic) NCEs reaching the market over the period from 1985 to 2004, with the balance of the total made up natural products isolated by extraction/fermentation and biomolecules.[4] A similar analysis by Grabowski and Wang[5] qualified the sharp decline in pharmaceutical productivity, from a peak in NCE numbers in the late 1980s, with the proviso that there had been an increase in quality over their period of study (1982–2003), based on a moderate increase in "first-in-class" agents over the study period. In the four years from 2005 to 2008 the average annual number of small-molecule NCEs reaching the market has been 19,[6] compared with ~45 in the late 1980s and 34 throughout the 1990s, suggesting that the downward trend is not reversing. Over the past dozen years the number of natural products launched has shrunk to <1 per year, while the number of biomolecules (vaccines, antibodies and peptide and nucleotide based agents) has slowly risen and now represents almost 20% of the total.

The great hope that the decoding of the human genome would usher in a golden age of drug discovery has largely faltered; so far, reading our DNA seems mostly to have taught us how little we still understand about our own biology. Many of the easier disease targets are now well treated, so researchers are left with complex problems like cancer, which is really not one disease at all but several thousand different ways that a human cell can go wrong, or Alzheimer's disease. The steady fall in the number of new agents reaching the market, and rise in associated costs, is illustrated in Figure 15.1.

The regulatory environment represents a third area of change that has impacted the industry, with regulatory agencies particularly focusing on patient safety and product quality issues. Since 1980 the number of clinical trials required to support a new-drug application has more than doubled, while almost three times as many patients are needed in each clinical trial.

Factors such as this have resulted in a greater than four-fold inflation-adjusted increase in the costs of the clinical-trial stage of drug development.[7] The expensive late-stage failures of agents in phase III (and even post-launch), which are becoming increasingly commonplace, are frequently the result of demands by regulatory authorities for tighter safety requirements, and for clear product

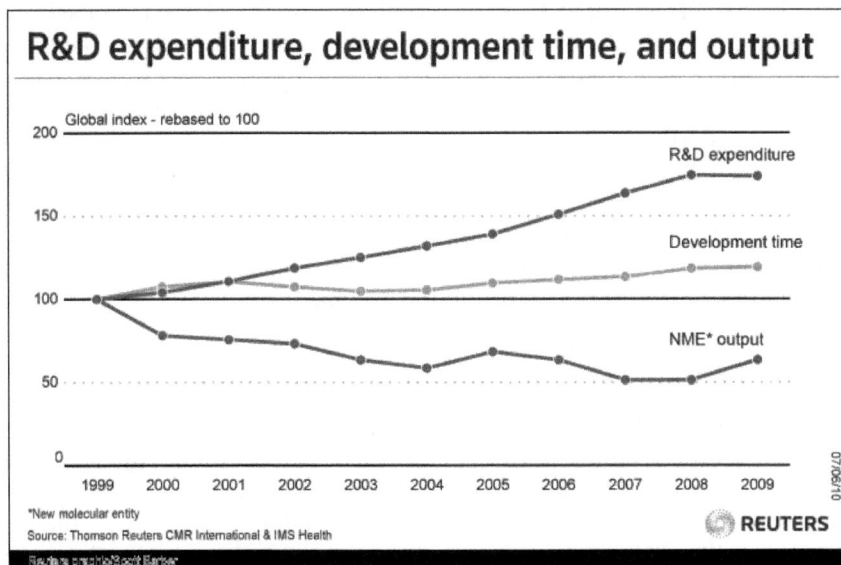

Figure 15.1 Pharmaceutical R&D expenditure, timelines and output (1999–2009) (reproduced by permission of CMR International, a Thomson Reuters business).

differentiation from existing therapies (in many cases a result of tighter regulatory scrutiny). The FDA has strengthened powers to assess already marketed drugs.[8]

For example, there is a greater focus on the risk of increased occurrences of rare ventricular arrhythmias associated with prolongation of the cardiac QT_c interval that has led to the withdrawal of several previously approved agents from the market. Analyses by the Centre for Medicines Research International in 2010 and 2011 showed that the number of phase III projects killed in 2007–2009 doubled the rate seen in the previous three years, while the average for the combined success rate at phase III and submission had fallen to $\sim$50% in recent years.[9] A separate analysis showed that since 1990 the rate of attrition has been rising in all phases of drug development, so that the overall survival rates of drug candidates has fallen markedly over that period.[10]

Even if the product is approved for marketing, there are still the pricing and reimbursement hurdles to be negotiated, and these issues are no longer restricted to the developed Western economies. Governments around the world are trying to contain costs and, as healthcare budgets constitute a very significant part of governmental spending, these costs are the subject of intense scrutiny.

15.2 The Impact of Changes on the Conduct of Process R&D

As discussed in Chapter 1, the pharmaceutical process R&D scientist continues to serve three disparate customers, providing the bulk active pharmaceutical ingredient (API) to fuel the development process, a robust manufacturing process for the production organisation, and contributing to the chemistry, manufacturing and controls (CMC) section of the filing to the regulatory agencies. Despite the above changes across the industry, the basic mission of the process R&D scientist is still to provide safe, economic, controlled and environmentally acceptable API synthetic processes, suitably documented and under regulatory guidance. However, the environment in which this work is carried out continues to change, and with it the conduct of process R&D is evolving. Some of the changes, challenges and opportunities facing scientists working in this area are examined in this section.

15.2.1 Lowering Early Development Costs

Chapter 4 discussed the "fit-for-purpose" early development paradigm, which is increasingly being used to minimise the chemical development costs incurred in taking an NCE to its clinical proof of concept (POC). These efforts hold down the costs associated with the failures, those NCEs that fail to achieve their POC and whose development is terminated. For the agents that achieve their POC, the minimisation of early work leads to an inevitable deficit in the knowledge about them. Once the POC hurdle has been surmounted, and the odds of the agent reaching the market have increased, the desire to progress the agent as quickly as possible through the late development programme leads to a rapid acceleration of the post-POC chemical programme. This frequently causes conflict between the desire for speed to market and the desire to identify the best route for filing and marketing. In turn, this leads to the risk of the developer getting "locked into" an acceptable, but not optimum (from a cost or environmental standpoint, for example), route that is optimised and validated for the filing.

Scientists working under this new regime, in which as much work as possible is deferred until after the drug candidate has successfully passed key attrition points, will notice a number of changes. First, the proportion of work in the unit devoted to early candidate work will decrease. Secondly, the need to make up for the pre-POC knowledge deficit, and to move the NCE as quickly as possible through the later stages of development, will intensify the level of chemical development effort applied to late stage programmes. Thirdly, the proportion of products filed with sub-optimum routes will increase, leading to the option of continuing the search for the optimum route after the route for the main filing has been decided. There are an increasing number of examples of "second-generation synthetic processes", with innovator companies refiling improved synthetic routes and processes once their products are marketed. A recent analysis of eight industry examples found that the median refiling approval time was merely 136 days.[11] Important examples of second-generation routes are provided by sertraline,[12] taxol,[13] sitagliptin[14] and pregabalin[15] (Figure 15.2), and the whole area of post-filing changes now seems to be set for rapid growth. This field thus provides increasing opportunities for the process R&D chemist and chemical engineer to work on products that are already on the market.

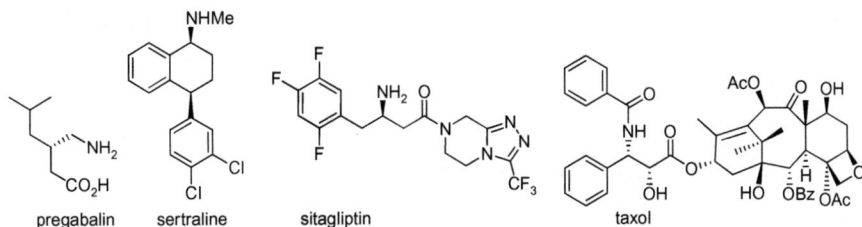

Figure 15.2 Major products produced by second-generation routes.

15.2.2 Outsourcing to Contain Rising Costs

The outsourcing of the manufacture of key pharmaceutical intermediates has been increasingly used to help cope with rising cost pressure. In recent years this trend has accelerated, with many Western pharmaceutical companies aggressively seeking low-cost country sources for pharmaceutical manufacturing and moving significant activities, such as the manufacturing of APIs, to India and China. For example, in 2009, AstraZeneca announced that it will move all of its current API production in the UK to China. Also in 2009, Novartis announced a $1 billion investment that will create China's largest pharmaceutical research plant yet, and Eli Lilly axed 5500 jobs in the US and added 2000 in China.

An estimated 30–50% cost savings has been the main driver for sourcing starting materials, intermediates and APIs from Asia. Cheaper labour, tax advantages, undervalued currency and lower capital, and overhead costs have all contributed to this. However, in China, for example, all of these advantages are expected to diminish in the coming years as inflation rises, currency appreciates and tax rebate structures start to evolve. In addition, a series of fraud and adulteration cases (including the falsification of clopidogrel batch documentation by an Indian supplier, and the adulteration of heparin from China) has eroded confidence in the quality of supplies from Asian sources.[16] The combination of concerns about security of product quality, erosion of the cost advantages, and realism about the levels of Asian investment needed, has recently led to some sourcing being pulled back to the West.[17]

Despite the inevitable flow of some jobs to Asia and other low-cost regions, the outsourcing trend will present some opportunities for process R&D chemists:

- The expansion of the contract manufacturing sector, as some pharmaceutical companies decide to outsource a greater proportion of their bulk synthetic operations, is producing some jobs in fine chemical supply companies. There are also opportunities for experienced Western pharmaceutical chemists in rapidly growing Chinese drug firms, with non-Chinese speakers focusing on interacting with clients and overseas collaborators.[18]
- Sourcing specialists with chemical development expertise will be needed to manage the placement and tracking of business with contract research and manufacturing organisations.

- Specialists will be required to audit suppliers, to "police the supply chain", and ensure that outsourced work is carried out to the required standards.[16]

15.2.3 Technology to the Rescue?

Some of the emerging trends in automation, process analytical technology (PAT) and process intensification were highlighted in Chapter 11, while the growing use of calorimetric investigations of reaction mechanisms was discussed in Chapter 7. Despite the increasing deployment of productivity enhancing technologies over the past decade, Figure 15.1 suggests that these technologies have to date contributed little to the goal of an overall reduction in development timelines; they have merely helped us to cope with increased workloads and lower project staffing levels (that is, to run faster to stay where we were). In the future, more will doubtless be done across the industry to routinely use these tools to more efficiently select, develop and optimise processes.

Exciting developments are taking place in the solid form area, which occupies the crucial interface between the drug substance and the formulated product. Computational tools are emerging to predict the probability of new polymorphic forms emerging, and modelling is increasingly being used to identify the effects of H-bonding patterns on the relative stability of polymorphs. In cases where the API does not have sufficient crystallinity, solubility or stability, the salt formation option can now be augmented by the exploration of cocrystals.[19]

However, the biggest opportunity to use technology to effect a transformative change appears to be provided by continuous processing (including for second-generation processes to approved products), which is still heavily under-utilised in pharmaceuticals, compared to other chemical manufacturing areas. Recent examples of the use of continuous processing in the pharmaceutical industry have been presented, including its application to APIs such as celecoxib and naproxcinod,[20] but the pharmaceutical industry has barely scratched the surface of the opportunities that continuous processing offers. According to an analysis of the kinetics of reactions carried out in the fine chemical and pharmaceutical industries, up to 50% of these reactions have the potential to be advantageously carried out using continuous processes.[21] Broader implementation of continuous processing seems inevitable, but will require close collaboration between chemists, engineers and, ultimately, a mindset change amongst the chemists devising early synthetic routes. The application of PAT to the understanding and control of manufacturing processes is well suited to continuous processing, and the views of the FDA on the potential impacts of quality by design (QbD) and PAT on waste reduction and the creation of more benign processes have been well articulated:[22]

"The agency is fully supportive of the industry moving in the direction of continuous processing...The principles of QbD and the implementation and use of PAT are inherent in the design and development of a continuous process."

Applying continuous processing to pharmaceutical syntheses thus offers many possibilities to improve both their cost effectiveness and their environmental performance.

15.2.4 Moving Away from "Blockbuster Dependence"

The shift from the blockbuster mindset to focusing on developing treatments for smaller patient populations is beginning to take hold, as there have been significant shifts in the corporate culture of both large pharmaceutical and biotechnology companies to begin investing in drugs for rare diseases, and in personalised medicine.

Since its passage in 1983, the US Orphan Drug Act has led to the approval of more than 350 drugs for around 200 rare diseases (defined in the US as diseases/disorders affecting fewer than 200 000 US patients), mostly thanks to small biotechnology start-ups looking for a unique marketplace niche. A number of other countries have subsequently adopted legislation designed to encourage the development of orphan drug products, including the EU, Japan, Australia and Singapore; the definition of an orphan product varies slightly from country to country, but the principle is the same. Following the problems experienced by big pharmaceutical companies with their traditional business model, some of the world's largest drug makers are aggressively entering the rare disease sector. The global market for orphan drugs was worth almost $85 billion in 2009; according to the US National Institutes of Health, around 25 million US citizens are affected by one of some 6800 rare diseases. Furthermore, the market is predicted to grow at a compound annual rate of around 6% over the next few years, which is at least as good as the growth achieved by some of the top companies in recent years.[23]

Personalised medicine is the translation of the revolution in the understanding of health and disease (brought on in large part by the sequencing of the human genome and the creation of a map of human genetic variation) to patient care by using genetic and genomic information in diagnosis, prognosis and treatment. The goal of personalised medicine is to provide the right diagnosis and treatment to the right patient at the right time at the right cost. The global market for a more personalised approach to medicine and health is growing at 11% per annum, and is expected to reach $452 billion in 2015 according to PriceWaterhouseCoopers' projections.[24]

Oncology leads other therapeutic areas[25] in the number of personalised medicines on the market as well as in the pipeline, with the expectation that within a decade all oncology drugs will have a related diagnostic. Other key therapeutic areas in which personalised medicine is making headway include cardiovascular, central nervous system and immunologic therapies, whereas personalised medicine development is just getting started for metabolic and respiratory therapies, as well as virology.[26]

The nature of healthcare is changing dramatically, and personalised medicine is now playing an important role in the drive to achieve better outcomes for patients in a cost-efficient manner. A recent survey by the Tufts Center for the Study of Drug Development (CSDD) found that between 2006 and 2010, drug companies increased their investment in personalised medicine by an average of over 70%, and that in the next four years these firms expected to further increase their investment in the area by more than 50%.[27]

The realisation by large pharmaceutical companies that peak sales expectations of products need to come down, should lead to more "niche" products which might otherwise have been neglected. There will, of course, be the need to develop these agents for smaller markets for far less than the current average $1 billion level, and the continuous processing option mentioned above could facilitate this, especially for smaller volume products. This trend to increase the effort devoted to rare diseases[28] and personalised medicine has the potential to start increasing the number of products reaching the market back to levels seen in the 1990s. This in turn will ensure that development scientists will have plenty of substrate to work on, but will need to redouble their efforts to increase efficiency and hold down development costs.

15.2.5 Drug Discovery and Development from Outside "Big Pharma"

In the past, the pharmaceutical sector was dominated by "big pharma", those fully integrated companies who carried out their own discovery, development, production, marketing and sales for the majority of their products. Most new NCEs approved for the market are still both discovered and developed by big pharma companies, though a recent review identified that a significant number of the 252 new drugs approved by the FDA between 1998 and 2007 were originally discovered within universities or biotechnology companies.[29] The "biotechnology" descriptor has become a potentially confusing one, as it encompasses both those companies who discover and develop biologically derived macromolecular drugs, and also start-up or "small pharma" companies, who develop platform technologies and/or use "biology" to discover small-molecule drug candidates. The issue has become further blurred, with companies bridging the divide; many big pharma companies now work in-house on macromolecules, and some of the larger established biotech companies such as Genentech and Amgen are now diversifying into small-molecule NCEs.

In recent years, large pharmaceutical companies have realised that they cannot survive on their own research and pipeline of innovations, and are becoming much more open to collaborating with smaller biotech firms, with universities, and even amongst themselves.[30] An increasing number of new drug candidates are being discovered outside big pharma, taken through early development, and if promising are then licensed to pharmaceutical companies for the later and more expensive stages of development. The trend is now towards more complex partnering relationships between biotech and pharmaceutical companies, with deals now involving more resource and budget sharing, and more deals being made at an earlier stage in drug development. While the small and nimble biotech model has proved its worth in the drug discovery arena, during the development process (and especially for later development) organisational scale still has significant advantages.

At a time when the number of jobs for development chemists in big pharma has been contracting, employment opportunities are being created in other settings:

- Some of the more successful small biotech companies are increasingly investing in their own development capabilities. Even an investment in 20 L glassware and the ability to prepare early API supplies on the 100 g to 1 kg scale can give the company improved control over its preclinical development timelines, as well as valuable experience to allow benchmarking of subsequent outsourcing partners.
- There have been similar developments in academic units that until now have only been involved in drug discovery, but are now also moving downstream into process development capabilities. Notable examples include the Institute of Process R&D (iPRD) at the University of Leeds in the UK (which has contract process R&D capabilities) and Vanderbilt University's Program in Drug Discovery (VPDD) in the US, which is growing its capability to develop agents it has discovered.[31]

Since 2007 a range of notable major alliances has been entered into between big pharma companies and academia.[32] Government responses to this trend have been very positive, including sponsorship of industry/academic partnerships such as the EFPIA Innovative Medicines Initiative in Europe.

15.2.6 The Rise of Macromolecules

The last quarter of the 20th century saw the rise of large organisations focused on the discovery and development of biologically based therapeutic agents, which can be composed of sugars, proteins, nucleic acids or complex combinations of these substances. An increasing number of biological agents, such as monoclonal antibodies and peptides, has been reaching the market over the past 20 years, and biological agents now make up about 20% of new drug approvals. However, these are expensive agents, with nine biologics in 2010 costing between $200 000 and $410 000 per patient year of therapy. These macromolecules are thus often targeted for acute life-threatening conditions, and require parenteral delivery.

Because of the molecular weight cut-off of around 550 for oral absorption, the majority of agents in clinical use will continue to be small-molecular entities, prepared synthetically. Although biological agents require rather different skill sets, there are opportunities for development chemists experienced with small-molecule agents to expand their skill sets to include peptide- and polynucleotide-based agents.[33]

15.2.7 Rising Regulatory Expectations

Environmental agencies are now far more stringent in their requirements, and Chapter 6 highlighted the progress that has been made in addressing the environmental issues associated with the development and manufacture of

APIs. Process chemists will be increasingly expected to design synthetic routes using catalytic processes wherever possible, rather than use older, well-tried stoichiometric synthetic methods. As much of the waste associated with API manufacturing is associated with work-up and isolation processes, these processes will become areas of increased attention. There has also been a closer focus on the potential for synthetic steps to generate genotoxic impurities.

Over the past decade, regulatory agencies have shifted their focus from compliance and ensuring that quality is assured through appropriate testing, to a paradigm centred upon enhanced process and product understanding and appropriate risk assessments.[34] With this move in philosophy from "quality by testing" to QbD, regulators are increasingly favouring filings in which knowledge, control and robustness are core values.

15.3 Conclusions

The global demand for better healthcare will continue for the foreseeable future. So despite the fact that big pharma has been through a difficult past decade, the industry as a whole is not in decline. It seems likely that in future an increasing proportion of the industry's drug candidates will be discovered by, or in collaboration with, smaller "biotech" units or academic laboratories. However, because of the advantages of scale during the expensive development phases, most drug candidates will probably continue to be developed by process R&D scientists in large companies.

A decline in industry productivity has been a concern since the 1990s, but there are indications that the long awaited impact of the genomics revolution is imminent. This, together with the growth in orphan drugs and personalised medicines, and a resurgence of anti-infective research due to problems with continuing bacterial and viral resistance, should ensure a sufficient flow of drug candidates to be developed. However, a key challenge will be to find ways of bringing the next generation of NCEs (especially the niche products) through to the market more cost effectively.

References

1. (a) Market growth forecast, *IMS Health*, 20 April 2010; (b) E. Sukkar, *Scrip World Pharm. News*, 20 April 2010.
2. (a) J. A. DiMasi, R. W. Hansen and H. G. Grabowski, *J. Health Econ.*, 2003, **22**, 151; (b) C. P. Adams and V. V. Brantner, *Health Affairs*, 2006, **25**, 420; (c) C. P. Adams and V. V. Brantner, *Health Econ.*, 2010, **19**(2), 130.
3. T. Stanton, *FierceBiotech*, March 26, 2010.
4. H. Murakami, *Top. Curr. Chem.*, 2006, **269**, 273.
5. H. G. Grabowski and Y. R. Wang, *Health Affairs*, 2006, **25**, 452.
6. "To Market, to Market", *Annu. Rep. Med. Chem.*, 2006–2009, **41–44**.
7. M. McArdle, *Atlantic*, June 20, 2010.
8. H. Ledford, *Nature*, 2010, **466**, 677.

9. (a) *The CMR International Pharmaceutical R&D Factbook 2010*, Thomson Reuters, London, June 2010; (b) J. Arrowsmith, *Nat. Rev. Drug Discov.*, 2011, **10**, 87.

10. F. Pammolli, M. Riccaboni and L. Magazzini, *The Productivity Crisis in Pharmaceutical R&D*, Working Paper Series, University of Verona, April 2010.

11. S. K. Ritter, *Chem. Eng. News*, 2010, **88** (Oct. 25), 45.

12. G. J. Quallich, *Chirality*, 2005, **17**, S120.

13. P. G. Mountford, in *Green Chemistry in the Pharmaceutical Industry*, ed. P. J. Dunn, A. S. Wells and M. T. Williams, Wiley-VCH, Weinheim, 2010, pp. 145–160.

14. J. Balsells, Y. Hsiao, K. B. Hansen, F. Xu, N. Ikemoto, A. Clausen and J. D. Armstrong III, in *Green Chemistry in the Pharmaceutical Industry*, ed. P. J. Dunn, A. S. Wells and M. T. Williams, Wiley-VCH, Weinheim, 2010, pp. 101–126.

15. P. J. Dunn, K. Hettenbach, P. Kelleher and C. Martinez, in *Green Chemistry in the Pharmaceutical Industry*, ed. P. J. Dunn, A. S. Wells and M. T. Williams, Wiley-VCH, Weinheim, 2010, pp. 161–177.

16. R. Mullin, *Chem. Eng. News*, 2010, **88** (Sept. 6), 32.

17. M. McCoy, *Chem. Eng. News*, 2009, **87** (April 27), 16.

18. J.-F. Tremblay, *Chem. Eng. News*, 2011, **89** (Feb. 14), 47.

19. C. Frampton, *Chem. Ind.*, 2010 (March 8), 21.

20. L. Proctor, P. J. Dunn, J. M. Hawkins, A. S. Wells and M. T. Williams, in *Green Chemistry in the Pharmaceutical Industry*, ed. P. J. Dunn, A. S. Wells and M. T. Williams, Wiley-VCH, Weinheim, 2010, pp. 221–242.

21. D. M. Roberge, *Org. Process Res. Dev.*, 2004, **8**, 1049.

22. C. Watts, presented at the ISPE conference on Continuous Processing in the Real World, Vienna, September 2006.

23. P. Charlish, *Scrip Mag.*, 2010, Sept 6.

24. *The Science of Personalized Medicine: Translating the Promise into Practice,* PriceWaterhouseCoopers report, London, December 2009.

25. A. Thayer, *Chem. Eng. News*, 2010, **88** (Apr. 26), 8.

26. *Tufts CSDD Impact Report*, Tufts University, Boston, MA, November/December 2010.

27. C.-P. Milne, presented at the conference on Personalised Medicine: Impacting Healthcare, Harvard Medical School, November 2010.

28. S. Borman, *Chem. Eng. News*, 2010, **88** (Apr. 26), 36.

29. R. Kneller, *Nat. Rev. Drug Discov.*, 2010, **9**, 867.

30. R. Mullin, *Chem. Eng. News*, 2010, **88** (Feb. 22), 12.

31. (a) R. Petkewitch, *Chem. Eng. News*, 2009, **87** (Apr. 20), 28; (b) A. M. Thayer, *Chem. Eng. News*, 2009, **87** (Jun. 8), 23.

32. L. M. Jarvis, *Chem. Eng. News*, 2010, **88** (Nov. 8), 14.

33. H. Carmichael, *Chem. World,* 2009 (April), 70.

34. *Pharmaceutical Quality for the 21st Century; a Risk-Based Approach*, http://www.fda.gov/oc/cgmp/report0507.html (last accessed 12 February 2011).

Subject Index

Page references to figures, tables and text-boxes are shown in *italics*.